AF540119

PLATE I

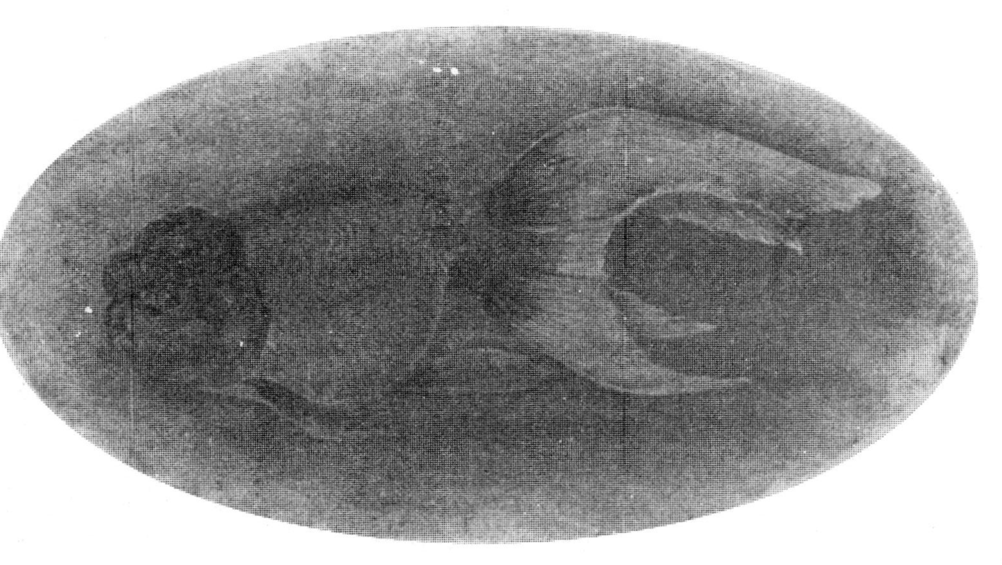

SHUKIN. JAPANESE GOLDFISH.

(By courtesy of the U.S. Bureau of Fisheries, Washington.)

[*Frontispiece.*

The BIOLOGY OF FISHES

New Introduction by Dr. Vijay Dev Singh

M. HARRY KYLE, M.A., D.Sc.

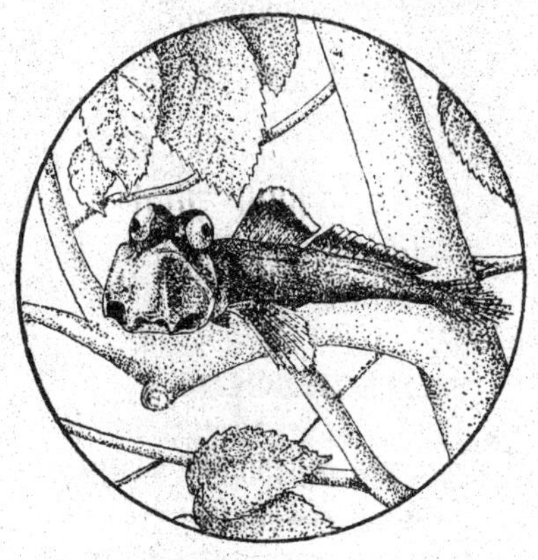

The Walking Goby (*Periophthalmus barbarus*) on the stem of a bush (from Jordan and Seale)

2017
Biotech Books
Delhi - 110 035

First Indian Impression, 2017

ISBN 81-7622-184-8

Published by : **BIOTECH BOOKS**
1123/74, Tri Nagar,
DELHI – 110 035
Phone: 27382765
e-mail: biotechbooks@yahoo.co.in

Showroom : 4762-63/23, Ansari Road, Darya Ganj,
NEW DELHI - 110 002
Phone: 23245578, 23244987

Printed at : **Chawla Offset Printers**
New Delhi – 110 052

PRINTED IN INDIA

WHATEVER GOOD THERE MAY BE IN THIS BOOK

IS DEDICATED TO

MY FRIEND

ERNST EHRENBAUM

INTRODUCTION

Reading a new book is like taking a new road for a journey. The traveller expects the road to be smooth and well laid out as well as interesting enroute to make the journey a pleasure. The biology of fishes by Harry M. Kyle is similarly both full of facts about the mysterious life of fishes and contains details of their biology as well. Unlike the present day publications on fishes which merely record facts and figures, reading this book is like discovering an old gold casket left buried in the depths of the ocean for half a century. The book deals with fishes in a much wider environmental context and introduces us to each new facet in the life cycle of fishes with such ease that even a layman would enjoy exploring the world of fishes. The author has described the various inter-linkages which must be kept in mind while undertaking any study of a living creature. Even though most of the fishes discussed in detail are European in distribution and certain data/information needs to be updated since the book was first published in 1926, it still proves to be a very comprehensive, systematic and naturalistic presentation of the biology of fishes.

This book contains 14 chapters. In the first chapter on the general characteristics of fishes, the author has described what makes a fish a fish, and how these characteristics have been adapted to enable it to live in an aquatic environment. The second chapter on the habits of fishes in general gives a very fine account on the movement of fishes in nature, their feeding and breeding habits. This is followed by the chapter on migration of fishes containing not only a detailed description of the phenomenon but also exploring the plausible reasons for doing so.

The chapter on The Development of Fishes covers the entire process from the egg, to the embryo to the development of the various tissues and organs of the fish. In the fifth chapter the author discusses the influence of balance and movement on the formation of the body structure and head in fishes. The next two chapters deal with the metabolism of fishes. Thus, the sixth chapter covers the functioning of the digestive, circulatory, respiratory and excretory system of fishes and the next chapter contains details of the rudimentary sensory, muscular and regulating systems present in fishes. A study on variations and differentiation of fishes complete with examples of various experiments makes up the eight chapter of the book. The author then traces the genealogy of fishes, both backwards and forwards, and the effect of the drifting of the continents on the distribution of fishes. This is followed by a chapter on the distribution of fish species in the bygone ages to their modern day spread.

A very detailed description of the growth of adaptations in fishes in relation to their mode of life in an aquatic medium, resulting in a specialized mode of respiration is the topic of the eleventh chapter. Common diseases, parasites and reproduction in fishes are dealt with in a separate chapter. In the chapter "the food question" the author first discusses the diet of fishes followed by the valuation of the sea and its resources. In the concluding chapter on the mental life of fishes some concepts like tropisms, reflex actions, intelligence, reason and parental care in fishes are presented in a remarkable manner.

The style of presentation of facts in the book remain as interesting and relevant today as before, giving credence to the belief that a good book is one which withstands the test of time. I, therefore, feel that all students and scientists of fisheries would enjoy and be greatly benefitted and enriched in their field of study by reading this very interesting and well written book.

DR. VIJAI DEV SINGH

PREFACE

In studying the life and characters of fishes one must face the problem of interpreting the phenomena. They may be taken as evidence of this or that theory, or they may be described as hard, cold facts without reference to any theory. A middle course may be suggested. One may form tentative deductions based on the facts and contrast them as impartially as possible with the conventional theories, so that the " fittest " may survive.

The possibility that characters have arisen at variance with the innate tendencies of " Heredity," and persisted without the aid of " Natural Selection," is a thesis which requires some courage to state. Yet this is one of the deductions that come, it seems to us inevitably, from a study of fishes, and a few words of explanation may be given here.

Fishes occupy a peculiar position in the hierarchy of animal life. We cannot be sure whence they came, but many of the important, even the essential, characters of Higher Vertebrates make their first definite appearance among them, and it is certain that fishes were the forerunner of all higher forms of life. We have these important Vertebrate characters, then, at the upper end, and nothing but a hypothetical form—very different from a fish and possibly a mutation from some larval form of lower degree—at the beginning. How did such characters first arise ? Which means, how can we conceive their origin ?

An answer may be approached from two different sides. In the development of forms at the present day each individual, to begin with, is little more than a muscular body with nerves. The earliest fossil remains of fishes, dating back some 400 million years, are of the same nature. The unfolding of the characters follows the same general lines in the development of the primitive forms of the present day and in the phylogenetic succession in the past history. But, whilst we may say that the unfolding of the characters in development is due to the inherited tendencies, such an interpretation would not apply to their original appearance unless we widen very considerably our ideas of heredity. As an example to be given presently will show, the reactions of an organism under the stress of new conditions are more elemental than specific. And it is just because they stand at the beginning of a new series, that fishes present us with more examples of this elemental power than any other group of bodied animals.

In the making of form and structure we can recognise a trinity of forces, inheritance, activity, and environment. For the reason stated, it is probable that the inheritance is of less importance among the fishes than in higher animals. The greater variability of their characters also indicates this. And experiments show that the characters are very sensitive to changes in the external conditions. The constancy of a species, in fact, depends upon the persistence of a normal environment. We might almost include the latter under the inheritance, were not this usage confusing to the student.

The activity of the individual, the link between inheritance and environment, is probably the most important factor in the making and remaking of characters. Under normal circumstances its importance is masked, but under altered conditions there is nothing left but this activity to account for the appearance of new characters. The plastic material that contains the traditions of the past responds in an elemental manner to the new impressions, and the resultant characters may be, have frequently been, of such a totally different appearance that the genetic relation cannot be discerned.

One has also to face the question, whether Natural Selection, the survival of the fittest and elimination of the unfit, has been the directive principle in differentiation and evolution. Does this theory give a reasonable explanation of the facts?

The author's attitude towards the theory may be illustrated by quotations from the writings of Charles Darwin. "The theory of natural selection implies that a form will remain unaltered unless some alteration be to its benefit." Again, "when we descend to details, we cannot prove that a single species has changed, nor can we prove that the supposed changes are beneficial, which is the groundwork of the theory. Nor can we explain why some species have changed and others have not."

Most biologists of the present day will probably agree with these words of Darwin. The theory of Natural Selection cannot be demonstrated as an interpretation of evolution in the past and we find ourselves unable to make use of it in unravelling the sequence of detailed events which lead up to the formation of characters in the present. Darwin has also admitted the possibility, that "through the nature of the organism and of surrounding conditions, but not through natural selection," new groups of animals have been formed. These statements apply with special force to the fishes and it is not difficult to point to numerous cases, where changes have been quite the reverse of beneficial, and yet the species have survived and multiplied.

One of these may be noted here by way of illustration. The Sun-fish when a tiny larva goes through a condition which may well be compared to dropsy. The natural course of development comes to a stop, the ordinary muscles and the nerve cord degenerate, at least do not develop, and the larva rolls about in the water for some time like a helpless drop of jelly. But the life is not extinguished; the form changes again, new muscles replace the old, new nerve-connections are established, and the Sun-fish persists to become one of the largest of fishes with a potential fecundity greater than anything known elsewhere in the animal kingdom. Examination of the characters of the adult Sun-fish affords no clue to its ancestors.

The deduction noted above is strengthened by this and many similar instances. If the activity of ill-balanced fishes has enabled them to escape from a fate to which they were apparently doomed both by heredity and by natural selection, may we not believe that the living organism possesses a power which is superior to either ?

How this deduction enables the phenomena of life to be interpreted in the case of fishes, will be illustrated in the following pages.

For many helpful criticisms the author tenders his sincere thanks to Professor J. Arthur Thomson, LL.D., of Aberdeen University.

H. M. K.

CONTENTS

LIST OF PLATES

FIGURES IN THE TEXT

THE BIOLOGY OF FISHES

CHAPTER I

THE GENERAL CHARACTERS OF FISHES

If we describe a fish as a slippery animal that lives in the water, the definition would be quite sufficient for practical purposes. Yet every one knows, that many other kinds of animals live in the water and all of them are slippery, that is, if they can be felt and handled. Then we may remember that fishes breathe by means of gills, which excludes most of the higher animals, and again that they possess some kind of backbone, which shuts out the lower animals. Hence a more precise definition would be, that fishes are gill-bearing vertebrates living in the water.

But the few words of a definition give only the barest outlines of the picture. Fishes have a structure that is eminently suitable to life in the water. We must know something of the former to understand the latter. And what sort of life do they lead? Is it all a monotonous drinking and feeding, or do they have some kind of sport and psychical life among themselves? How far away or how near do they stand to ourselves? To try and answer these questions, will be the aim of the succeeding pages. Meantime, we may consider briefly, how the fish has come to be what it is.

1. Origin and Nature of a Fish

To say that a fish must have learnt to swim before it became a fish may seem a truism, but there is a meaning behind the statement. The whale and the seal are not fishes; they have entered the sea from the land at a different level of evolution, with very different structures to begin with—nose opening into the mouth, and lungs, but no gills. Their outer form may change to that of a fish and even the legs disappear, but they have not retrograded to such an extent as to lose their internal characteristics.

The ancestors of the fish swam in the water before the structures distinctive of a fish developed. The earliest fish record of the rocks, perhaps 400 million years old, shows only the impression of muscles grouped as in a fish, nothing solid about it. But, it may be said, the worm is also a muscular body; how then could a distinction be made?

If we examine the description of a worm in text-books, we find that the muscles are principally of two kinds, one in the form of circles round the body, the other running along the body. Within these we find a central space occupied chiefly by the digestive canal. Now, when such a body moves, it does so by intensive wrigglings with almost as much slip as progress. It cannot present an effective biting surface anywhere along its body, and thus its advance is comparatively slow in spite of prodigious efforts.

In a fish, however, the space with the digestive tract occupies only the lower half of the body and even only a part, the front part of the lower half. The rest of the body is all bone and muscle. The digestive tract is, as it were, shoved out of the road to give the muscles freer play. Further, the muscle segments are inclined towards the tail, just as the blades of a propeller are inclined backwards, so that the surface of the body literally strikes the water with each movement.

When a fish is flattened out, therefore, even when no bone is present, the muscles have quite a different arrangement from those of a worm. In fact, there is a very wide gap between them, and if fishes came from the worms, as many suppose, several important changes in structure had to be made before the new arrangement could be attained. In the simplest "fish" there is very little appearance of backbone or ribs, yet some kind of strengthening rod runs down the mid-line of the body and above it lies the nerve cord, just as in higher vertebrates. The brain and sense organs are not much to speak about—certainly insignificant by comparison with what are found in insects and other invertebrates, but the median rod with the dorsal nerves makes all the difference. A worm, it may be said, is too flabby to have any balance in the water, nothing to hold the different parts firmly together; the fish has obtained rigidity whilst retaining flexibility.

But if we try to picture a muscular body trying to move through the water of its own accord, we can readily see that the head and body must be somewhat compressed or flattened, otherwise it will roll and twist in an uncomfortable manner. Now some of the forms immediately preceding the fishes (Appendicularia) display such a twisting, and we may believe that the fishes have passed through and to a great extent overcome this difficulty. Having done so, as we shall see later, the fins developed out from the flattened sides and at the end as a natural consequence. In the development of fishes at the present day we can still see the process being repeated.

Other differences from the worm may be noted. The heart is a powerful, muscular organ and lies ventrally just behind the head. To some extent it looks as if the nervous and the vascular systems had changed places in changing from the worm to the fish condition. Some think even, that the body has not only turned upside down but goes in the reverse direction to what it did as a worm. These changes, however, will be discussed in more detail later.

The worm is able to move forward in a definite direction and thus may be said to have a head. Further, it may have tiny eye-spots, ear-stones, and tactile or taste organs, but these are so minute a microscope is required to see them. In fishes, however, there can be no doubt about the eyes, the nose, and the large size of the head. It is plain that "brains" must lie inside the latter, and this represents the cardinal difference from the worm. Its organising and controlling faculties are centralised instead of diffuse, and it consequently has more mastery over its surroundings. Yet it has to be noted, that the simplest form of "fish" known—though it may be a degenerate—has little more brain or sense organs than a worm and lives almost like a worm.

The older, and still the popular, idea of a fish as a being that moves through the waters is the one that should be kept most in mind. Just as flight is the outstanding characteristic of birds, swimming is what we associate most naturally with a fish. We shall be obliged to modify this idea somewhat as we go on, for most fishes, so far as number of types or species is concerned, have eased off in their swimming inclinations and a few hop about on land—reminding us that in the distant past some kind of frog emerged from the water and learnt the lessons which made higher vertebrates possible; but of the total numbers of fishes most remain, as they were in the beginning, creatures that live and move in the sea.

2. Form and Movement of Fishes

The ideal fish is, then, one that cleaves the water. In shape it is like a spindle, or rather a torpedo, with the sides com-

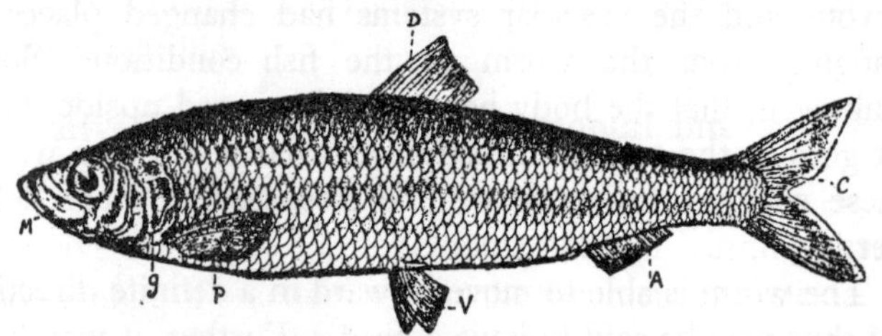

Fig. 1.—The Herring (*Clupea harengus*). M., maxillary; *g*., gill-cover; P., pectoral fin; V., ventral or pelvic fin; A., anal fin; C., caudal fin; D., dorsal fin. (From Ehrenbaum.)

pressed. Pointed in front under the pressure of the water it swells out across the shoulders or pectoral region, then the depth and thickness gradually taper towards the tail. The slenderness of the body at the root of the tail is a sure sign of a good swimmer. Then follows the broad and long caudal fin, as it is called.

Even this torpedo form, however, shows many variations, as we can see at once from the shape of the tail. In the Cod (Fig. 2) the tail is flabby and square-cut at the end. In moving, this kind of tail beats from side to side and the whole body undulates, even the head. The resistance the Cod presents to the water must be considerable, and it is not to be classed among the most rapid swimmers. In the Herring (Fig. 1) the tail is forked, and from this as well as the position of the fins we may conclude that the fish swings or twists its body in moving. In the Mackerel family (Fig. 62) the tail fin is more deeply forked and looks for all the

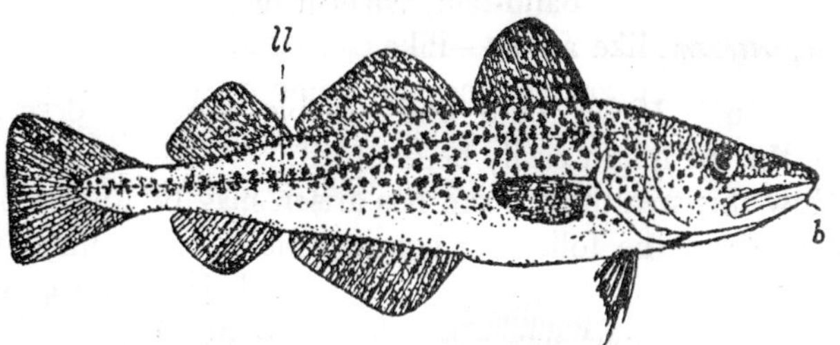

FIG. 2.—Young Cod (*Gadus morrhua*). *ll.*, lateral line; *b.*, barbel. (From Ehrenbaum.)

world just like a propeller. We find on dissection also, that the different vertebræ are so linked together, that the whole apparatus forms a shaft and screw operating directly on the skull and driving the fish forward with great velocity. Judging from the ease with which the Bonito keeps up with ships, its speed has been estimated at 16 to 20 knots per hour. In the Sharks and Dog-fishes we find another sort of tail. The upper lobe of the fork is very much larger than the lower and moves like a scull from the stern of a boat. Here also the body and head must rock and swing from side to side, and the speed of most sharks is probably considerably less than that of a Bonito. As will be mentioned later, even smaller fishes than the Bonito are able to keep up with and attach themselves to a shark.

The compressed torpedo form is only one of many we find among the fishes. A partial list of the others with some examples would appear as follows :—

Anguilliform, shaped like an Eel—Eels, some Blennies, Lycodidæ, Lung-fishes ;

Depressiform, flattened from above downwards—Angler, Rays ;

Disciform, flattened from side to side—Flat-fishes, Scorpididæ ;

Globiform, rounded or spherical—Sun-fish, Porcupine fish ;

Aculeiform, needle-shaped—Pipe-fishes ;

Taeniform, like a ribbon or very elongated—Oar-fish, Band-fish, Macrurids ;

Sagittiform, like a dart—Pike ;

and so on. If we could find suitable words to describe them, the list might easily be doubled.

How this wealth of form has arisen may be understood perhaps from the following considerations. The aim of a fish is to move, and as quickly as possible. To this end it has developed the muscular region of the tail, thus restricting more and more the space for the digestive organs. Let us suppose that the centre of balance is about the middle of the fish to begin with, then these muscle-changes will bring it further forward near the head. This in itself means a change of form. The tail may become very long or asymmetrical in the effort to raise and push forward the heavier front end of the body. The latter will wobble from side to side and instead of going faster, the fish will

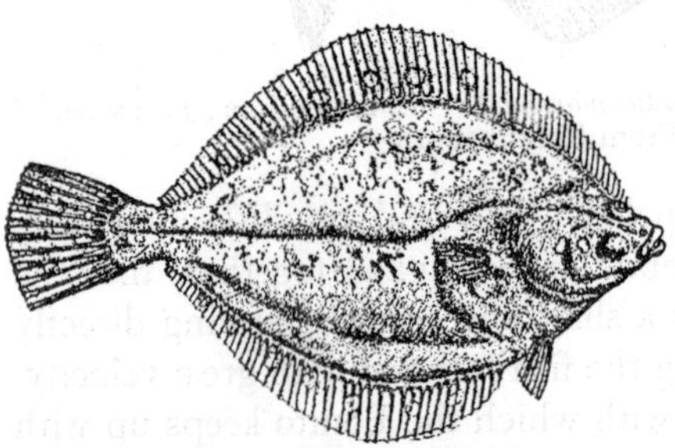

FIG. 3.—The Plaice (*Pleuronectes platessa*), a fish that lives on the bottom, in sand or mud. (From Ehrenbaum.)

make slower progress. Then it gives up the effort to swim faster and takes to wriggling along the bottom, where it finds plenty of scope for its restricted abilities and changes its shape in various ways according to its mode of life. Or it manages to obtain a stronger framework with a more compact body, and is thus able to drive its way through the water, in spite of the greater weight being at the front end.

Various accidents from time to time have also helped to change the form of fishes. It is highly probable that the earliest fishes had an air-bladder, for example, and that this had an opening into the gullet or stomach, as in the Herring and Salmon of the present day. But the restriction and consequent twisting of the digestive organs led to the loss of this opening, sometimes even of the whole air-bladder. The fishes could no longer rise and fall in the different layers of water, as the Herring does, and were obliged to restrict their activities to particular zones, where less movement again modified their structure.

FIG. 4.—Young Frog-fish or Angler (*Lophius piscatorius*), developing its broad head and mouth. (From McIntosh.)

An interesting thing is, that the great majority of the marine fishes have lost this open communication to the air-bladder, whilst the freshwater fishes have retained it, with but few exceptions. And monstrous forms, such as we find in the depths of the ocean, are not to be found in freshwater. The conditions in the sea, especially of pressure and oxygen-supply, present greater extremes, which would account for the greater variety of forms living there.

Some idea of this variety may be obtained from the accompanying figures and others elsewhere in the book. A few more particulars may be given of the different parts of the fish.

In the commonest fishes, Cod, Herring, etc., the mouth is anterior, at the front end, but sometimes it is drawn out from side to side as in the former, in others pointed as in the latter. It may be pointed upwards as in the Star-gazers (Uranoscopus) or downwards as in the Mormyrs of the Nile. Sometimes the jaws protrude forward in a snout with a tiny mouth at the front end as in the Pipe-fishes, sometimes they are drawn backwards until the head seems all mouth, as in the Angler or Frog-fish (Lophius) and Gastrostomids (Figs. 4 and 10).

The natural position of the mouth is in front, as it was in the earliest fishes, but in many the jaws have been pulled down and backwards and a snout has grown out above. Thus, in the Sharks and Rays the mouth is always ventral and drawn out sideways in a crescent. As will be seen later, we are able to connect this queer shape with the mode of life and movements of the early embryo. These are cartilaginous (gristly) fishes with an impressionable structure, but even some of the bony fishes have the mouth in a ventral position, the Sturgeon, for example, the Macrurids, allies of the Cod, and the Common Sole. In the Plaice (Fig. 3), the mouth opens sideways, a peculiarity shown by some other flat-fishes but not by all.

Round the mouth there may be fringes of skin or barbels (*b* in Cod, Fig. 2), which have a tactile function, that is to say, without seeing their food these fishes can " feel " for it. This is specially noticeable in the case of the Cat-fishes (Siluroids), which have a number of sensitive barbels round the jaws, and can be seen to make use of them when living in aquaria. The Red Mullet has a couple of barbels on the throat, whilst other fishes, like the Blennies and Gurnards, use their fins for that purpose.

In the common fishes the nostrils lie above the snout, each with two openings as a rule, but sometimes only one. In the Sharks and Rays, however, the nostrils are the deep openings on the ventral surface just in front of the mouth. There is no internal opening in these, but in the Lung-fishes (Dipnoi) the nose communicates with the mouth.

The eyes in the common fishes are usually very large, larger than in the Sharks, Rays, and Lung-fishes, and this can also be connected with the mode of life of the young stages. Sometimes they are close together on the one side, as in the asymmetrical flat-fishes (Fig. 3), or on the top of the head, but as a rule they are separate, and the fish may be said only to have " monocular " vision. In the fishes that live in caves or at great depths in the ocean, the eyes have degenerated, partly from want of use, but more especially because they require the rays of light to develop the pigment behind the retina, without which they cannot see.

At the side of the head, behind the mouth, is the large and movable gill-cover (*g*, Fig. 1). On the inner side of this lie the gills in a large chamber below the hinder part of the skull. The gill-cover is supported by a number of bones and moved by muscles which cause it to open and shut as the water passes through the mouth and over the gills in breathing. Sometimes the external opening is reduced to a small slit above (Pipe-fishes) or below, as in some Eels. In the Sharks and Rays there is no gill-cover, and the gill-slits or openings can be seen externally just behind the head.

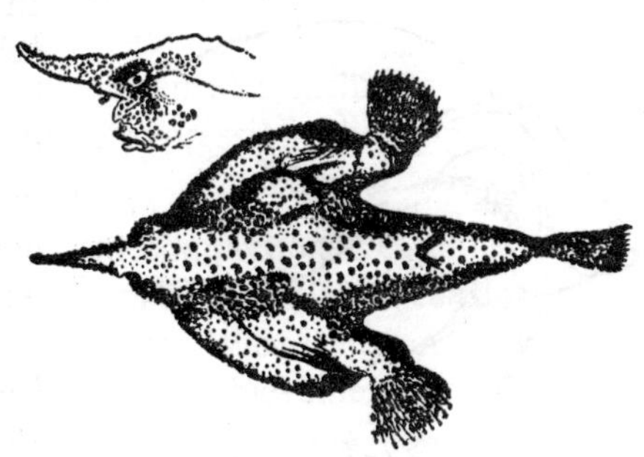

FIG. 5.—The Bat-fish (*Malthe vespertilio*) belongs to a deep-sea group of fishes, but lives in shallow water about the West Indies. (From C. et V.)

The head itself is the most variable part of the fish; naturally, for it comes under more varied influences. The body moves it forwards or swings it from side to side, whilst the body muscles also act more directly upon the jaws, and thus on the front part of the skull. And as the head moves it comes under the pressure of the surrounding water, which has helped to mould the shape according to the rate of movement. What remarkable heads can arise under these influences may be seen from several of the figures.

In the body the most characteristic structures are of

course the fins. These may be all round the body, from the back of the head or even starting on the head at the snout, along the back (dorsal fin) to the tail (caudal fin), then along the lower edge (anal fin) almost as far forward as the head. These are called the marginal, median, or vertical fins. Potentially, every muscle segment of the body may contribute to the support of the marginal fins. But, owing to the movements of the body and the interference of the abdominal cavity below, only parts of the fin develop. Thus, in the Herring there is only a short dorsal fin about the middle of the body, whilst the Cod has three, and the Plaice has a single dorsal extending the whole way from head to tail.

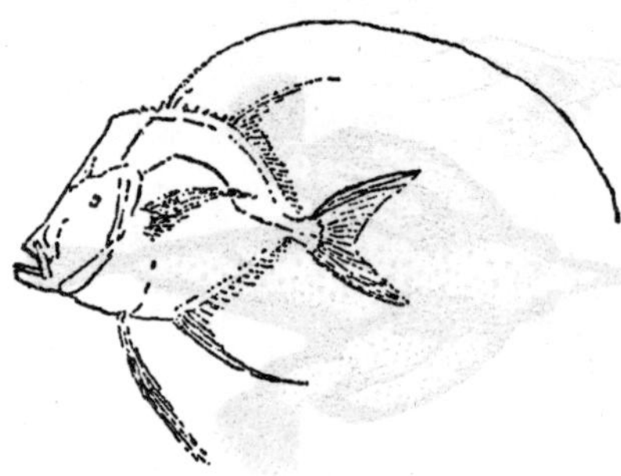

FIG. 6.—A Carangoid (*Argyreiosus vomer*) from the West Indies, with the head depressed downwards. (From C. et V.)

In the great majority of fishes the root of the tail is left free; owing to the strong movements there the fin rays are not able to develop. The anal fin is usually single, but again, as in the Cod, there may be two fins, each opposite a corresponding part of the dorsal fin. The caudal fin may be quite absent, as in some of the Pipe-fishes.

In addition to the marginal fin or fins, the fishes also have what are called the paired fins; the pectoral fins (*P*, Fig. 1) on the shoulder girdle just behind the head, sometimes low down, sometimes high up on the side, and the ventral or pelvic fins, as they are called by many authors (*V*, Fig. 1), anywhere along the ventral margin from about halfway to near the mouth, but most often below or behind the pectorals. These paired fins in fishes have arisen just like the marginal fins from buds pressed out from the muscle segments, but their development is more restricted owing to the interference of the abdominal cavity. The ventrals especially suffer a great deal from the changes in the abdominal cavity and frequently, as in the Eels, never develop at all. The pectorals are more constant in their appearance and as they extend up the sides with increasing growth of the caudal region, they sometimes become of huge dimensions, as in the Flying-fishes.

Apart from the caudal fin, the service rendered by the fins is for the most part a passive one. They help to maintain the vertical balance when the fish is resting or not swimming rapidly. When the Tunny is moving, the fins are folded

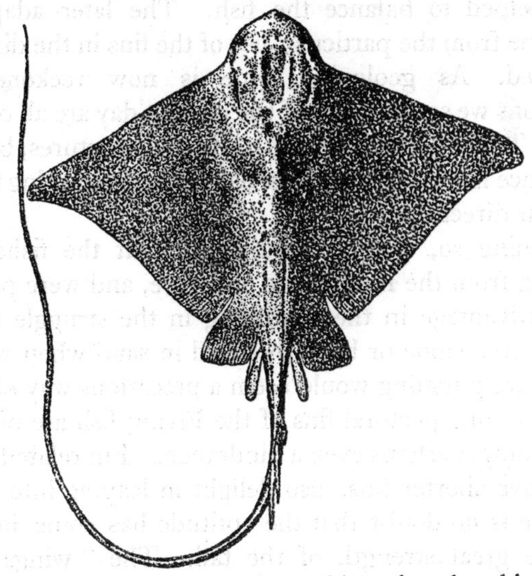

FIG. 7.—The Eagle-Ray (*Myliobatis aquila*) reaches a breadth of 10 feet; it is said to fly rather than swim in the water. (From Smitt.)

back out of the way. The pectorals and ventrals also, though they look like and function as side-keels, are not altogether essential. The fish can manage to balance itself without them so long as the lateral line is intact.

Owing to this moderate usefulness, so far as the balancing is concerned, the paired fins have been adapted to a number of other uses. The pectorals are employed not only in flying or parachuting, but for paddling or scrambling, feeling the ground, holding on to stones, keeping the water in motion over the eggs and even to go backwards. The ventrals, being nearer the ground, are still oftener used as tactile or holding organs.

These adaptations will be described more fully in succeeding pages. The significant point is, that the fins were not made for these purposes. They arose from the beating of a soft, muscular body against a resisting medium, first to the one side, then to the other. The skin-fold or membrane thus pressed out along the margins then became occupied in various parts by horny fibres or scaly rays, which helped to balance the fish. The later adaptations have come from the particular use of the fins in the directions mentioned. As geological time is now reckoned, the adaptations we see in fishes of the present day are all of recent origin. These fishes have not made the structures, but their intelligence has been at work in using and developing them in particular directions.

In doing so, we can readily see that the fishes were departing from the normal use and type, and were probably at a disadvantage in the beginning in the struggle for life. To cling to a stone or bury the head in sand when watchful enemies are pursuing would seem a precarious way of escaping. The long pectoral fins of the Flying-fish are of no use in swimming, perhaps even a hindrance. But related species which have shorter fins, also delight in leaping into the air, and there is no doubt that the aptitude has come in reality from the great strength of the tail. The "wings" have been developed in the air, from the efforts of the fish, and not in the water.

When one part changes, others change with it. If one fish takes to swimming rapidly or flying, another to clinging on to stones or crawling on the bottom, it is clear that the moulding pressure of the water must react on the form in many different ways. To be able to grasp the complexity and variety of the forms, let us think of an intermediate type, a moderate swimmer of comparatively simple body structure, like the Sprat or Herring, and endeavour to picture how, by coming under different conditions, fresh or salt water, pressure, etc., or by accidents, and then by using the structures in different ways, such a moderate swimmer became modified and gave rise to all the different forms of the present time. It is fairly certain that these pelagic fishes of moderate swimming powers have remained more constant and similar to the earliest, most primitive fishes than any others.

3. Skin and Coloration of Fishes

Apart from the shape of a fish, the characters that impress us most are its slipperiness, the bright scales, and the coloration. Taken together these are more than a simple covering, they make up a dress. Every fish has a covering of some sort but there is an infinite variety in the make-up. Still more, what is good enough for one sex or season does not do for another. Some fishes are drab all their lives, others become brilliant at least once a year, and others again are brilliant all the time. In beauty and play of colours, and in the control of colour, the fishes are superior even to birds and insects, yet it must be admitted that our common fishes are not of this kind ; to confirm the picture one must go to the tropics or visit a well-equipped aquarium.

The slipperiness of fishes comes from a secretion of the skin, which varies greatly in quantity and probably in quality in different species. A single Hag-fish placed in a bucket of water soon turns it into a thick jelly of a whitish colour, beneath which it hides itself. The mucus is secreted by cells of the outer layer, or epidermis, which form pits or glands opening to the surface. It is a curious thing, worth noting, that all the sense organs have been derived in the same way from the epidermal layer. They are in fact glorified mucus glands. The latter may be distributed all over the body and head with a mass of nerve or sense cells at their base, as in the Sharks or Rays.

In this way has arisen the characteristic lateral line system of fishes and frogs, which was formerly thought to be just a connected series of mucus glands. The lateral line can readily be seen on most fishes (*ll*, Fig. 2) but not in the Herring or its allies, where it is absent except on the head. It frequently runs along the mid-line of the side, but by no means always, and sometimes there are two or more lines. The line really consists of a canal or tube sunk into the skin, and opening to the exterior by a series of pores, which can be seen when a scale is examined. There is a group of sensory cells beneath or near each pore, and these serve to give the fish impressions of minute differences in the pressure of currents of water. It is believed, for example, that the blind cave-fishes can tell when they are approaching a rock or wall, simply by the wave of their own movement reflected from the stone.

The "hearing" organ is a more complicated development of the glandular sense-organs of the skin. It responds to more violent shaking of the water, but it is not believed that the fishes can really "hear" as we do. Like the lateral line system it is of use in the balancing and orientating of the fish. Inside the organ are several "ear-stones" or otoliths which probably serve as dampers on the waves set up in the endolymph, or secretion of the ear, by disturbances from the outside. By means of these otoliths it is possible in many cases to determine the age of fishes, if they are not too old (Pl. II). The hearing organ (statocyst) lies on the side of the head behind the eye and has no external "ear."

The olfactory organ has been formed in the same way from the glandular organs of the skin; perhaps the eye also. In some mysterious manner the glands fold or sink into the underlying tissues, then become shut off to varying degrees whilst remaining or coming into contact with the central nervous system, which of course is also formed from the epidermis, likewise in the same mysterious way. The development of the various parts depends to a large extent on the permanence of the surrounding influences. It has been found in the case of the eye, for example, that if magnesium salt is mixed with common salt in the water containing developing embryos, the lenses and other parts of the two eyes tend to fuse together into a single median or cyclopean eye.

The scales are likewise products of the skin's activity, and are formed in two different ways. In the Sharks and Rays the skin is, as it were, blown out into minute papillæ or cones. The outer layer or epidermis of the cone becomes hard like enamel by the deposition of chalk, whilst the dermal part inside forms dentine with a central pulp cavity containing the blood-vessels and nerve. These give rise to the rough denticles or placoid scales, which are not pleasant to meet with on a Ray or Dog-fish. The teeth are formed in a similar way. The base of the cone spreads out in the shape of a plate, but leaving a hole in the centre for the blood-vessels.

A few of the bony fishes have similar denticles, and these may also be present in the larval stages, but as a rule scales are formed as simple plates in the inner layer of the skin, the dermis. That is to say, the protruding denticle or spine is not formed. Some of the earliest known fishes (Dipnoi) had scales of a similar kind. As the plate grows, it may form roughly four-sided scales (Ganoid) fitting into each other, but in most of the bony fishes of the present day it is circular or oval (cycloid scale).

The difference between the rhombic and the cycloid scale most probably represents a difference in the movements of the fish. The scales are closely connected with the underlying muscles and the contractions of the latter have naturally a tendency to wrinkle the skin into circumscribed areas. A change in the disposition of the muscles, for example, if the fish travels faster, would alter the shape of these areas. In this way it is supposed, the rhombic scales gave place to the cycloid scales, which permit of greater freedom of movement. At the same time the unequal pull of the different parts of a muscle would lead to the scale or plate lying obliquely, and thus, as it grows, to the hinder end appearing through the skin. Sometimes the scales lie embedded in the skin ; at other times, as in the Herring, the covering of the scale is very thin, and the scale seems to be lying on the surface.

As the scale grows, it assumes various shapes from almost

circular to a long oval. Further, the edges of the scale may grow unequally, and rows of small points then project from its surface, which when they harden are sharp and rough to the touch (ctenoid scale). This roughness sometimes affords a ready means of distinguishing different species; thus the Dab is rough, where the Plaice is almost always smooth.

Since the scales grow with the fish, and the fish grows unequally at different seasons of the year, the markings on the scales have been used in many cases to determine the age of the fish. When the fish is growing rapidly, as in the spring and summer, a large number of rings or circuli are laid down in the scale, but as the growth slows down in the autumn or ceases in the winter, the circuli become fewer in number and closer together (Pl. II). These winter rings or annuli may thus be said to mark off a year's growth, though, strictly speaking, they only represent a certain physiological condition of the fish.

In favourable cases it has been possible to count up to fifteen of these annuli; but it must be admitted that there is often a great deal of uncertainty in the matter owing to lack of definiteness, which again may be due to unequal growth even in the summer months. But on the whole, the scales have provided a valuable means of comparing the rate of growth in different years and also the composition of catches from different areas, especially in the Herring, and the method has been largely employed in the international fishery investigations. The scale is a definite, small proportion of the total length of the fish; if the fish in any one year grows, say, half its former length, the scale will show an annulus about half as broad as the total of the earlier annuli. Hence from the breadth of the annulus, it is possible to say whether the fish has had a good year or a bad year of growth at any time in the past. This method has been developed especially by the Norwegian observers (Hjort and Lea).

In some cases the scales are never developed (Silurids) or they may remain hidden in the skin, as in the Eels. In the

PLATE II

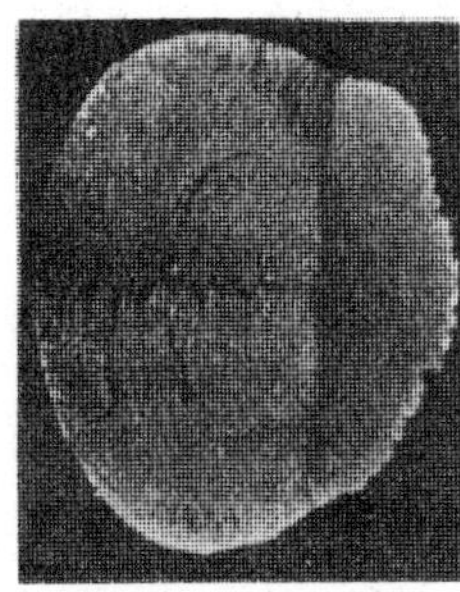

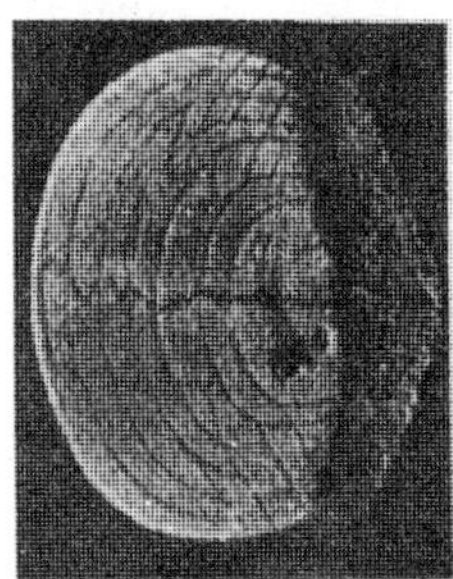

Scales of the Herring.

Photographed to show the annuli; the scale on the left is from a small Herring 18 cm. in length and has only one winter ring; that on the right is from a large Herring 31 cm. in length and has eight winter rings. (From Lea.)

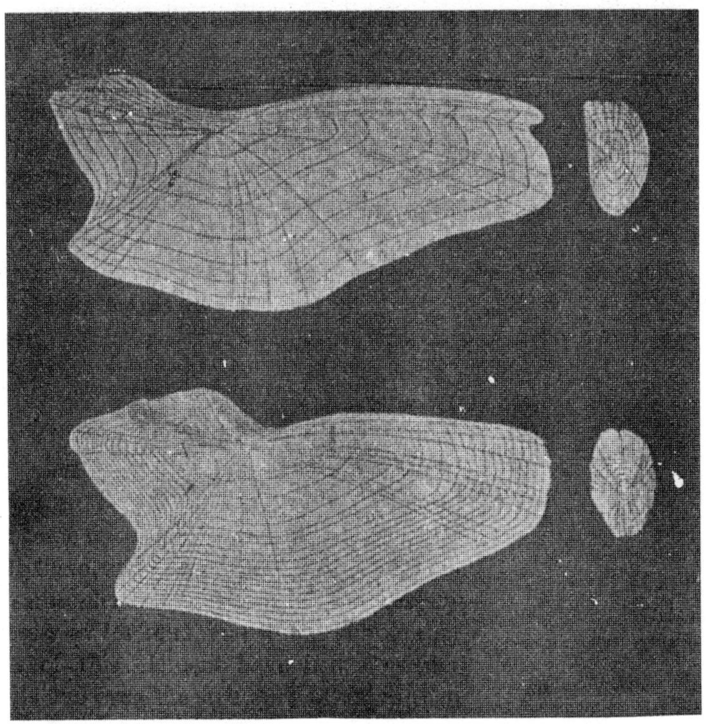

Interopercular Bones and Ear-stones (*otoliths*) of Plaice.

Showing lines of growth. (From Heincke.)

latter they do not develop until the seventh or eighth year (Ehrenbaum and Marukawa). But even when scales are absent, and the otoliths do not show the annual rings very clearly, it is still possible to tell the age of a fish from other structures, the vertebræ in some cases or bones of the gill-cover (Pl. II). The different parts of a fish are not equally developed in all species, and one has to search for the most suitable structure.

Coloration.—The scales, being formed of a bony substance, do not contribute much to the colour, as we can readily see on rubbing them from the sides of the Herring. Yet by reflecting the rays of light at different angles, especially when they are large as in the Salmon, they add to the silvery or glassy appearance of the fish. The large scales, about two inches across, of the American Tarpon are made into ornaments.

The real colouring matter lies beneath the epidermis in two layers, an outer and an inner or deeper layer (Fig. 8), each consisting of a large number of pigment-containing cells or chromatophores. These are formed from the connective tissue cells of the dermis and are usually branched. The colouring matter within may be red, yellow, or black, and perhaps blue (v. Zeynek); each cell keeping to a particular colour. When blended in different proportions other colours result, such as green or brown, and various markings are formed by differences in the amount and position of the pigment in the two layers. These chromatophores have also the peculiar power of concentrating or spreading out the pigment within the cell, apparently by stream movements (Fig. 8), and this in itself means a change of colour. For, if at one moment the chromatophores are expanded and then the pigment contracts, the whole appearance of the fish changes, a black spot becomes a grey and the red or yellow spots become fainter or almost disappear. This power is under the control of the nervous system, so that many fishes can change their colours to suit their surroundings. How this is accomplished will be described later.

Tropical fishes owe their wealth of variable colours to

these chromatophores, but even in the temperate zones we can find some fishes with brilliant colours, for example, the Wrasses and others which live among the seaweed, and the Gurnards which live in deeper water. In the summer months also, the Red Mullet appear among the bluish-grey Whiting. On the whole, however, the temperate forms do not possess such quantities or variety of chromatophores

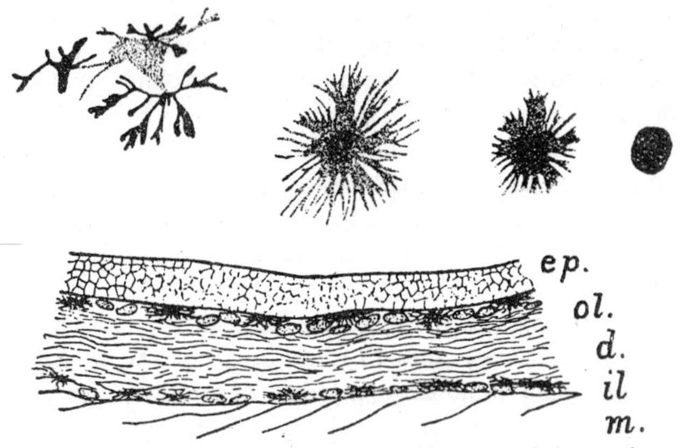

FIG. 8.—Pigment cells and layers. Above, a chromatophore of the Herring in four stages of expansion and contraction; below, a section (semi-schematic) through the skin of the Turbot (*Rhombus maximus*); *ep.*, epidermis; *ol.*, outer layer of chromatophores and iridocytes; *d.*, connective tissue of dermis; *il.*, inner pigment layer; *m.*, muscles. (From Schnakenbeck.)

as the tropical forms. Their colour comes from a different source.

Among the chromatophores in both layers of the skin are a number of small grey plates, which are usually of a rounded or polygonal shape. These are very opaque, and have a strong reflecting power; hence their name iridocytes. According to the way the light is reflected from them they appear white or a bright silver. They are composed of guanin, a waste product that should escape from the body like other urates by means of the kidneys. But the latter organs are not so well-developed in the fishes as in higher vertebrates, and thus the guanin and lime become deposited in the dermis. In the beginning bone was also a waste product, that is, the lime salts were deposited in various places of the body instead of being excreted to the exterior. And in common with guanin lime is often present in the internal tissues of the fish, as in the retina of the eye, peritoneum, air-bladder, etc. We thus see how the waste products of metabolism have been made use of by the fishes in building up their structure and producing their colours. The pigment of the chromatophores had probably a similar origin, and we may note that when a part of the flesh is injured, either from without or by the unequal working of the muscles on the skin or other structures, black pigment assembles round the parts. The leucocytes or scavengers of the blood are removing the noxious particles, converting them into substances that appear to us as pigment and placing them in the skin where they can do no harm.

The distribution of the chromatophores and iridocytes in the skin of the fish is very significant. Thus, in the common fishes like the Cod, the darker colour of the back is due to the great abundance of the chromatophores, which of course come under the influence of the rays of light; the sides are more greyish or silvery, the chromatophores becoming less abundant whilst the iridocytes form a more connected layer. On the belly, lastly, the chromatophores are quite absent and the iridocytes form a dense, continuous layer, called the argenteum. This is in the deeper layer of the skin and gives it the dead white or silvery appearance, while the iridocytes of the outer layer produce the iridescence.

This distribution is in keeping with the view, that the pigment is essentially a waste product stored up in the cells of the dermis. Where the rays of light are able to operate on the cells the pigment stuff becomes transformed, just as the green colouring matter of plants requires sunlight for its development. Where the rays are more vivid, as in the tropical and shallow waters, the coloration is more brilliant; where the rays do not operate, as on the under side of fishes, especially flat-fishes and Rays, there is nothing but the dead whiteness of the argenteum. The blind cave-fishes have also no "colour."

Whilst the origin of the colours might be thought to be a sufficient explanation of their existence, many authors have gone further and ascribed to them a purpose or usefulness of great importance to their possessors, either for "concealment, aggression, or protection." Of recent years, however, a strong reaction has set in against this somewhat too easy method of interpreting the colours and structures of fishes, indeed of all other animals as well. It is very doubtful if we have any logical right to ascribe "purpose" to these phenomena; all we can say is, that in many cases the fishes seem able to make use of their particular coloration to adapt themselves to the surroundings.

4. Size and Age of Fishes

The smallest fish known, and probably the smallest vertebrate in the world, is a tiny Goby which lives in Philippine waters. It is only 12–14 mm. long, about ½ in., and has a name longer than itself, *Mistichthys luzonensis*, Blk. Some other Gobies, *Aphya* and *Crystallogobius*, which live in European waters, are but little longer, the latter being about 1½ ins., the former 2 ins. And a number of the tropical Carplings (Cyprinodonts), which live in tropical countries and feed upon the larvæ of dangerous mosquitoes, are just about the same length. Many of the brilliantly coloured fishes that live among the coral reefs are only 3 to 4 ins. long.

The largest fishes are found among the Sharks and Rays The largest is perhaps the Basking Shark, which lives in sub-Arctic waters, but may wander south into the Atlantic and Pacific. Its length is about 40 to 45 ft. and the more dreaded Carcharodon of tropical seas, Australia, and New Zealand, is said to reach the same length. But the most extraordinary monsters are the Eagle-Rays. The Vampyre Ray of the West Indies may reach a width, across the wings, of 20 to 25 ft. (Fig. 65). If we can judge from the size of the teeth, however, some of the sharks that lived in earlier days (Chalk Period) must have been giants in comparison with which even the Basking Shark is a dwarf (Pl. III).

PLATE III

Mouth of Gigantic Fossil Shark.

(From the *Journal* of the American Museum.)

The Bony fishes or Teleosts never attain such dimensions. The swift Tunny may reach 10 ft. and the Sword-fish with its sword about 12 ft. The freshwater Catfish (Silurus) of the Danube also reaches a length of 10 ft. or more, and a weight of 400 lb. The largest salmon on the official records had only a length of 4½ ft., and a weight of 70 lb. On the other hand, the Arapaima gigas of South America reaches a length of 15 ft. with a weight to correspond.

The most remarkable fish in many ways is Regalecus or

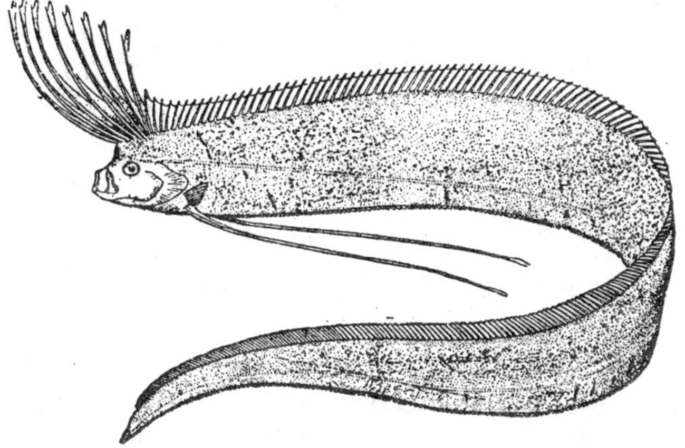

FIG. 9.—The King of the Herrings (*Regalecus glesne*) has perhaps been taken for the sea-serpent. (From Jordan and Everman.)

the " King of the Herrings," which occasionally drifts into the littoral waters of Northern Europe, and has been caught close to the coast of North-East England. It reaches length of 20 ft., one example is said to have been over 25 ft., but the peculiarity is that its thickness does not exceed 4 ins., whilst its height or depth is only about a foot. Hence the fishermen have called it the Oar-fish, and we need hardly wonder that it has been seen lying on its side. When it wriggles through the water with the first dorsal fin or crest on its head above the surface, it may have given rise to at least some of the stories of the Sea-Serpent.

The age of fishes is as diverse as their size. The small

Aphya is believed to live for only one year, and a small Scopelid living in the Mediterranean is said by Tåning to live no longer. Some of the small flat-fishes (Arnoglossus and Bothus) do not seem to live more than four years. On the other hand, Heincke from a study of the bones of Arctic Plaice came to the conclusion that they may live at least sixty years and possibly longer. The Plaice of the North Sea seldom get the chance nowadays of living to thirty years. Under natural conditions there seems no reason why some fishes, like the Sharks for example, should not live a hundred years or more, but we do not know what puts a limit to the age. Old Cod frequently develop a crooked mouth, and that may put a stop to their feeding, but there may also be some kind of senility among fishes. The Herring, in spite of constant persecution from all sorts of animals, has recently been found to reach the quite respectable age of twenty years.

5. Organisation

The belief that fishes are cold-blooded is only comparatively true. We can hardly believe that the energy spent in moving and working the muscles does not produce a considerable amount of heat. And as a matter of fact, the temperature of the blood in fishes investigated has always been at least half a degree above that of the surrounding water. In some cases a much greater difference has been found; in the muscles of the Tunny temperatures of 10–12° C. above that of the water have been recorded.

The fishes, as a rule, do not have to work hard for a living, and this may account to some extent for the comparatively low temperature, though it should be noted that we have but scanty information on this difficult subject. The scaly covering or, where this is absent, the thick, leathery skin, is a sufficient protection against the loss of heat or the penetration of water into the tissues. The muscles especially are so protected, and this makes an essential difference from the lower animals. The fish is not permeated by the environmental influences, it stands apart or above them and controls its own life-processes on a higher level.

Apart from the temperature, the constitution of the blood, that is, the quantity of water it contains in addition to salts and organic compounds, is regulated mainly from within and not from without. Thus, it has been found that the blood-pressure, though equivalent to 8 to 10 atmospheres, is far below the pressure of the surrounding water, except in the cartilaginous fishes, where a peculiar condition prevails. This blood-pressure is much the same as that found in man and the higher vertebrates generally, and depends upon the power possessed by living membranes of selecting and excluding the various ingredients, gaseous or otherwise, in the water.

The processes of life, metabolic changes, that go on within the fish are thus distinct from the surrounding processes. Each fish is a little world encompassed by a different world. The food that it takes in has to be transformed by the digestive changes to suit its needs in quite the same way as in higher vertebrates, but the details of effecting this may be very different. The digestion of food proceeds along the same lines, but the apparatus, the retorts for distilling, circulating, and separating the ingredients are in an elementary condition. The heart with its two or three chambers, though a great advance on anything seen among the invertebrates, makes little or no distinction between venous and oxygenated blood. The purification of the latter takes place in the gills after it has passed through the heart, and it is only in the lung-breathing fishes and those with an additional breathing organ—apart from the gills—that the head and brains receive a purer supply than the rest of the body. The brain power and psychic life of fishes, as of higher animals, has evidently depended on the distribution of oxygen by way of the blood to the central nervous system.

In one other respect also the fishes are far below the higher vertebrates. The organ for the excretion of waste liquid products, the kidneys, is at a much lower stage of evolution and is possibly less efficient. Compared with amphibians and reptiles it has not the blood-vessels, known as the renal-portal system, which convey the used-up blood from the hind limbs or caudal region through the kidneys, and the result is that the blood is not cleansed of the urates and lime salts to the same extent. As mentioned previously, these are deposited in various parts of the skin and body, and it is probable that bone has arisen in the same way. This is not very remarkable in itself, when we remember how low in the scale of vertebrate life the fish stand. They are not far removed from the invertebrate series in which the kidneys were of a different and more primitive nature. In fact, as we shall see, the excretory system of the fishes remains in some ways in the same condition as in the earliest stages of animal life.

With the imperfect oxygenation and the clogging of the blood through the waste products, we may connect the low level of psychic life shown by the great majority of fishes. This can be done with greater assurance, since we find that the forms which have made advances in the one direction or the other, display a higher level of intelligence and, apparently, reasoning powers. At least, their actions reveal many of the mental attributes we associate usually with a much higher stage in evolution—higher so far as the structure is concerned.

CHAPTER II

THE HABITS OF FISHES IN GENERAL

If peace and quiet rumination mean happiness, we should follow the example of Isaak Walton and spend some time occasionally in observing the lazy Brown Trout watching us from the bottom of a deep pool or the Salmon in an eddy waiting for its leap. There is a fascination in the ways of animals which, if we allow it to sink in, takes us out of our everyday selves, perhaps because we see in them our own reasoning powers. And the different medium in which fishes live is not a hindrance to our understanding them. In an aquarium we can watch their ordinary movements, how they breathe, feed and rest, even test their intelligence by experiment, tame them or encourage them to fight. From this it is but a short step to try to understand how they live under natural conditions.

1. Haunts of Fishes

Fishes are adapted to live in the water, and the great majority of the twelve to thirteen thousand known species do so, but what this means can be understood best by considering the contrast, that some fishes live quite a long time without water surrounding them, and many others take excursions on the land. The same element, oxygen, is as necessary for their living processes as for ours, and it is probable that the varying amounts of this element have had a greater influence on the habits and even the structure of fishes than we yet know. Deep-sea fishes which live in waters poor in oxygen manage somehow to fabricate a store for themselves, and keep it in their air-bladder, whereas the fishes that live in freshwater, with ready access to the surface and the air, have no store of oxygen. Some come to the surface, like the mammalian porpoise, to get rid of the used-up gases and take in air, others take in gulps and blow them out as bubbles, and others again must have access to the surface waters, where there is more oxygen, or they perish, even in the water.

We can thus understand why many fishes can hop or squirm about on mud and land; one or two even climb branches of bushes. So long as they can retain moisture to cover their gills—and our lungs also require moisture—they can breathe quite well in the air. On a damp night the Eel has been followed many miles over land, travelling so fast in the grass that the observers could hardly keep up with it.

It may be that this greater thirst for oxygen, and ability to satisfy it, accounts for the greater liveliness in general of the freshwater fishes. But if they live amongst the mud and reeds, as some do, they become too large and lazy to be very active. The inshore fishes that live among the seaweed are also lively, moving about with quick intelligent movements, and here again they enjoy a bountiful supply of oxygen given off from the plant life. We have but to compare the behaviour of a little Shanny or Crested Blenny with that of a Cod in an aquarium to note the difference.

But some of this liveliness comes, of course, from the different nature of the food they eat. A Blenny has to deal with the jerky shrimps, and has learnt to keep still among the seaweed until one spring is sufficient to settle with the shrimp. But a Cod grubs along the bottom and picks up the mussels or pursues smaller fishes than itself, like the Herring and Sand-Eel, and takes them in by the gross. Not much intelligence is required for this kind of hunting.

The inshore waters between tide marks are populated mainly by the small fishes which can remain in pools and between the rocks, such as the Blennies mentioned, the Cottoids, Pogge, Needle-fishes, and the Butter-fish or

Gunnel. The young of the larger fishes also play about among the seaweed in their first year, and the larger fishes, three to four year old Cod or Rock Cod, Whiting, Plaice, and others, frequent the shallow water, outside the range of the tides, in the summer time. Where the water is deep enough, for they are too wary to risk less than a fathom of water, the Grey Mullets will enter and leave harbours with the tides. And the Bass in the autumn pursues the Sand-eels right inshore. There is in fact plenty of food in these inshore waters, the richest zone lying outside the ebb marks, but the water there may not be deep enough to suit the bigger fishes.

The pelagic fishes like the Herring, Mackerel, and Tunny, which can swim up to the surface, are of course independent of the depth; but even these live for the most part in deep water from ten down to over a hundred fathoms, and most of the fishes in our temperate zone live at the bottom. The reason for this behaviour may be the food supply. For three parts of the year the upper waters have abundance of the lower forms of life that go to form the food of fishes, but there is a scarcity in the winter. And it may have been just the necessity to adapt themselves to the winter conditions that led fishes to try and keep to the bottom, which again reacted on their structure to such an extent, that most of them cannot now rise to the surface even if they try. Their internal organisation will not permit it.

On the bottom they find a great deal of food, Molluscs, Star-fishes (Ophiuroids), and Crustacea, but there is probably not enough to go round. At any rate they feed for a great part on one another, and the poor Herring especially has a hard time of it. It certainly gets the most and the best of the food, for it is a rover into all layers, but the hungry crowd below, Cod, Green Cod, Angler, and all sorts are in pursuit. In the Norwegian Fjords the Herring are sometimes literally hemmed in by their enemies, and so crammed together that they build up a mountain in the water, "Sild-bjerg," rising to the surface. And there the Sharks and Killer Whales claim their tens of thousands. Man of course takes not a few, but there is some justification—apart from feeding his kind. If the Herring, in spite of all this persecution, is able to live for more than twenty years, the surplus that escape each year must be greater even than the immense quantities taken. If man did not help to keep down their numbers, the Herring would have to turn upon themselves; indeed, they feed upon their own eggs and fry at the present time, as well as consuming large quantities of Sand-eels.

In older days, before the intensive herring fisheries had begun, it used to be said that a man standing on the rocks by the shore could fish up as many herrings as he wanted with a bucket at certain times of the year. The fjords of Northern Iceland until recently were also filled periodically to the same extent. On the western side of the Atlantic, Newfoundland, Nova Scotia, and the States, the same story is told. The quantities of the Herring in the seas are beyond anything we can imagine, and this abundance comes not from their fecundity, which is relatively low, but to their ability to change from one layer of water to another. Other members of the same Clupeid family, Sprat, Pilchard, and Anchovy, have the same ability, which comes from a peculiar disposition of the air-bladder. Having this structure and these habits they have naturally the pick of the food at all seasons of the year.

An interesting thing about the Herring is the effect upon them of the food they eat. If they are feeding upon the oily, vitamine-rich Copepods and Euphausids, they are in splendid condition; but if they take to Pteropods (*Limacina*) their flesh runs to "water" and they are of no use as food or for smoking or salting. Other fishes are also said to become poisonous when they feed upon small jelly-fishes—but perhaps these have been mistaken for Pteropods. The Norwegian fishermen, who know the Pteropod symptoms quite well, pen up the herring shoals until they have got rid of the contents of their stomachs.

The greatest wealth of fishes lives within the hundred fathoms zone, and the great fisheries of the world, in the North Sea, Newfoundland Banks, Sea of Japan, and elsewhere, are prosecuted within these limits. The reasons for this are fairly clear. On the one hand, the coastal waters have a rich supply of plant life, not only in the form of enormous forests of seaweed, but also in the form of floating organisms or phytoplankton—all converting the refuse or waste products from the land and rivers into useful food and oxygen. The result is that the animal life or zooplankton of the littoral waters is exceedingly rich, far richer than out over the oceanic depths, and the young of many fishes as well as the Herring profit thereby. On the other hand, the littoral waters are the meeting-place of diverse currents, from the ocean, from the north, and from the rivers, and the meeting-places seem to be specially favourable to the production of plant and animal life.

In addition to the fishes that move actively about, there are many others that lead a more easy life, resting for the most part on the bottom. Four different kinds can be distinguished. Some are like the normal fishes in shape and yet bury themselves in the sand or mud. Such are the Weevers (Trachinus) with poisonous spines, and the sharp-pointed Sand-eels (Ammodytes) which provide more food to fishes than any other forms. A second group tends to be broadened out or flattened, like the Three-bearded Rockling, the Sea-snails (Liparis), Lumpsucker and Toad-fish (Batrachus), from crawling as close to the bottom as possible without actually burying themselves. And some others are still more broadened out until they seem a cumbersome mass, all mouth and stomach with very little tail, like the Frog-fishes (Figs. 4, 5); or again, all head and side-fins, like the Monk-fish or Angel-shark (Rhina). The last leads to a third group, the Rays, which have the broad side-fins attached along the side of the head, and the mouth under the snout. These all keep the normal position, and have the belly undermost. But the last group, the so-called Flat-fishes like the Plaice and Sole, lie on one side of the body with the two eyes together on the upper side; most of these also bury themselves in the sand or mud.

As a general rule, saltwater fishes keep to the sea, freshwater fishes to the rivers and lakes ; to change them suddenly is to condemn them to death, for their delicate gill-membranes are so nicely balanced to the outer conditions, that a change of water acts like poison. Some fishes, however, like the Smelt, young of the Herring and Sprat (Whitebait), some of the American Killifishes (Fundulus), and many more, live in brackish water, and some, like the Eel, Stickleback (*Gastrosteus aculeatus*) and one Fundulus, are able to tolerate salt or fresh water with indifference. But the most extraordinary thing is, that a few fishes, like the Salmon, make a practice of changing regularly from the one to the other every year when they grow up. A few more, like the Flounder and Eel, spend their youth in the fresh-water, but descend later into the sea. This ability to stand sudden or even slow changes in the salinity is a mystery which has not yet received sufficient attention.

Beyond the continental plateau the water deepens rapidly from 100 fathoms down to 1000 fathoms, and then down to the great depths of the oceans, 3500 fathoms and more. Formerly, it was thought that no fishes, indeed no life, existed at these great depths, but this at any rate was disproved by the *Challenger* and more recent expeditions. It is not always certain that the depths from which fish have been recorded were the actual depths where they lived, as they might have been taken in higher layers when the net was being hauled in. On the other hand, we may be sure that more escape the slow-moving net than are taken in it. From an examination of the records Hjort has come to the conclusion that some fishes (Macrurids and deep-sea Blennies) really live at depths greater than 2000 fathoms. On their expedition across the Atlantic in 1910, Murray and Hjort obtained considerable quantities of fish down to depths of 1500 fathoms. Those from the deeper layers were black in colour (Astronethes and Gastrostomids, Fig. 10), whilst by contrast the numerous deep-sea prawns obtained—some of extraordinary size—were red. But most fishes were obtained in the intermediate layers, that is, living pelagically and not at the bottom (Scopelids, Alepocephalids, etc.).

An interesting thing proved by these investigators was, that many of the fishes from the depths ascend in the water during the night-time and descend again during the day. This shows that they are tuned or adapted to the light as well as to the depth. The light rays descend to considerable depths, more than was formerly thought; the ultra-violet and blue rays being present in abundance at more than 300 fathoms, and even present down to 500 fathoms, but the red rays come to an end at about 60 fathoms.

The fishes that live in the intermediate and upper layers over great depths are very silvery in colour (Scopelids, Argyropelecus, etc.), and the young forms that live near the surface (Leptocephalids, Bothus, etc.) are like transparent

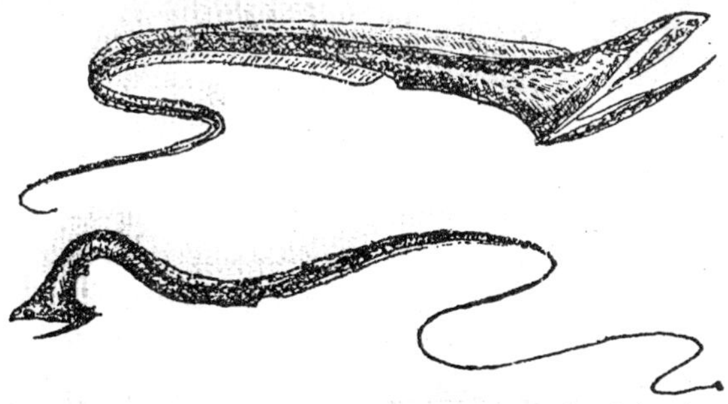

FIG. 10.—Two Gastrostomids from the depths of the Atlantic. (From Murray and Hjort.)

pieces of paper. On or near the surface, too, live larger fishes, the Basking Shark which seems almost the only fish to revel in the sunshine, the Pilot-fishes with their comrade the Sand-Shark, the "Dolphins" (Coryphæna) which pursue the Flying-fishes, Sun-fish (Mola), and many more. On the whole, indeed, there would seem to be no lack of fish life in the ocean, even though it may be scattered.

The dwelling-places of fishes have thus a wide range, from 11,000 ft. above sea-level to at least 14,000 ft. below, a greater range than any other group of living forms, except perhaps Bacteria and Crustacea. To this bathymetric range we may add the horizontal distribution from the Arctic Ocean across the tropics to the Antarctic, and from east to west round the world—in all possible seas and lakes.

The significance of regarding the haunts of fishes in this broad perspective lies in the conclusion it indicates. The difficulties encountered by the fishes in becoming adapted to their surroundings have arisen, in the first place, from the physical conditions, temperature, light, pressure, and the more subtle chemical conditions, salinity and gaseous contents of the water. The habits of rising to the surface to obtain more oxygen, of swimming fast or slow or resting on the bottom, of shunning the light—few fishes like to bask in the sun like lizards—changing from fresh to salt water and so on, are elementary adaptations of even deeper importance than the biological conditions, that is, the relations to the animate surroundings.

2. The Wanderings of Fishes

Among the habits of fishes which are of great importance to an understanding of their biology we must include their wanderings. The population of the sea is mobile in more senses than one, and we can find no parallel among higher animals, not even among those that live in the sea. For this reason, perhaps, it is not easy to interpret the meaning of these wanderings, though the facts have become well-known through the intensive investigations of recent years.

By way of introduction, it may be said that the eggs are spawned at one place, the fry grow up at another or several other places, and the adults wander to and fro. Some of this movement is passive, some of it active and deliberate, and some of it active, it may be, but compulsory, as if the animal were in no way a free agent. Difficulties arise when we endeavour to determine the causes behind these wanderings. To simplify matters, the migrations of the adults for the purpose of spawning may be kept apart and treated separately. During these the fishes are not quite normal and the causes operating are of a different nature from those seen in the ordinary movements.

On the east coast of Scotland the passive drift of the eggs of Plaice and other fishes has been studied by Fulton by means of floats or drift-bottles set free at the grounds where the species were known to be spawning. These bottles passed down the coast from off Aberdeenshire towards the south; some of them stranded on the coast, but quite a number drifted across the North Sea, in its northern part, and landed on the Danish coast, some even reaching Norway. The Plaice eggs, however, take only about three weeks to hatch out, and in that time the drift bottles had only reached as far as the southern counties of Scotland, and there could be no question of the eggs or fry crossing the North Sea. In fact, the young Plaice are found, some three to four months after the eggs are spawned, all along the margin of the shore in shallow water.

There is a certain correspondence therefore between the prevailing currents and the drift of the earliest stages, but when we look more closely into details the correspondence becomes less clear. Thus, other fish eggs are also in the same drift—Cod, Haddock, Long-rough Dab, etc.—yet the young fry of these species are never found on the shore. It follows, that at some time during the early stages the different species separate from one another and go their several ways. When this separation is effected and by what means are problems still to be solved.

Several factors are concerned in the process. Firstly, the eggs, as the embryo within absorbs the lighter yolk, increase in density and tend to sink down from the surface; hence, they are no longer under the influence of the surface drift. A great deal, however, depends on the prevailing temperature; if this is high the eggs seem able to expand and thus do not sink even when the density of the water decreases (Johansen, Strodtmann). Thus the rate of development of the eggs, the nature of the yolk, and the external conditions of temperature and density, all come into play in the separation of the developing eggs of different species, and even of the same species.

Secondly, when the young fry escape from the eggs,

we come to a different set of phenomena. How soon they choose their own way is uncertain, but we cannot regard them any longer as quite helplessly drifting with the currents. And they are now able, by swimming, to overcome the difference in specific density between themselves and the surrounding water—to some extent. Many of the young larvæ swim actively up to the surface, but when they cease to swim, they gradually sink. They must keep on the move to remain on the surface. We may say that on the whole they remain in the body of water surrounding them and are thus carried along by the currents, but their own activity in choosing the body of water has to be taken into account. For example, the fry of the spring Herring make their way from the bottom up to the surface and take about three weeks in their ascent ; the fry of the autumn Herring, on the other hand, do not come up to the surface. They remain near the bottom throughout the winter. In neither case can we explain the phenomena as wholly due to the fry being adapted specially to particular conditions of temperature and salinity or density.

This is also seen clearly in the case of the Plaice. When the fry sink down to the bottom, they ought to remain there, whatever may be the depth. Yet they only occur near the shore, right on the beach. Redeke, it is true, found some one-year old Plaice some distance from the coast off Holland ; but the water there is very shallow and the specimens may have wandered out from the shore. It may be, of course, that there are under-currents in the sea which bring in the young fry to the coast and that the fry of the other species do not come under the influence of these under-currents for some reason or another. But for the present, it seems more reasonable to believe that they actively find their way there, although in size they are only about half an inch long. The fry of the other fishes also choose their own particular region or zone.

Some would say, that the fry are adapted to particular conditions and perish if they do not come there, and for this reason we do not find them except under these conditions.

But even if this explains anything, which is doubtful, there are still some difficulties in the way of accepting it. The Plaice eggs from the Danish Belts carried into the Western Baltic, to conditions quite different from those of the North Sea or the Belts, still manage to develop and the fry appear to grow there remarkably well.

The young Plaice remain for a year or more in the neighbourhood of the shore, and then they begin to wander outwards into deeper water and along the coasts. In some cases, as in the North Sea, this wandering takes them back in the direction towards the spawning grounds of the adults, and in this way they complete a cycle in the course of three to four years. In other cases, however, as in the Channel, this cycle is not so obvious, and the wanderings seem to be directed simply towards where the fish can find the most suitable food. It would appear that some ancestral memory guides them, for we cannot imagine that they leave one bay and travel ten to twenty miles to another knowing that better food awaited them there, when they themselves have had no experience of the fact.

This example will perhaps suffice to indicate, that the psychical element cannot as yet be neglected in endeavouring to understand the wanderings and other habits of fishes. And it does not matter much whether we speak of intelligence or instinct in connection with them; both imply something that cannot be reduced to mechanical laws. One might say that the food attracts the Plaice to particular grounds, and in the case of the old Plaice which return yearly, after spawning, to the same bays, the word has a definite meaning, but when the young have never been there before it is difficult to speak of attraction.

A further interesting example of this psychical element may also be mentioned. After the Plaice have spawned and are out of condition, it has been discovered that they do not return at once to their ordinary haunts and food (Molluscs), but seek out the softer muddy grounds where they can feed on worms; or it may be, that they choose these grounds to suit their colour.

The extent of the wanderings of the Plaice in the North Sea and Channel is very limited. In the latter a few small bays have a population all to themselves ; the Plaice seldom wandering beyond. In the North Sea the areas appear to be greater, owing probably to the wider expanse of shallow grounds. Thus there appears to be only one race or population for the whole of the southern North Sea and another for the northern parts, but even among these there may be local communities.

In Icelandic waters, where the depths near the coast are much greater than in the North Sea, the wanderings of the Plaice appear to have a much greater range. A number of specimens marked by Johs. Schmidt on the east and north coasts were retaken later up to 240 miles away, moving towards the west and south, and the distance actually travelled was probably much greater. In these waters the Plaice spawn, it is said, only on the south and west coasts, and the fry appear to be carried much longer distances round the north to the east coast, probably owing to the slower rate of development in the colder waters. A curious thing, if this interpretation is correct, is that on returning to the west and south coasts to spawn, the Plaice of the north coast travel backwards along the north coast, whilst those of the east take a shorter route along the south. Thus the latter appear to make a complete circuit of Iceland. Whether the adult Plaice make annual wanderings of this extent—just as they wander out and in yearly in the North Sea and Channel—is not clear. So far as the observations go, it would appear that the Plaice of Iceland have adopted quite different habits from those of the North Sea.

In the Norwegian waters the Herring and Cod have also been found to make extensive wanderings. Thus, whilst the spawning grounds are restricted to definite areas of the coast, from Stavanger to Romsdal for the Herring, and the Lofotens for the Cod, the fisheries at other seasons extend along the whole coast from Finmark in the north to the Skager Rak in the south. The eggs and youngest stages are naturally carried away by the currents, some over the deep water in the neighbourhood and many into the fjords (Damas); but the Herring seem to collect into enormous shoals when about two years old, and then make their appearance far away from the spawning grounds. This concentration cannot be wholly explained by reference to the currents. The Cod, Green Cod, and other fishes have a good reason for collecting together, for they pursue the Herring; but what causes the Herring to wander together in such quantities and to such distances?

In the North Sea the Herring has also presented many hard problems to the scientists. It is no longer believed that they make extensive wanderings—at least, not as far as the ice-floes of Greenland as was formerly thought. Nevertheless, countless millions disappear from sight, almost as if they did not exist, and some have imagined that mysterious influences may be behind some of the phenomena, the change in the fishing grounds from bygone years and the seemingly long-period fluctuations in some of the fisheries. For many years now the great spring and summer fishing along the east coast of Scotland and North England has been very constant with no reduction in the quantities. At this time the Herring have concentrated to feed, apparently on the plankton organisms brought in by the Atlantic currents, and it may be that they belong to one vast community or race (Johansen) or to several (Heincke). After feeding they spawn and then disappear once more, but within recent years large quantities have been taken during the summer and autumn months by trawlers fishing in the western and northern parts of the North Sea (Borley and Russell), and later in the winter in the Skager Rak. It appears therefore that these Herring live mainly in the North Sea region throughout the year.

Similar phenomena are displayed by the Herring on all the continental slopes of the Northern Hemisphere. They appear at certain times and places for the purpose of feeding, at others for the purpose of spawning, and it is not always easy to distinguish between the two. Some have thought that their movements are determined by the hydrographical conditions, temperature, and salinity of the water. But it has to be remembered, that during millions of years this pelagic form has become adapted to a much wider range of variations in these conditions than the North Sea and Norwegian Sea can now show, even from winter to the height of summer. And it seems more probable that the wanderings at any rate are simply dictated by hunger and need of food. Formerly it was thought that they were restricted in their diet to the plankton organisms; but it has been shown recently (Hardy), that a very large proportion of their food consists of fish, particularly young Sand-eels. In the winter the latter migrate out from the coast into deeper water to spawn, and here in the spring the young are pursued by the Herring, which again are pursued by the Cod, Haddock and Mackerel, etc. In short, it is probable that the biological conditions, food, and escape from enemies, are now the main determining influences in the wanderings of the Herring.

Many other fishes undertake wanderings, some of greater extent, others much less, but these examples will suffice to show the nature of the problems that scientists in many countries are endeavouring to solve. Final solutions, however, should not be expected; that has become very apparent from the intensive investigations of recent years. Even if all the factors determining the wanderings and abundance or the reverse were known and also their relative importance, they are so closely interwoven that only a close and constant study can tell which is most operative at any particular time and place. For example, it is probable that the removal of large quantities of Cod, Haddock, and other Gadoids by the fishing has had a great deal to do with the Herring's extraordinary abundance in the North Sea during the past thirty years. But in this as in so many other problems connected with the fisheries, we have no measure for comparison with the other factors influencing the quantities of Herring. Fluctuations have been traced right back to the spawning, that is, to good and to bad spawning years; but again what makes a good or a bad spawning year is still a matter of conjecture.

3. Feeding Habits

Those fishes that are constantly wandering about chase their food and are themselves chased; they are mostly flesh-eaters. Some of them are exceedingly voracious and seem to have a very elastic stomach. The deep-sea Gastrostomids (Fig. 10), allies of the Eels, and others can indeed extend their stomachs outside their body and are thus able to take in booty—sometimes of their own kind—larger than themselves. Some few—not only the Sharks—take a pleasure in killing far more than they can eat at a time, and one species combines this lust with a morbid appetite to such an extent, that it is said to eject the contents of its stomach when full in order to go on with its feeding.

Few fishes take time to taste or chew their food, though the Grey Mullets seem to do so, and more than two thousand years ago Aristotle remarked, that a species of Scarus was the only fish that "chewed the cud." As a rule, the booty disappears as rapidly as possible, probably to prevent rivals having a share, with curious consequences in some cases. Eels have been taken with another Eel in their mouth which had wriggled its way between the gills and thus suffocated its captor. And live fishes have been taken in the stomachs of others. It is on record, indeed, that the live booty has on occasion bored its way through the lining of the stomach and sometimes perhaps escaped in this way; and in other cases remained embedded in the tissues.

Some of these active fishes have taken to a less strenuous mode of hunting and feed on the smaller animals of the sea, either altogether or at times. To judge from its size the Basking Shark does remarkably well on the floating and swimming organisms (plankton) near the surface of the sea, and the Herring and Salmon are at their best when they have been feeding on the Copepoda. Deeper down, the Cod and many other fishes pick their food from the bottom, Crustacea, Molluscs, Worms, and some kinds of Star-fishes.

But some fishes do very little hunting. They either lie in wait in a secluded nook until their special food comes along and then pounce upon it, like the Blennies in the rock-pools; or they bury themselves and let the prey come to them. It is a peculiar thing, for example, that the Greater Sand-eel, a sluggish fish, is almost invariably taken with a Lesser Sand-eel in its stomach. More extraordinary still, the latter is always in head-first; when disturbed, it dives seemingly into what it takes to be an opening in the sand and finds too late that the opening is the mouth of its larger relative. The Anglers have improved upon this method. By means of odd flaps of skin and tentacles they rest on the bottom and attract the curiosity of other fishes, which pounce at what seems to be a tempting bait and disappear into a capacious stomach.

Fishes are not always feeding however; some take long rests between meals, and it may be that the rapidity of digestion varies a great deal. The Plaice, which feeds mainly on shell-fish and is of a restful disposition, digests slowly and thus always has something inside it; but the active Cod and Herring are frequently empty. The best feeding times for the fishes of temperate climates are when the tides are changing.

When resting, it is possible that the nerves relax and fishes sleep, but they have no eyelids and it is not easy to say whether the blank stare of a Cod in an aquarium means sleep or not. The Wrasses, however, undoubtedly sleep. This was known to the ancients, according to Smitt, and more recently Möbius and Heincke kept a Wrasse (*Ctenolabrus rupestris*) in a tank and observed that it retired regularly at night-time and went to sleep lying on its side. One of the mouth-breeders tucks its young into its mouth at night-time, and even if they do not sleep, for we cannot well observe what they do, we may consider that they ought to, with such careful mothering.

Among freshwater fishes some peculiar modes of obtaining food have been developed. That many should grub among the mud and eat debris, is not altogether

PLATE IV

TOXOTES JACULATOR

Earns its name and its living from shooting down flies. (From Stansch.

wonderful, for quite a number of the marine forms—Red Mullet, for example—do the like. But a great many have acquired a fine taste for insects and insect larvæ, and this has meant a new kind of hunting for the fishes. If the insects lay their eggs in the water, like the mosquitoes, the small fishes find a rich harvest in the developing larvæ, and in tropical countries have proved of inestimable value in keeping down and even exterminating those dreaded evils, malaria and yellow fever. But the taste for flies has become so great, that one fish has developed into an expert sharp-shooter in stalking and smothering flies—with a drop of water and mucus (Pl. IV). Others again have taken partly to a land life, even to climb bushes, in search of their favourite food.

4. Breeding Habits

The most interesting and important habits of fishes, as of other animals, are those which centre round the propagation of the species. Among the active forms it is the rule in the great majority of cases, that the production and setting free of the sexual products are merely passing incidents in the yearly round, impelling them in many cases, it is true, to undertake extensive migrations, but otherwise of no interest to them. The body becomes charged with these products at certain definite seasons, and they get rid of them under particular conditions of ground, temperature, and depth of water. It is apparently their selection or need of these conditions that makes them undertake the journeys and not any care for the offspring. And we may note, that whilst the same instinct affects both sexes and brings them to the same grounds—thus, it is often said, caring for the good of the species—yet it is not quite true, for the males usually ripen earlier and arrive on the grounds before the females. This is probably the main reason why a large percentage of the eggs spawned in this way have been found to be unfertilised.

A number of fishes, again, lay their eggs on the bottom, on the sand or seaweed, like the Herring and Sand-eel, and having done so pay no further attention to them. It is not certain whether the males follow and attend on the females, but they pass over the same ground as they shed the milt, and it appears that both sexes may partake of the eggs around them. Freshwater fishes like the Salmon and Trout go through a similar but more elaborate performance. The choice of ground is very carefully made, the sand and gravel hollowed out and the male guards the bed from other males and intruders. With the deposition and fertilisation of the eggs, however, the duties of the parents end. But in some other cases the parents cover over the eggs with sand and stones.

" Parental care " is most evident among the sedentary fishes—perhaps as the result of their more sluggish metabolism and temperament. The eggs are shed in a crevice of the rocks, under shells or in hollows made in the ground, and then the male, seldom the female, broods over them and guards them as well as it can against enemies. Then some build nests of various kinds—again almost always the males. In one kind, the seaweed is carefully plaited together by means of mucus threads, secreted from the mouth or the kidneys, and the male then invites or entices or—if his solicitations are not favourably received—drives a female to lay her eggs within. If she escapes through the other side, as the female Stickleback sometimes does, he finds another or several more until he has obtained his quota, then he inseminates the eggs, drives the females and other intruders away, and remains for many days jealously guarding his possessions.

Mating of the sexes occurs in many cases and results in some in the eggs being laid after they have been fertilised, as among the birds, or the whole embryonic and even larval development is passed through within the female. Internal fertilisation is the rule, with only one exception perhaps, among the cartilaginous fishes or Elasmobranchs. And many of the bony fishes or Teleosts have acquired the same habit, we may believe, quite independently. The probable origin of this habit will be discussed later.

Most species spawn only once a year and this may be taken as the primitive condition. Crystallogobius and some of the small Scopelids (Tåning) appear to spawn only once in their lifetime, when they are one year old. The Eels also spawn only once, but they are then much older. The Common Eel, which lives in freshwater whilst immature, may remain there from ten to fourteen years before the ovaries become mature. The reason why these forms spawn only once appears to be, that the whole organ ripens at one time and the strain on the fish is so great, that either the tissues disintegrate or the fish is too exhausted to recover. Many of the Salmon perish in this way, and one American species, that runs up the rivers on the Pacific coast for a long distance, is said not to return.

In most cases, however, the eggs ripen gradually, one batch after another ; otherwise the fishes would all probably suffer the same fate. It has been calculated by Fulton, for example, that the weight and volume of the whole mass of eggs in the Plaice when fully mature would greatly exceed those of the fish itself. This succession in the ripening of the ova and eggs is also seen in some of the viviparous fishes. Most of these are also annual fishes, but in the diminutive Carplings (Cyprinodonts) an extraordinary condition of things has arisen. Some of these are able to produce up to seven clutches or broods in the year and one mating with the male is sufficient for the lot. We may refer their small size (1½ to 2 ins.) to this excessive reproduction and the manner of it.

The number of eggs laid by fishes varies in a remarkable degree even within the same species. The smaller or younger specimens have fewer eggs than the older and the eggs are also smaller in size. The active fishes that pay little attention to their offspring usually have a very large number of eggs, up to many millions, whereas the sedentary and viviparous fishes have only a few hundreds. There is thus a general agreement between the number of eggs on the one hand and the size of the fish and its mode of life on the other.

Some fishes lay their eggs together in a string or mass

covered with mucus, derived probably from the cells lining the oviduct, and in this way the eggs or egg-masses become adhesive to whatever they may fall upon. The Perch and other Freshwater fishes lay strings of eggs which become entangled on the water-plants. In other cases the eggs simply remain in a clump together on the bottom, but in the Frog-fish the mucus mass floats on the surface. Sometimes the outer layer of the egg-shell (chorion) bursts on coming into contact with water and rolls up at one end, as in the Smelt and Gobies, to become an organ of attachment. Or the outer layer wrinkles up in various ways and may be thrown out into long filaments, which become attached as they harden to the drifting or fixed seaweed, as in the Flying-fish and Garfish (Fig. 13, 3). In Dog-fishes and Rays (Pl. V) the fibrous capsule is formed already in the oviduct, like the egg-shell of the hen, and the long tendrils at the corners curl round the seaweed, or the egg-capsule is simply buried in the sand (Rays). Sometimes these capsules appear on the beach as mermaids' purses, but as a rule they are quite empty.

These are sometimes called adaptations, in their various ways, and it is significant that the egg-capsules of the Chimæridæ laid in deep water, and of most of the Rays, do not have the attaching tendrils. But the purposes usually ascribed to the special adaptations become mutually contradictory when we take all cases into consideration, so that the supposed advantages are purely hypothetical. Thus the mucus strings of the Frog-fish are a hindrance to the free development of the young fry and probably account for its peculiar structural abnormalities. In any case, where a definite advantage can be detected, we need not fail to admire the ability of the fish in using it.

Fishes as a rule are unisexual, but some very interesting exceptions occur. Even normal fishes like the Cod, Ling, Mackerel, and Herring have occasionally been found with both kinds of sexual organs in one individual. Some of the Sea-Perches (Serranidæ) are constantly hermaphrodite and self-fertilising. In the Daurade (Chrysophrys) the male

PLATE V

Female Dog-fish (*Scyllium canicula*) and its Egg-capsule.

(From a drawing by H. Varges.)

and female organs ripen alternately in succession. The Hag-fish (Myxine) is apparently hermaphrodite to begin with, then a male, and later a female, but the changes have not yet been clearly proved.

Summing up with regard to habits and trying to interpret them biologically, we may say that the most salient features are perhaps the ancestral pelagic and demersal (or bottom-living) tendencies and the greater diversity among the latter. Primitively and primarily fishes are pelagic, with a tendency to become sedentary or more sluggish as they grow older. This is the basis from which fishes have diverged. The sedentary fishes, more sluggish physically, display on the whole more mental development, variety, choice and workmanship in their habits, and some of them have in early days emerged partly from the water to become lizards and frogs. It is as if, even in those early days, the worth of an organism was not to be determined solely by the work of his fins or hands, and it is of interest to note, that the pairing, nesting, carrying of eggs and young seen among the frogs, are already present among the fishes. The care for the young, when properly understood probably the main factor in progress, has thus been learnt already in the waters, near the shores and in the rivers apparently, where the variety of conditions has given more intricate problems for the mind and brain to solve.

We must not conclude, however, that the pairing and nesting habits have been undertaken deliberately by forms with sufficient intelligence to perceive that their offspring would benefit thereby. Far from it ; it would be too much to credit even the highest animals with so much foresight. The result, success or failure, usefulness or otherwise, of their habits did not concern them—that is only the way we look upon them—but we can see that the self-seeking tendencies have in some changed to self-forgetting, just as a structure developed for one function may change to another.

CHAPTER III

MIGRATION OF FISHES

THE term "migration" is frequently employed to express three different kinds of movements. In the early stages of life the eggs and fry may be carried by the currents some distance from the place where they were born. This helpless drift is at first involuntary, but later the young fry are quite able to make their way in and out of currents or remain stationary. Then the young fishes and the adults spend most of their time seeking food; they scatter about within their area; but good feeding grounds or incoming currents rich in plankton food bring many together in a concentration or shoal which persists for some time. These may be called voluntary movements on the part of the fish.

Rightly considered, the true migration is something very different from the drift and wanderings. Here the fish is urged to leave its ordinary hunting grounds and even to stop feeding, in order to undertake sometimes long and perilous journeys. The call of the blood, a deeper necessity common to all living things, takes possession of the fish, and the thousands and millions in like case forgather year by year almost at the same place and time—a circumstance of great importance both in the past and in the present.

Most of the great fisheries are based on the spawning migrations of fishes, and at the present day we can observe how fishing populations move from place to place and even from one country to another in pursuit of their calling. History tells us also, how the New Englanders would hardly have surmounted the hardships they encountered in the first years of their endeavours to settle in America but for the rich fisheries of the coast; and again, how the Dutch fishing population spread from one country to another in the seventeenth century, and how England and Holland went to war on a small matter of the fishing industry. In ancient times also the Greek and Phœnician colonists chose their settlements where the fishing was of importance, and the nature of the fishing is sufficiently indicated by the representation of the Tunny on the old medals of Carthage and Cadiz. These under-currents in the life of nations are not apparent in the surface histories, yet Edward Forbes was justified in his quaint comment, "the existence of Alexander may have been determined by the migration of a shell-fish." It is a question whether the leading or the feeding of peoples is of the greater importance.

The Tunny (*Orcynnus thynnus*) is indeed the most famous example of a migratory fish; from Aristotle onwards classical literature makes frequent mention of the ways and character of this magnificent fish. The watchman on the hill overlooking the sea, mentioned in the Bible, is possibly another reference to it, and the methods adopted at the present day, the outlook and even the fishing nets, are the same as those used at least two thousand years ago (Pl. VI).

Aristotle and the ancient writers presumably echoed the common belief of those days in saying, that the Tunny migrated from the Atlantic into the Mediterranean, following the current along the north coast of Africa to Egypt and Palestine and thence into the Black Sea. It was supposed to spawn in the Sea of Azov, thence returning along the northern coasts of the Mediterranean out into the Atlantic. Such wide-reaching ideas with regard to the migrations of fishes are the common property of fishermen of every country, for other fishes as well; it is their tribute to the mystery of their calling. But the scientific investigations of the past century have greatly reduced the limits of the migrations, except in one remarkable instance to be noted later.

The presence of the young in all stages and the appearance of the adults in all parts of the Mediterranean at practically the same time (summer), are among the phenomena which led Pavesi to believe that the Tunny lives for most of the year in the deep water off the coasts. As the spawning time approaches, the fish first of all make a vertical migration, from the depths to the surface. It is said that new arrivals in the surface waters can be distinguished from the earlier by their darker colour. Then they move towards the shallower waters to spawn. It is the latter, the horizontal movement, which has given rise to the idea of immigration from the Atlantic and which accounts for the elaborate preparations made by the fishermen for the reception of the Tunny.

It is not only the Mediterranean, however, that the Tunny frequents. The greatest fishery indeed lies outside, in the Bay of Cadiz, and here as in the Mediterranean the concentration of the fish on the coast is due to the spawning migration. It is also taken on the western side of the Atlantic. On the other hand, it has been found much further north, in British waters, North Sea, Scandinavian coasts, and the skulls have even been taken in Icelandic waters; within recent years quite a number have been captured in the North Sea.

These are not spawners; they are the larger fish which, after spawning in the summer on the Spanish or Portuguese coasts, follow the Sardine and Mackerel shoals northwards, and later feed on the autumn Herring of the North Sea. When the trawl with its load of Herring is hauled up, the Tunny follows and snaps at the protruding heads and tails. Possibly it was just as common in earlier years; there is a record by Schönevelde that it was fairly common near Eckernförde in the Western Baltic as far back as the beginning of the seventeenth century, also in connection with the presence of Mackerel. The Porpoise may often have been blamed for its depredations, but until the trawl was used for the capture of Herrings in large quantities, its occurrence was regarded as a rarity.

As with the Tunny, so with the other members of the Scombroid group of fishes. They are all strong swimmers

PLATE VI

Thynnoscopi of the Adriatic or Ichthyoscopi of the Greeks.

Watchmen on high ladders to give notice of the arrival of the Tunny. (From Faber.)

of the ocean and they all, so far as we yet know, migrate towards the shallower waters in order to spawn. In British and northern waters, as well as in American waters, the commonest form is the Mackerel. As is well-known, the Mackerel arrives off the coasts in the summer months in immense shoals or schools. Spawning over, they also go in pursuit of the Herring, and in winter retire into deep water.

This same tendency to migrate from deeper to shallower waters in order to spawn, is also shown by the Clupeids, Herring, Menhaden, Pilchard, Anchovy, etc. As in the case of the Tunny long migrations used to be credited to the Herring, as far as the Greenland ice, for example; but recent investigations make it doubtful whether the migrations are of any great extent; the Herring frequent the deeper waters off the coasts and forgather towards the shallower waters when the impulse to spawn comes upon them; but, as mentioned in the previous chapter, the wanderings along the coasts may be very extensive.

A curious thing about the Herring, not known in any other species, is its ability to spawn at all seasons of the year and in fresh as well as salt water, thus under the most diverse conditions as to temperature and salinity. Its variability in this respect suggests that from the beginning it has been a shallow-water or brackish-water form; it can readily adapt itself to any conditions from 0 down to 100 fathoms. There are winter Herrings which spawn in the Firth of Forth and in the west; spring Herring in Norway, in the Schley, and north of Scotland; summer Herring off the north-east coast; autumn Herring spawning in the southern North Sea, and so on. But the Herring keeps to the temperate zone; in the tropical and subtropical regions the variability has changed it into other species.

Another interesting problem in distribution is furnished by its relative, the Anchovy. This is a southern, subtropical form, with its main spawning places in the Mediterranean and adjacent waters. Yet a northern branch appears regularly year by year, and that in great quantities, in the

Zuidersee and neighbouring brackish waters. In southern regions the species spawns in water of comparatively high temperature and salinity; here the water is brackish and cold. There is a wide gap between Spain and the Zuider-see; few specimens have been taken in the Channel or Bay of Biscay. Are we to suppose that it migrates this distance yearly, or that it remains somewhere in the Channel, where it eludes capture? In either case the origin of this branch remains a mystery; possibly it represents the original form of Anchovy, as we can more readily understand a migration to the pleasanter conditions of southern waters than the reverse. On the other hand, this particular branch may have suffered greatly from its enemies, Mackerel, Tunny, and others, and found a safe retreat from their pursuit in the Zuidersee. But no definite solution can be given to such a problem; of greater interest would be the investigation of the structural and developmental differences between forms of the same species growing up under such wide differences.

The principal thing about the species so far mentioned is, that the migrations are from deeper to shallower water in the sea. Some Clupeids, like the Shads (*C. finta*, *alosa*, *sapidissima*, etc.), go further and ascend rivers to spawn in fresh water. Some are permanent inhabitants of the inland seas (Caspian, North Italian Lakes, and Great Lakes of America).

The same phenomena are found among the nearly allied Salmonoids. Whilst some forms, like Argentina, spawn in deep water in the open sea, and others, like Mallotus (the Capelan), come close inshore to spawn, thus constituting a most important source of food, for example, for the Eskimos of Greenland, others like the Smelt enter brackish and fresh water, and the principal species of the group ascend rivers right to their sources, whilst whole families belong only to the fresh water.

The Salmon, it is now believed, does not move far from the mouth of the river to which it belongs; yet, like the Anchovy, it is rarely found in the sea except where it assembles to feed, as in the straits of the Pacific and in the

Southern or South-eastern Baltic. The latter case is of near interest; the Salmon are taken on hooks, but some escape and are caught later, with the hooks still attached, in the lakes of Finland. The Salmon is thus able to spend its whole life in fresh or nearly fresh water. In the Rhine again, when it was a clean river, the winter Salmon used to remain nearly a whole year in the middle reaches below Basel before proceeding to spawn.

It sounds almost incredible, therefore, that the Salmon does not feed after it has entered fresh water. Yet the evidence from all countries supports this view, and further, the digestive organs undergo a certain amount of degeneration. How then can we explain that the Salmon rises to the fly, and has even been taken with a Trout in its stomach? For one thing, it is certain that the Salmon of different regions, even different rivers, vary greatly in their powers and habits; for another, we have only to take a wider perspective to understand what happens. In the larger Norwegian rivers the Salmon rise to the fly in the lower reaches, but provide no sport in the upper. In the still longer American waters the Salmon provide no sport at all. They feed, therefore, or endeavour to feed only when fresh from the sea.

All the time they are in fresh water, and it may be a whole year, they live on the materials stored up in their body. This has been considered a wonderful provision or adaptation of nature. Regarded from the Salmon's point of view, it seems more like a tragedy; the Salmon wishes to feed but cannot! Certainly, its body must be an accumulator of energy of very good quality and enormous power. The fish ascends rivers against strongly running waters, jumps numerous falls (probably not more than 10 ft. at a single leap however); if a male it has to fight other males and keep off enemies, then help to prepare a bed in which the female lays its eggs. It is not to be wondered at, when the spawning is over, that the fish is completely exhausted and allows itself to drift tail-first down the stream.

Even in the short rivers of Scotland the mortality from

exhaustion and other causes is distressingly high. In the longer rivers of the Pacific, where the distance travelled upstream may be anything up to 1000 miles, the mortality reaches a maximum. Evermann has found that "in the head waters of the large streams, unquestionably all die; in the small streams and near the sea, an unknown percentage probably survive." [1]

From experiments made in Scotland and Norway the Salmon would appear to possess a homing instinct, and Buckland thought that the organs of smell, which are very strongly developed in the Salmon, have something to do with this (Young). In one case noted by Buckland and quoted by Day, some twenty or thirty salmon ascending from the sea were captured and marked, then taken round an island and set free in a lake not far distant from the original water, but separated by land. In the course of the same season some of these marked salmon were retaken in or near their own pool; to do which they must have travelled round the coast some forty miles, passing several streams on the way. But we cannot take this as a general rule. It may be that these particular Salmon have more intelligence; they may have been trained by many generations of persecution and learnt to make a finer discrimination of the waters; but in other waters under less strenuous conditions the Salmon is frequently at fault. Thus some have been found trying to enter the Thames, from which pollution has long since driven the spawning fish. And in the Pacific, where countless millions throng along the coasts, they frequently crowd one another into blind alleys with no access to the upper waters. Yet even there the fish displays a certain amount of selective power. It is said, for example, that the Sockeye or Red Salmon rarely or never ascends a stream that has not one or more lakes at its head waters, whilst the Humpback and the Dog Salmon seem to prefer the smaller coastal streams. Regarding the Humpback, Evermann

[1] The idea that Salmon change their quarters in order to get rid of parasites, has been forestalled by the Mediterranean fishermen, who used to believe that the Tunny came into the shallow waters in order to get rid of their parasites by scratching on the rocks.

PLATE VII

Salmon jumping Falls in Alaska.

(Photograph received through the courtesy of the Bureau of Fisheries, Washington.)

and Goldsborough write, "If they enter large rivers at all they are apt to run up the first small tributary stream which they reach."

On the whole, it may be said that the run of Salmon up the rivers is connected with spawning, yet some important exceptions have to be made. Many specimens will go some distance up a river only to turn back (and this may happen several times), or even remain months in definitely fresh water before proceeding to spawn. There may be a simple explanation of these eccentricities; it must be remembered that the Salmon is not far from being a freshwater fish. According to Paton and Newbigin (1900) the factor determining migration from sea to river is not the "*nisus generativus*, but the state of nutrition." The spawning comes later.

With the Salmon we may finish this account of the one stream of migratory fishes, those which journey from deeper to shallower and even fresh water in order to fulfil their life purpose. This is by far the most common form of migration; a very large number of marine fishes, even the Sharks, Skates, and Rays, follow in the main the same course. As suggested by Meek, the term "anadromous," formerly used only for those entering fresh water, may well be applied to all those fishes which follow the same general rule in their migration, namely, an ascent from deeper to shallower waters. To a very large extent, this at once indicates the nature of the problem to be solved and even suggests the solution.

But there are two groups of apparent exceptions. In the one the fishes exhibit but little tendency to undertake any spawning migrations. They spawn, as it were, where they find themselves. This applies naturally to the stationary fishes, though even among these a certain amount of migration can be noted. Thus, among the Blennies the Butter-fish or Gunnel has a short, anadromous migration of a few miles, whilst the Angler or Frog-fish shows a longer migration in the reverse direction. Among the Gadoids, again, most species prefer the deep water for spawning; but the Cod, if it has any preference, comes from the deep water to shallower banks.

These groups, in which the migratory movements are of little account or in which different tendencies are shown by different species, may be called neutral for the moment. The probability is, indeed, that they are not so much affected by the spawning process. The seeming migrations may be either a return to the natural home places, following upon the drift of the young away from these, or they may be mixed up with the movements in search of food. The Cod, for example, is a rover with a circumpolar distribution like the Herring, and it shows a decided preference for the latter as food. It pursues the latter at all seasons of the year and at the proper time it is able to spawn wherever it goes —within certain limits, of course, but they are very wide and correspond closely to those of the Herring.

From the biological standpoint these neutral groups are of great importance in any theory regarding the causes of spawning migrations. We may take it, that the species composing them are so well-adapted to their surroundings that they are not urged to make any great change when the need to spawn comes upon them. Viewing their life-history as a whole, they show a better balance between the inner and outer conditions.

The second exception is represented by the Eel group. With the Common Flounder the Eel constitutes the old group of katadromous fishes, those which migrate from the fresh water to spawn in the sea. The Flounder is certainly katadromous and like the Plaice seeks water of greater density than its ordinary feeding grounds for the purpose of spawning; but other Flat-fishes like the Soles show the reverse procedure, and it seems better to regard the group as neutral. There is in any case no comparison with the life-history and migrations of the Eel.

The Eel seems to belong to quite a different world from other fishes; it would be difficult to imagine a greater contrast. Whilst other fishes are content with a reasonable area of distribution, thus admitting the limits of the piscine nature, the Eel ranges between the most extreme conditions on the globe, from mountain stream to the depths of the Atlantic. It passes through two transformations in the course of its life, when it enters and when it leaves fresh water, other fishes have only one or none at all. The Eel takes some ten to fourteen years (in fresh water, plus three more in salt water) to become mature, other fishes everywhere only one to three.

Other Eels, Conger, Muraena, etc., do not show such a great contrast, so far as we yet know their life-histories. All seem to be katadromous fishes, migrating out from shallow to deeper water, but only the Anguilla genus enters fresh water.

From many years' study of the conditions in Northern Europe, C. G. Joh. Petersen came to the conclusion, that the eels of the Baltic and Western Europe migrate out to the Atlantic to spawn. Where they spawned was the mystery—many and varied have been the speculations regarding the solution of the mystery—and the matter was somewhat complicated by the conclusions of the Italian workers, Grassi and Calandruccio, that the Eel spawned in the Mediterranean. They had found the young fry and even thought they had found the eggs in the latter region. It was not known at that time that these fry were already three years old.

Petersen's work was taken up by Johs. Schmidt. With the Italian conclusions in mind, two years were spent in fruitless search of the Mediterranean. The Eel certainly did not spawn there, he discovered, and no young specimens could be found below the size of about 2½ ins. But many hundreds of the same size had already been taken by him out in the open Atlantic.

This young form, Leptocephalus, belongs to a well-known group of pelagic fishes found in many parts of the globe. At an earlier period they were considered distinct species by many authorities, and certainly there is not much resemblance between a Leptocephalus and an Eel (Fig. 12).

Convinced that the Eel does not spawn in the

Mediterranean, Johs. Schmidt extended the scope of his investigations out in the Atlantic. By this time he had found from careful examination and comparison of the adult Eels of the Mediterranean and Northern Europe, that no racial or specific difference existed among all the European Eels, whether from the Adriatic or Finland. Hence he

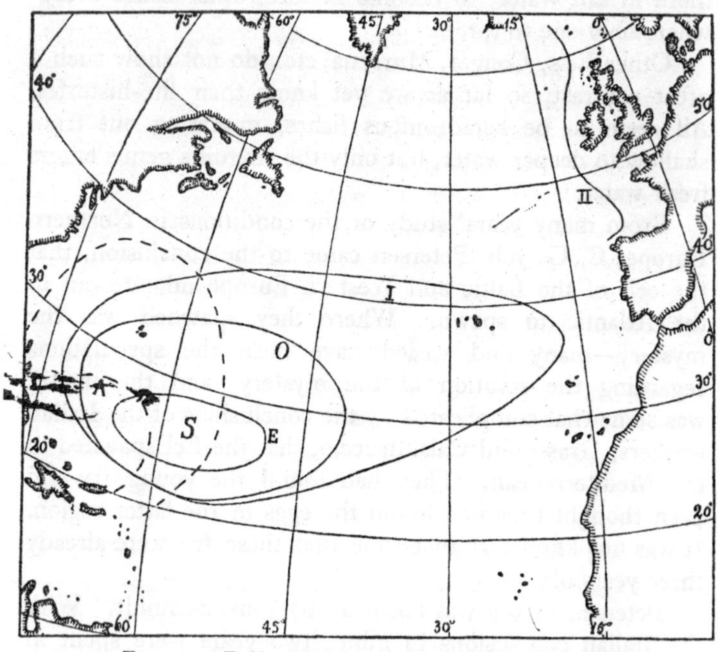

FIG. 11.—Drift of Eel Leptocephali in the Atlantic.

A. - - - - American Eel.
E. ——— European Eel.
S. spawning-ground near the West Indies.

O, 1-year old.
I, 2-years old.
II, 3-years old.

concluded that all the Eels of Europe must migrate out into the Atlantic when their time comes to spawn. It is a mass migration on a scale previously unsuspected and quite unknown for any other fish. None return and few of the countless millions have been taken on the way; where do they go?

By a series of brilliant observations and deductions

Johs. Schmidt has at length traced the Eel to its spawning places (1921), in an indirect yet quite convincing manner. The post-larval Eels (Leptocephali) were found in ever increasing quantities the further he proceeded westwards, but they became gradually smaller. Successive lines or zones could thus be marked off, according to the size, until at length he found the smallest larva with the yolk-sac still attached and even the eggs out over the deepest parts of the Atlantic, near the West Indies and nearer the American than the European coasts. These were taken in the middle and upper layers, and there is therefore some doubt whether the adults actually spawn at the bottom. The depths on the "spawning grounds" were about 3000 fathoms.

Two more of Schmidt's discoveries are of interest. At the same place where he found the larvæ of the European Eel he also took those of the American Eel. The spawning grounds of two forms, living thousands of miles apart, are practically the same, and the young of both species were occasionally taken in the same haul of the net.

The second discovery is even more extraordinary. The two streams of post-larvæ spread out and move northwards towards the Bermudas; there they separate, one going east to Europe, the other west to America. The latter form spends only one year in the Atlantic and is then ready to enter fresh water; the European form takes three years to cross the Atlantic (on Chart, O—first year, I—second year, II—third year) before it is ready to change its constitution and enter the rivers. How these tiny fishes can settle which direction to take, even allowing for a difference in constitution, is a profound mystery. In illustration of the immense quantities of these young Leptocephali it may be mentioned that Johs. Schmidt took as many as eight hundred specimens in one haul out in mid-Atlantic.

We have now to consider whether any explanation—something more, that is, than a simple reference to "instinct"—can be given of the wonderful behaviour of fishes on these occasions. First of all, a reference may be made to the corresponding mystery in the case of birds, another of the few groups of animals that show extensive migrations. In the words of Prof. J. Arthur Thomson, these are due to " constitutional rhythms and impulses " and have nothing to do with the movements in search of food. They have arisen from the difficulties presented by the seasons in certain areas and take place every year, even the yearlings taking part. From the experience of many generations birds have learnt to evade the dangers of the winter at their breeding places. They have adapted themselves to the prevailing

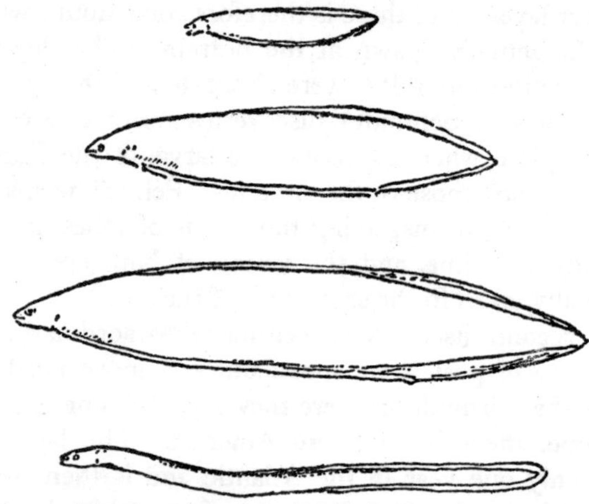

FIG. 12.—Atlantic stages of the Eel in the first three years of its life ; below a Glass-Eel in fourth year ready to enter fresh water. All natural size. (From Johs. Schmidt.)

conditions, let us say by means of their intelligence, and the main problem in connection with the birds is the manner by which they find their way over thousands of miles of land and sea.

Though the migrations of fishes may appear the same superficially, they are widely different in important matters. They are not seasonal migrations in the same sense and the young take no part in them. Fishes have nothing to fear from the winter, not those we are dealing with at any rate ; both the Salmon and the Herring spawn during the winter months. Further, the migration brings an advantage or benefit to the birds. On the one hand, they are urged on by the constitutional fever within; on the other, they are enticed forwards by what lies at the end of the journey. This is wanting in the fishes; the urge is there but not the inducement or reward. If an Eel could foresee what lay before it in the depths of the Atlantic, would it undertake the journey?

We may say, then, that the migrations of fishes cannot be regarded as adaptations, as we apply that term to birds. They are on a lower level, more primitive and crude. In all probability they are an inheritance from remote ancestors, which the fishes from lack of sufficient intelligence perhaps have not been able to overcome.

Again, it should be noted that the birds do not show contradictory phenomena; the migrations are always to avoid the severity of the external conditions. We might say this of the summer spawners among fishes; but all sorts and conditions are to be found, and the extreme cases, Salmon and Eel, seem wilfully to be seeking the most dangerous and most deadly conditions to be found on the globe.

Many efforts have been made to connect the migrations of fishes with the external conditions, temperature, salinity, light, etc., and it is probable that each has some influence, but the problem has another side. In the first place, it is essential to distinguish the food concentrations and wanderings from the spawning migrations, a matter not so easy as at first appeared. It would seem, indeed, that a feeding concentration, in many cases if not in all, precedes the actual spawning, and it may be that the storing of food-stuffs leads directly to the internal changes in metabolism which give the stimulus to change of grounds and conditions (Paton, 1900).

Physiologically the feeding and the migratory impulses are very different. In the one the fish is seeking to obtain something, in the other to get rid of something. The latter may be called a morbid, even pathological condition, affecting every part and sufficiently powerful to make the fish ignore terrible injuries and even to overcome the impulse to feed. The digestive organs partially or wholly degenerate, and the fish is then urged to seek relief by unusual or extraordinary means, that is, to find if possible external conditions in better agreement with its changing or changed constitution.

One cannot say definitely what happens in any particular case, owing mainly to the protean nature of fishes. A Dog-fish, for example, is more in harmony with the surrounding medium than a Salmon, and a Stickleback will survive under conditions that kill most fishes. These phenomena depend apparently on their ability to alter the blood-pressure, or concentration of salts, as they pass from one set of conditions to another. And this ability, again, depends upon how far the gill-membrane and the skin permit of the passage of water outwards and inwards, a question that has not yet been sufficiently investigated.

It is possible, therefore, that the internal stimulus to migration comes from some internal change in the blood; probably in the blood-pressure. This is always much less than the pressure of the surrounding water, but it varies. In the Salmon, Greene found that the concentration of salts in the blood was far below that of the salt water in which the fishes were feeding voraciously, and gradually altered as the fish entered fresh water. It may be that a slight reduction in the pressure within the blood is a source of discomfort to the fish and urges it to enter the rivers. In the case of the Eel, on the other hand, there would seem to be a great increase of the internal pressure accompanied by an increase in the amount of carbonic acid at the time when the fishes are beginning to migrate from fresh to salt water. If they are detained on their journey in any way, the body thickens and swells and even increases greatly in weight, though the fish are apparently not feeding. Petersen has shown that the digestive tract is in process of degeneration at this time and even the air-bladder decreases in size. The average Silver Eel in its breeding dress is about 70 cm. (27 to 28 ins.) in length and weighs about 1½ lb. The large specimens referred to were from 110 cm. and more, and weighed from 4 to over 8 lb. Most remarkable of all is the great increase in the size of the eyes. These become more than four times their former size and weigh twice as much.

Without being able to say positively that such is the case, one may conclude that these phenomena point to a great increase in the internal concentration of salts and absorption of water. And this increased pressure within seems to be the cause of the Eel leaving the rivers to seek the greater pressures of the salt water. When the impulse to migrate comes upon them, the Eels are restless and ill at ease, moving up and down for a time in a manner quite unlike their usual sluggishness, then definitely turning down the rivers. At every stage of their descent they come under increasing pressures without. And it may be recalled here, that the Eels are well-known to be extremely sensitive to the differences of pressure produced by thundery weather.

In the case of marine fishes like the Plaice or Cod, we know that when the spawning fish are brought up in the trawl, the spawn continues to run even when the fishes are still alive. This indicates that the internal pressure persists for some time at the level of the deep water whence the fishes came. Since the low pressure inshore would help in getting rid of the spawn, the migration of the Plaice from shallow to deep water would be unnecessary for that purpose. Hence we may conclude, that an increase of pressure within has occurred with the maturation of the sexual organs, and the resulting discomfort has urged the fish to seek deeper water and greater pressures. Similarly in the case of other fishes. Each species is urged to seek the external conditions, whether of lower or greater pressure, which corresponds best to the change in its own constitution.

How far the fish in its migrations is guided in any way by its own sense organs, we cannot say. The external conditions—temperature, salinity, dissolved gases, etc.—all make for differences in pressure in diverse ways. Possibly, as Hofer has suggested, the extreme sensibility of the lateral line system to slight variations in the strength of currents, may enable the Salmon to detect the entrance of a river or side-stream. But on the whole, one may doubt whether the senses play much part in the directing of the spawning migrations.

CHAPTER IV

THE DEVELOPMENT OF FISHES

As a general rule, the phenomena presented by the growing organism are interpreted in terms of the adult form and structure. The development is regarded merely as the unfolding of the adult characters, and from this standpoint the early stages are of very subordinate importance. It is possible, however, to take another view; namely, that the structure of the adult is to a large extent determined by the life and movements of the larva and the conditions under which the embryo develops.

The second view will be illustrated in the following pages, after some information has been given regarding the eggs and general development of fishes. And from this it will be seen, that the recapitulation of the past ancestral history by the organism can be read more clearly in the general form and mode of movement of the young stages than in details of the adult structure.

1. Eggs of Fishes

It would be only partially correct to say that all fishes begin at the same starting point. One cell is the beginning, it is true, but different substances, organic and inorganic, have been used in the making of that cell; its contents vary as to quality and quantity, and the place where it finds itself greatly differs. On the one hand, it has an inherited constitution, on the other an inherited environment.

We can readily distinguish these two factors in most cases. The freshwater fishes generally lay their eggs on the bottom, either rolling on the ground, or among the vegetation, or in nests; most of the marine forms have pelagic, floating eggs, but many also fix their eggs to the bottom, Herring, Blennies, etc. The constitutional element is clearly that the eggs are heavier or lighter than water. And to emphasise this we find that the Gouramies, which live in fresh water, have eggs that are lighter even than fresh water and float to the top.

The size of the egg varies a great deal. Most of the pelagic eggs are about 1 mm. in diameter, but some are smaller, while those of Halibuts may be 6 mm. or more. The demersal eggs, those that stick fast to the bottom, are in general larger than the average pelagic eggs, but the chief characteristics are the greater thickness of the shell or chorion and the larger amount of oily ingredients. The eggs of the freshwater fishes are usually much larger. Thus, the small Bitterling, Rhodeus, though only about 70 mm. altogether, lays eggs of 3 mm., and some of the Siluroids have eggs of 17–18 mm. The largest eggs are those of the Elasmobranchs, Sharks and Rays, which are laid in a hard shell or capsule.

The number of eggs also varies in a remarkable manner. Forms with pelagic eggs usually have an extremely large number. Fulton estimated that a Ling contained from 14,000,000 to 60,000,000 eggs, and Day gives no less than 160,000,000 for a large specimen. The Cod and Conger are content with about 8,000,000 to 10,000,000. Until recently these were considered the records, but now Johs. Schmidt has found that the Sun-fish has about 300 million eggs in its ovaries.

The fishes which look after their eggs usually have only a few hundreds, the viviparous fishes still fewer. But it is astonishing what a number they can pack into a small space; the Viviparous Blenny is recorded to have had 300 young in its ovary at one time, though the usual number is much less.

An interesting thing is, that the Herring though it lays demersal eggs has a considerable number, from 30,000 to

60,000. In this and many other ways it occupies an intermediate position between the freshwater and the pelagic and demersal marine fishes.

When the eggs are ripe or mature, a small opening appears in the shell, the micropyle, through which the polar bodies escape and water enters. In this way the pressure within is regulated to the pressure without, and in many cases the outer membrane swells out, sometimes increasing the size of the egg by as much as four times (Shad and Long-rough dab). A perivitelline space is then formed in which the embryo is bathed during its development. The eggs of quite a number of fishes, particularly of those living in deep water, display the same phenomenon. In the Herring, on the other hand, the reverse condition prevails ; the capsule seems to be inelastic and does not expand. Instead, the yolk contracts or is compressed and the small perivitelline space is formed in this way. In most cases the water mixes directly with the yolk at the time, and causes the egg to become transparent.

This phenomenon is due to the fact, that the pressure of the blood and body fluids (concentration of salts), in the Teleosts is much lower than that of the surrounding water. But the strange thing is, that after the egg has been fertilised by the spermatozoon and the micropyle has closed, no further exchange takes place. Gases are able to pass through, oxygen inwards and carbonic acid outwards, but when living the membranes are impermeable to water and its dissolved contents. Dakin has found that the developing eggs come to have the same specific density or partial pressure of salts as the parents, though they begin with the same pressure as the environment.

It is thus not only the substance but also the essence or regulating influence that is passed on by the parents to the offspring. As will be seen, this is a new and important fact in connection with the constitutional inheritance. Further, if the environment be altered, if potassium or magnesium salts replace the sodium, the partial pressures to which the embryo is accustomed are altered and the whole course of development is changed. Hence the inherited environment is also of fundamental importance.

Regarding the chemical changes that go on within the egg, we know very little. When the egg is laid free in the water, the shell at once becomes hard. This hardened chorion, which is absent in the eggs developing internally, serves as a protection to the embryo; but in other ways it is a disadvantage, since it restricts the growth of the embryo. It has been observed by Ehrenbaum, however, that the egg-shell becomes thinner as development proceeds and the egg then increases slightly in size. Probably, as in the case of crabs and lobsters, part of the substance of the shell is withdrawn and absorbed by the embryo, and this seems to occur to a greater extent in the demersal eggs, where the embryo is more advanced on hatching, than in the pelagic eggs.

Whilst the embryo seems to be immune to the chemical influences, thus to the salinity, it is very sensitive to changes of temperature. When Herring eggs were placed in a temperature of about 12° C. the fry hatched out in eight days; but in temperatures about 4° C. the period has been prolonged to forty days. Observations on other species have given the same result, and there can be no doubt that under normal circumstances this speeding up of the metabolism within the egg is one of the main factors in causing variations.

The period of incubation under normal conditions seems also to be regulated by the inherited constitution in the wide sense. It varies extremely in fishes which seem to be under the same external influences. The embryos of many Elasmobranchs take only a few months to develop, but Acanthias takes two years. Among the Teleosts the demersal eggs usually take a longer time than the pelagic eggs, but here again the Herring takes a short time. The Agonus embryo spends a year in the egg, the young Stickleback emerges in 25 to 40 days. The Pipe-fishes take only about two weeks to develop, the Viviparous Blenny about four months. How these differences have arisen, we can but faintly imagine, but it is of some importance to remember that they have come from some previous condition; they have not been separately created.

The principal difference between the eggs of fishes is one of size, and this is mainly due to the varying amount of nutriment or yolk. Amphioxus has a very small egg with

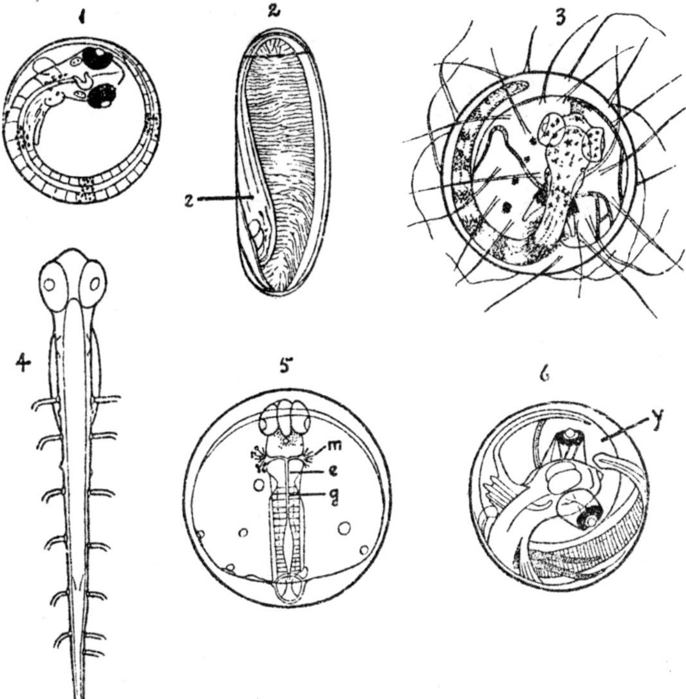

FIG. 13.—Eggs of fishes. 1. Cod; 2. Bdellostoma; *e.*, embryo; 3. Belone; 4. Anchovy larva; 5. Weever; *m.*, mouths of funnel; *e.*, ciliated portion of *g.*, gut; 6. Trachypterus; *y.*, yolk. (1 and 3 from Ehrenbaum, 2 from Bashford Dean, 4 and 5 from Williamson, 6 from Lo Bianco.)

little yolk, and the living protoplasm, when it passes from the one-celled to the many-celled stage, is able to include the whole of the yolk within its cells. Development is then rapid and the young embryo assumes the adult form within a few hours. In the true fishes this *holoblastic* mode of segmentation is rare, yet we find it in a slightly modified form in some important types ; thus in the Lampreys at one end, Sturgeon and Dipnoi at the other, the segmentation is complete but not quite equal, just as in the frogs. In the Teleosts and Elasmobranchs, on the other hand, the greater portion of the living protoplasm of the egg collects into a disc at the one end. There it divides up and only gradually manages to develop a layer of cells round the yolk. This *meroblastic* form of segmentation is undoubtedly the more specialised. In some Teleosts, however, especially the Clupeids, we find a superficial segmentation of the yolk, distinct from that of the disc, and this is probably due to a large quantity of the living protoplasm remaining in the substance of the yolk. The same phenomenon appears to occur in the egg of an ancient shark (Cestracion) described by Bashford Dean. This seems to be an intermediate stage between the more and the less complete mastery of the yolk by the formative protoplasm.

From the biological standpoint, we may question whether this accumulation of yolk has been an advantage. More advanced forms are able to do without it, perhaps by improving the quality of the transmitted materials, but more particularly by bringing the egg and embryo into a different kind of environment. Thus, we may connect the fact, that the fishes have remained stationary at a comparatively low level of vertebrate development, with the conditions under which the formative protoplasm has to work in the earliest stages.

2. Embryo

In the beginning the cleavage and increase in number of the cells are the only appreciable signs of the life within. We can hardly speak of an embryo until some structure is apparent ; before that it is simply referred to as the blastodisc or blastosphere. This works itself into a gastrula, the two-layered stage, in which the outer layer gradually extends over the surface of the yolk and goes to form the sensory and nervous systems, whilst the inner layer proceeds with the absorption of the yolk-food, and the formation of the muscles and internal organs.

Specialisation or organisation has thus begun, and it cannot be without some pregnant meaning that the first part or organ to be formed is the central nervous system. In some mysterious way a furrow appears along the top of the blastodisc, and the sides swell up and approach each other until they meet over the furrow, then fusing to form a canal or it may be at first a solid axis. At the anterior end several swellings soon appear, denoting the vesicles of the brain.

The stage now reached may be said to be the possession only of the fishes and higher animals (Craniata), yet signs of it are not wanting amongst still lower forms (Cephalochordata and Urochordata). And going back still further, the two-layered stage or gastrula is common to many Invertebrates. In this way, therefore, the past history is being repeated, but the gap is very great, and some zoologists believe that here at any rate nature has broken with tradition and completely reversed the order of earlier proceedings. The larva of the worm, it is supposed, has turned upside down and develops forward instead of backward, the nerve cord becomes dorsal instead of ventral, whilst the heart comes to lie below the nerve cord and alimentary canal instead of above. Further, an entirely new mouth is formed. As we shall see, the formation of a mouth has been the most difficult problem for the fishes.

This revolution theory cannot, of course, be proved, but it provides us with the simplest explanation of the immense change in form and structure we find among the fishes and Vertebrates. The organism has come, as it were, into a new world, found it suitable, and has then expanded in many different directions.

Whilst the dorsal nerve cord is forming, the lower layer of cells is engaged in laying down a mass of muscle cells along each side and these soon come together in groups, probably already under the directing influence of the nervous system, to form the muscle bundles or segments so plainly seen in the young embryo (Fig. 13, 5). The embryo is now distinctly raised above the yolk, and the lowermost layer of cells, generally distinguished as a separate layer, arches upward in a groove below the nerve cord. With the downward growth of the muscle bundles at the sides this groove is pinched off and becomes the notochord, the forerunner of the vertebral column. Underneath the notochord the lowermost layer now forms the upper wall of the alimentary canal, and the extent to which it remains in contact with the yolk becomes a distinctive mark of the different groups of fishes.

Viewed from the outside the most important signs of life up to this stage have been the gradual uplifting of the embryo from the yolk and its growth backward to form a tail. The form or shape of the future fish depends greatly on the amount of space the embryo now has in which to expand. In all forms the embryo is curved round the yolk to begin with. In the Hag-fishes (Cyclostomes) it remains in this position pressed tightly into the yolk until it escapes, whilst the head is compressed against the egg-capsule (Fig. 13, 2). From this we can understand three important features in the adult Hag-fish, the feeble development of the brain, the rounded shape and the simple condition of the gut and kidneys. This condition may be contrasted with that of the Dog-fish (Scyllium), which has almost the same amount of yolk, but more space in its capsule, and thus reaches a much higher grade of development. In many Teleosts a similar compression of the head can be noticed, especially where the yolk completely fills the egg.

In the Elasmobranchs, on the other hand, the embryo soon becomes quite free from the yolk except for the placental cord (Fig. 14), and this freedom results in the greater development of both the internal organs and external structures. For example, the gills are already present, the mouth is open, and the alimentary canal has its spiral. The young Lamna or Herring Shark, which is a giant among embryos, takes advantage of this freedom to feed on the yolks intended for other embryos. What appears to be the yolk-sac in the figure, is really its bloated and protruded stomach.

As the embryo separates from the yolk from before backwards, a space appears below the head, and it is here

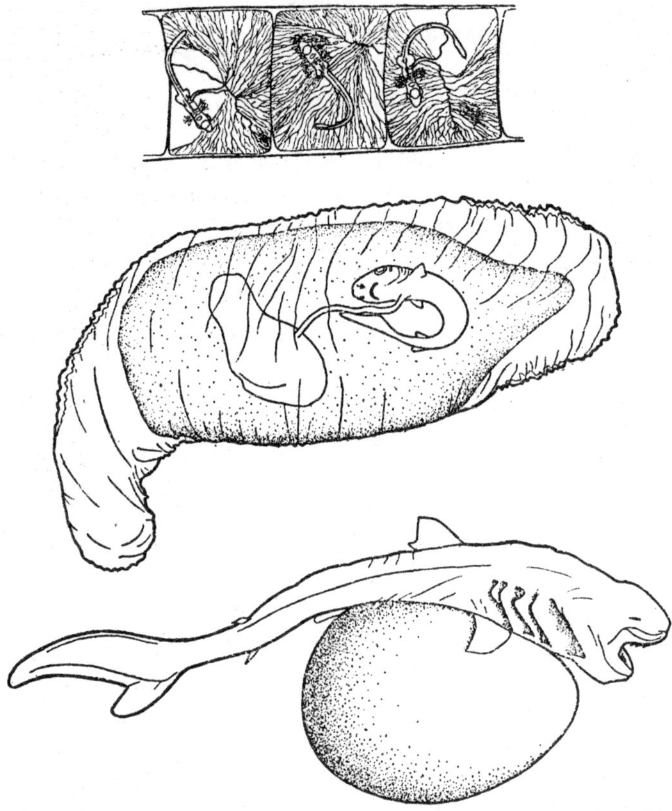

FIG. 14.—Embryos of Sharks and Dog-fishes. Above: Acanthias (from Graeffe). Middle: Mustelus (from Joh. Müller). Below: Lamna (from Lohberger); actual length 22 inches, weight 5½ lbs.

that the heart is formed, first as an indefinite group of cells, then as a ring, and finally a tube, which may remain straight but usually bends into a horseshoe. It begins to pulsate almost at once and as it pulsates, corpuscles within the upper margin of the yolk begin to vibrate to and fro, and then some separate from their fellows and race towards the heart. It is as if the heart were pulling them along, and to some extent this is the case, a slight vacuum being formed in the wake of the rhythmical contraction from before backwards. This process leads to channels being formed over the surface of the yolk as fresh contingents of corpuscles are drawn forward, until a complete network of vessels is formed over the yolk. The corpuscles drawn through the heart are forced along a narrow space between the gut and the notochord into the still narrower spaces in the head and between the muscle-bundles. Then the channels over the yolk extend inwards towards the body, first of all in the pectoral region then at the tail, and connect up with the internal channels to form a complete circulation.

Definite blood-vessels are not formed, however, until the circulating corpuscles become red, and the stage at which this occurs varies greatly. The Elasmobranch and most of the Teleost demersal eggs soon develop the red blood; but the Herring and most of the Teleost pelagic eggs do not show any sign of red blood until the larvæ are hatched. The Ganoids, except perhaps Polypterus, are like the latter in this respect.

Williamson (1897) has shown that the vitelline circulation in the pelagic eggs of Teleosts is essentially the same as in the demersal forms, but the red blood (or rather hæmoglobin) and definite walls to the channels are wanting. Consequently, the embryos of these forms cannot receive their oxygen through the vascular system. Further, it is remarkable that, whilst the Elasmobranch and the Teleost demersal embryos with red blood have the gills developed and the mouth open before hatching, this is not the case with the Herring group and the pelagic forms. For some considerable time after hatching the larval Herring has no definite blood-vessels in the body except the dorsal aorta.

The aeration of the tissues in the Herring and the pelagic forms appears to come through the gut and not through the vitelline circulation. The embryonic gut here communicates directly with the perivitelline space by means of two canals in the pectoral region (Fig. 13, 5 *m*). These have funnel-shaped outlets, or rather inlets, and Williamson has observed the passage of small bodies through these pores into the gut. The latter possesses a number of fine cilia, which vibrate from before backwards and thus convey a fine stream of the perivitelline fluid through the gut and out by fine pores in the region of the future anus. Williamson has also observed in one form (Trachinus), that in the later stages there is a prolongation of the gut forward which divides into two branches, one opening dorsally on the head, the other ventrally.

As the fluid thus obtained from the perivitelline space soaks through the embryonic tissues, the embryos probably obtain most of their oxygen in this way. It is possible that the yolk of these pelagic forms contains a large proportion of dissolved organic substances, and that these reach the tissues by means of the heart and loose vascular system instead of through the gut. In the later embryonic stages the latter is shut off from the yolk (except perhaps in the Sturgeon).

Another important feature in these embryos is the presence of what appears to be an external excretory system. The kidney (pronephros) does not develop until the embryo hatches out, but long before this small papillæ with projecting tubules have appeared along each side of the body (Fig. 13, 4). These were formerly taken to be mucus canals, but they are in all probability excretory organs.

These curious phenomena in the early life of the Teleosts can perhaps be interpreted in the following way. The master activities, muscle and nerve, are well-developed, but there is confusion in the sustentative. What should be the respiratory system is dealing mainly with the food-supply, the digestive gut is respiratory, and the excretory organs are of a special kind, opening to the exterior instead of the interior. That these are very primitive features is apparent, and we may conclude that the regulatory system is imperfectly developed. There is a lack of differentiation and co-ordination. An interesting thing is the light they throw on the future differences between the various groups of fishes. The nervous and vascular systems of the Teleosts are inferior to those of the Elasmobranchs, and the digestive tract of the former retains throughout life the tendency to be a respiratory organ. The Ganoids and Dipnoi are the same in this respect, whilst the Elasmobranchs show no sign of it at any time.

3. Larva and Postlarva

With the termination of the embryonic life the dominance of the constitutional influences comes to an end. The larva now comes into direct contact with the inherited environment and its movements henceforth are responses to stimuli from the surroundings. Its further growth is therefore determined as much from without as from within.

The free-swimming larva of the Teleosts is still embryonic in its character and carries about the yolk-food given it by its parent for some time, quite as long as the period within the egg. As it is a free-swimming form, however, we must call it a larva and a new word has been coined for the later stage, after the yolk has been absorbed. This is the most important stage in the life of the Teleost, when bony substance appears and many transformations take place.

Some authors prefer to use the term " prelarva " for the yolk-bearing stage and " larva " for the transformation stage; but it seems better to follow established usage and call the latter the postlarval stage. Further, it emphasises the unique character of the Teleosts.

The degree of development reached by the embryo on hatching depends, apart from the temperature, on the amount of yolk and also the amount of freedom for movement in the egg. The development reaches its highest point where both are at a maximum, as in the Elasmobranchs ; it is incomplete to a varying extent in the Teleosts.

That the accumulation of yolk has had its disadvantages, when it limits the space for growth, is vividly shown by the fate of the Hag-fish Bdellostoma (Figs. 13, 15). In spite of the embryo's rich supply of blood and food, the brain (cerebellum) and nervous system do not, it seems cannot, develop in the normal manner. The materials are there but they are too crowded together and the tissues remain in an elementary condition; in a sense one may call them degenerate.

In the opposite extreme, plenty of space and little or no yolk, the embryo is complete but small and stunted. In Amphioxus and Appendicularia the brain is not developed at all, nor is there any head or definite vascular system, but the muscular system is well represented.

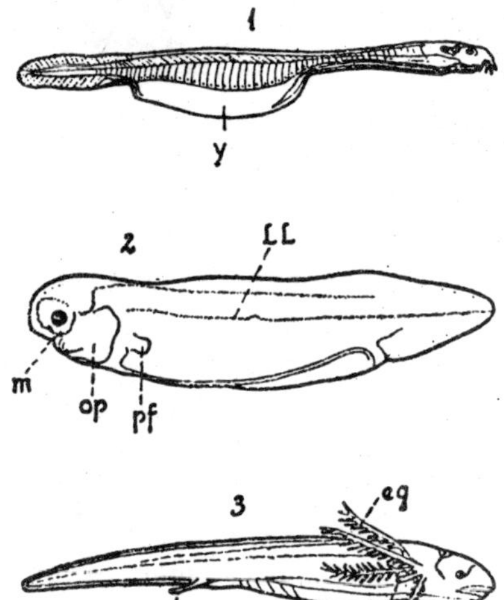

FIG. 15.—Larval fishes. 1. Bdellostoma; *y.*, yolk. (From Bashford Dean.) 2. Neoceratodus; *ll.*, lateral line; *m.*, mouth; *op.*, operculum; *pf.*, pectoral fin. (From Bridge, after Semon.) 3. Lepidosiren; *c.*, cement organ; *eg.*, external gills; *p.*, pectoral limb; *v.*, pelvic limb. (From Bridge, after Graham Kerr.)

Similarly in the Teleosts, little yolk with a perivitelline space, or even without the latter, means rapid but incomplete development, as in the Clupeids; a sufficiency of yolk and space gives more fully-formed larvæ, as in forms with demersal eggs. Here the mouth is open and the sustentative organs, blood, gills, and gut, are well-developed when the larva emerges from the egg. Such a degree is not reached by the Sprat larva, for example, until nearly a week after it has hatched.

If we think now for a moment of the conditions under which the embryos of Higher Vertebrates develop, we see that progress has consisted in the improvement of the inherited environment. In the Mammals the egg receives but little yolk, but it is kept in intimate connection with the nourishing fluids of the mother. And the "larva," when it emerges, is complete in all its parts and organs; what follows we may call growth rather than development, such as we see in the Teleosts.

The development of certain deep-sea Teleosts, which have come from advanced or specialised types, is worthy of a special note. Trachypterus, for example, lives at considerable depths, but the eggs when spawned rise into the upper layers, though not to the surface. The egg then swells, a perivitelline space is formed and the embryo thus obtains more space. Hence the development is rapid and the muscular system again is early formed and comes into use (Fig. 13, 6); then the various organs follow in succession, head, vascular system, fins, even the peculiar coil of the gut, until the embryo is a complete miniature of the adult. The eyes of the embryo when first brought to the surface are large and protruding (telescopic), showing that the embryo reacts to differences of pressure, but they become normal after a few days.

The simplest and most primitive kind of fish larva can be pictured from a comparison of the larvæ of the Clupeids and their allies, for example, Herring, Sprat, Albula and Chirocentrus. The characters of such a larva are as follows (Fig. 18, 1; p. 85).

There is a uniform segmentation of the body (myomeres) from head to tail. The notochord is very large, occupying most of the breadth of the body, and is unicellular or bicellular (Fig. 17, 1) where in later forms it becomes multicellular. The nerve cord lies immediately under the dorsal margin of the body. The straight gut, closed anteriorly, extends the whole length of the body right to the tail, indicating the nearest known approximation to the neurenteric canal. The just hatched embryo has no mouth, no gills, and no kidneys; it cannot breathe or feed or excrete like a fish. The head is a membranous swelling in front of the myomeres and the eyes are large but unpigmented, thus without the power of sight. Other sense organs are as yet wanting. There is no pigment in the body. The blood is colourless (lymph) without hæmoglobin, and the vascular system is not developed.

From this simple organism all other fishes, past or present, can be derived, as will be seen from the following pages.

The change from the larval to the postlarval condition consists in the complete absorption of the yolk and the development of the various organs. The mouth forms as the head separates from the yolk; the head continues to swell, the yolk decreases and the membrane in between ruptures. In the Elasmobranchs, where this occurs early, the embryo is anchored for a long time to the heavy yolk-sac and the mouth takes the form of a crescent. The head can only move from side to side and gradually becomes flattened and pointed. In the Teleosts, such as the Herring, the head tilts upwards when released from the yolk connection and the mouth then becomes narrow and opens forwards. How the mouth grows will be considered in the next chapter.

The gills or breathing organs are forming in the pharynx before the mouth opens, and very frequently the outside membrane connecting the shoulder-girdle with the head is ruptured so that the gills are for a time free and unprotected; in other cases the membrane persists and a special opening has to be made, as in the Pipe-fishes, Plectognaths, and Eels. These phenomena are connected with the movements of the head on the shoulder-girdle.

As the yolk diminishes, the liver can be seen as a large dark-coloured organ lying around the blood-vessels leading from the yolk into the heart. A large vacuole in its substance indicates the gall-bladder and various lymph spaces. Behind this there is another lobed organ similar to the liver but slightly lighter in colour, the pancreas, and mixed up with it we find the beginnings of another organ, which has played a very important part in the life of the bony fishes, the air-bladder. This begins as a slight swelling like a small finger on the top of the gut (posterior end of the œsophagus), as if a globule of air or gas had been swallowed and stuck there. In some forms, Pipe-fishes for example, it swells up almost at once into a large, thin-walled sac ; in others, like the Herring, the finger pushes backwards above the gut for a short distance, then appears to bend round or send forward a branch, which gradually works its way onward to become connected with the statocyst (" hearing " organ) in the head. Its progress is rendered easy in these forms by the numerous lymph spaces. The posterior end meanwhile expands backwards until it occupies the whole space above the hind portion of the gut.

Some interesting facts with regard to the development and fate of the air-bladder have recently been discovered. Moser has described how the digestive tract twists round during the development, so that the position of the duct changes, travelling round with the gut. In this way we can understand how the duct disappears in the specialised Teleosts (Physoclisti), where the gut shows a varying amount of twisting. In the Pipe-fish Tracy found that when the swelling had reached a certain length, the duct leading to its hinder end from the digestive tract shrivelled up and disappeared. In many of the Clupeids the air-bladder obtains a second opening directly to the exterior, the " safety valve," as de Beaufort calls it, when the young fish sink to the bottom and come under greater pressures.

4. Origin of Ossified Structures

The principal feature in the postlarval stages of the Teleosts is the development of bone, and to understand its formation we have to borrow from what is known regarding the general processes of living organisms and the past history of fishes. Some people think that bone arose because of its usefulness, but this is putting the cart before the horse ; it was probably far from useful to begin with.

In the beginning the fish had no skeleton—no skull or backbone or fins. It was more or less simply a worm. But even a worm is a living being, with blood and muscles and nerves. Moreover, it feeds, and within the recesses of its lowly system are many chapters of organic chemistry.

One of these concerns the excretion and utilisation of by-products. In civilised countries the rank of a corporation can be measured by the use it makes of waste, and in this respect fishes mark a turning point in the evolution of animal forms of life.

In the lower half of the Animal Kingdom substances that are a hindrance to the vital activities are disposed of in various ways. Some of the worms simply sweat out the chalk from their systems and then settle down comfortably in the tube. The star-fishes carry about a houseful of chalk, and the corals are more chalk than animal.

The snails and mussels may also be called to mind, but the point is that they make an inferior use of this by-product. They get rid of it from their active cells, but place it where it becomes a nuisance to the corporation as a whole.

This is the problem that was in great part solved by the fishes. The calcareous deposits have been taken to various parts of the skin, especially of the head, and to internal spaces where they would not interfere with the working of the muscles. Sometimes they are left loose in narrow spaces, as in the vertebræ of the Elasmobranchs, but the proper method is when the chalk is enclosed in cells and these arranged in the meshwork of membranes.

This has been achieved by the Teleosts or bony fishes and, as it appears, by some of the earliest of fishes. The accompanying figure of the Sprat's skeleton (Pl. VIII) shows the result in a simple form. It shows a wealth of small curving bones in a parallel series decreasing gradually towards the tail. We have but to picture the muscles on each side of the bones to realise what happens. The active muscle cells do not want the calcium crystals, and these have been deposited in the passive membrane between. Or this deposition may be due simply to the varying dissolving power of the colloid substances in the muscles and the connective tissue. In the form pictured the ossification or calcification is of the slenderest description. The bones of the Sprat or Herring, though decidedly inconvenient, are not dangerous. It is not necessary to say how many there are, but the small figure shows the number that can be perched on one vertebra (Fig. 16).

The ossification of these flexible intermuscular bones of the Clupeids is so slight, that it cannot interefere with the movements, and the latter have evidently determined the arrangement of the bones. In the more strongly ossified fishes like the Perch, on the other hand, the bones control the working of the muscles. The Clupeids are thus of a more primitive nature in this as in other characters.

We can distinguish, therefore, between the origin of bone and its usefulness. In fact, there are two stages to be passed through before a structure becomes useful. It has to fit in with the parts already present, and then be moulded by the surrounding forces. For the most part, the structures of fishes are only in this regulating stage; very frequently they do not fit in well with one another, and some part disappears or cannot develop. And it seems necessary, for bone to develop, that the larva should be free and swimming in the water. Bone is a product of the metabolism of a moving body; the embryos or larvæ of the Elasmobranchs, which lead sheltered lives, do not develop bone though calcified rings may form round the vertebræ.

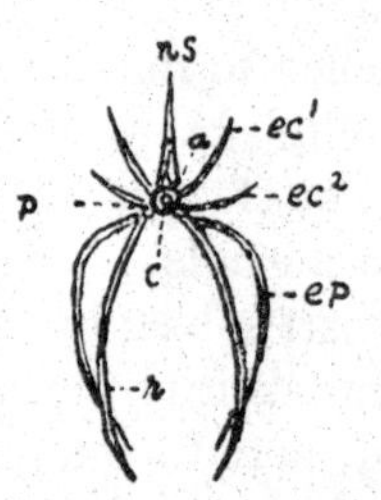

FIG. 16.—Vertebra of Herring. *a.*, neural arch; *c.*, centrum; *ec*[1], *ec*[2], extracostals; *ep.*, epipleural; *ns.*, neural spine; *p.*, parapophysis; *r.*, rib.

The vertebral column or central axis of the fish depicted consists of a long tube filled with a gelatinous substance, the notochord. Though like a jelly, it is by no means lifeless. It shows a cellular structure which varies in a significant manner in different groups of fishes. In the Clupeids two cells (or one whole cell above and two halves below) go to form one centrum of the vertebral column (Fig. 17). If these separated, each forming a cylinder or centrum, we should have two centra to each neuromere or

PLATE VIII

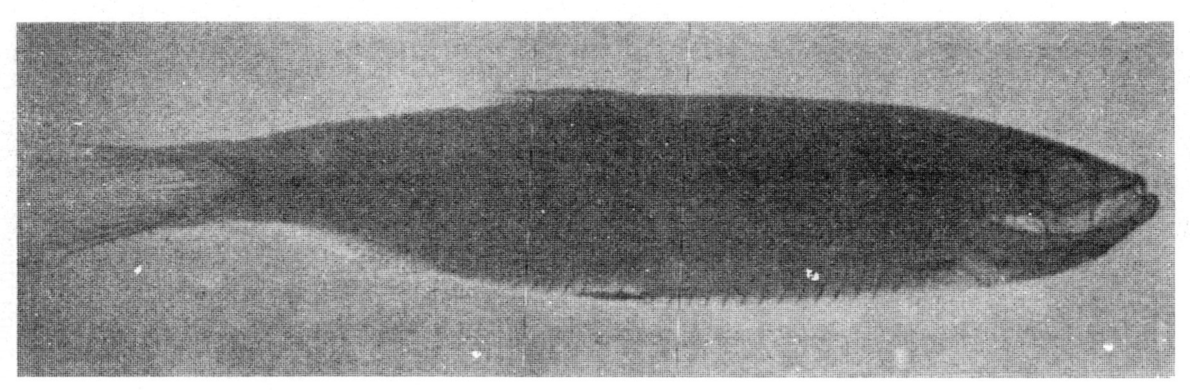

Skeleton of Sprat from an X-ray Photograph.

nerve element. This is the condition found in the Elasmobranchs—but only in the caudal region, which emphasises the fact that the segmentation of the vertebral column is effected by the muscular action. The same condition is found in the caudal region of Amia and its fossil allies (Fig. 17) as well as in some other important cases to be mentioned later. It is possible that the duplication of the vertebræ, especially the caudal, in the Eels, many Blennies and other elongated Teleosts, has been brought about in the same way. But in the specialised Teleosts (Percoids, Scom-

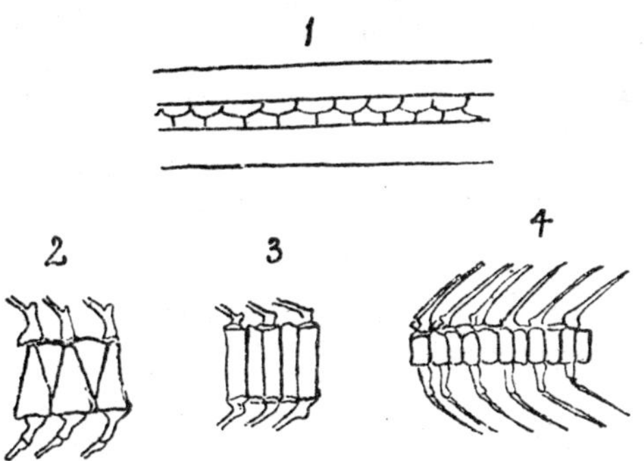

FIG. 17.—Segmentation of notochord and vertebræ. 1. Cellular structure of notochord in a Clupeid. (From Ehrenbaum.) 2. Precaudal vertebræ of Eurycormus. 3. Caudal vertebræ of Eurycormus. 4. Caudal vertebræ of Amia. (2, 3, 4, from Bridge, after Zittel.)

broids, etc.) the notochordal substance has become more concentrated and multicellular.

For a long time after the young Teleost is hatched, the notochordal tube is straight and its walls become slightly ossified in the same way as the small bones. Then the movements of the muscles lead to its breaking up into segments or centra, each between a pair of muscles. Ossification ceases in the parts between the centra where they are working on each other, and thus the whole axis comes to have the appearance of a long series of small tubes or cylinders, most clearly seen in any postlarval Clupeid. The number of these cylinders in any species is not determined until the little fish has been moving about for some time. In the

Teleosts, therefore, the number is determined under the combined influences of the movements and the outer conditions.

As the fish grows and the intermuscular bones develop, the arches curving over the nerve cord come to press upon the sides of the small cylinders and become cemented thereto, as it were, by cartilage. With the continued pressure of these arches against the sides, the cylinders take on the shape of an hour-glass, most typically seen in the Clupeids. On examining the centra of a Herring we find that they remain quite hollow, filled with the notochordal substance and without ossified bounding membranes.

This is the simplest form of vertebra found in the fishes, and may be added to the primitive characters of the Clupeids. Further progress or differentiation consists in the cartilage of the arches working its way round the cylinders and replacing to various degrees the original bony covering. Then the cartilage itself becomes calcified and we get the more diversified vertebræ we see, for example, in the Cod or Perch, with the bounding membrane also ossified and showing annual rings.

In the earliest fishes of the Palæozoic period the calcareous by-products were deposited more in the skin than in the vertebral column and more especially in the head region, at least in most of the forms preserved as fossils. These appear to have been slow-moving fishes living on the bottom and in fresh water. The probability is, that their constitution, and particularly their excretory system, was inferior to what we find in the fishes of the present day. Yet some of the early forms were undoubtedly active with very much the same structure as the modern fishes, and it is therefore not easy to say whether fishes have really made any " advance " since those early days.

CHAPTER V

REGULATION OF THE FORM AND STRUCTURES

FISHES have been living in the water for some millions of years and have learnt, or been forced to learn, how to adapt themselves to circumstances. In the beginning, however, they were simply innervated muscular bodies, and structures have arisen as a natural expression of the living processes and activity, without any reference to utility. In the same manner each individual begins as a muscular body and in its development presents us with a picture of the past history. From a number of these pictures we may abstract the essential features in the formation of fishes.

In certain ways, as by some authors, the regulation of the structures to one another and to the surrounding conditions might be called adaptation. But in considering the origin and making of things it is better to get away from the idea of utility implied in the latter term. The special adaptations of fishes will be discussed in a later chapter.

1. THE INFLUENCE OF BALANCE AND MOVEMENT ON THE FORMATION OF STRUCTURE

With the completion of the postlarval stage and ossification of its skeleton, where this occurs, the young fish becomes fixed in its characters for the rest of its life. It grows in length and thickness, but is no longer able to effect any noteworthy rearrangement of its structures. Its form, sense organs, digestive system, even the teeth, are already determined, and all it has to do is to find a suitable environment. Needless to say, the environment of the adult is not necessarily the same as that in which it was born. All fishes pass through several different environments in the early stages and the resultant structure is a summation of responsive co-ordinations. The most interesting period, and the one from which we may hope to obtain the best information regarding the past history, is therefore the early stage when the structures are forming.

It has been shown in the preceding chapter, that the vertebral column arises under the combined influence of the movements and the outer conditions. But the nature of the movements depends naturally on the balance or poise of the fish. If the fish wishes to raise its head, the muscles along the back are called into play. And if the fish is too heavy at the head end, there will be a pull on the tail end upwards. The latter phenomenon occurs in many fishes and leads to the formation of a characteristic structure in fishes, the tail. How the latter is formed in various ways under the influence of the movements, will be understood with the aid of the accompanying figures.

Taking one of the commonest types, we find that the notochord to begin with is quite straight and slightly ossified (Fig. 18, 1, 2). Then the muscle along the top seems to break away from the tip—apparently connected with the raising of the head—and crosses the notochord some distance down (3). Here the separated muscle cells proceed to form a number of cartilages, and with these as base, the muscle then bends up the tip of the ossified notochord (urostyle) into what is called the heterocercal position. Under the strain at the bend successive portions of the urostyle are broken off on each side and form the so-called uroneurals. Then the notochordal substance at the tip being constricted is withdrawn or atrophies, and the tip of the notochord also collapses to form a uroneural or two uroneurals. The number of cartilages below (hypurals) may vary from two to eight, as in the figure, and in the final stages several splint bones (epurals) develop above to support the upper rays of the caudal fin. It is in this way that the apparently symmetrical, homocercal tail is formed.

An interesting correlation, first pointed out by Schmalhausen, lies in the fact, that the homocercal tail is only found where an air-bladder occurs or has been, and one can readily understand why this should be. The air-bladder buoys up the fish, so that the continual use of the tail to raise the head is unnecessary. The Elasmobranchs, however, have lost the air-bladder and the tail remains in the heterocercal condition.

If the body is long and thin, the tail is less under the

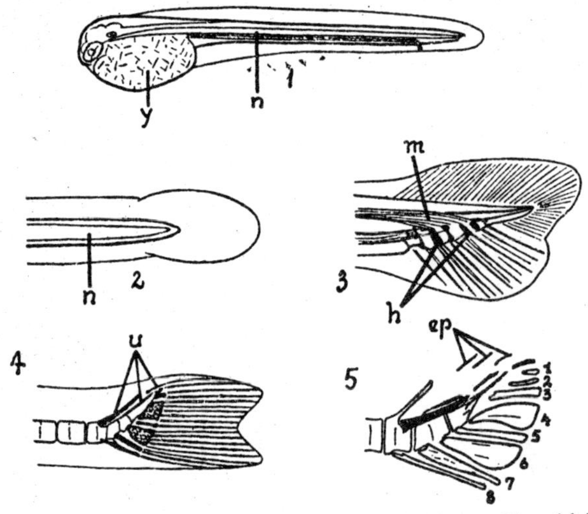

FIG. 18.—Development of the homocercal tail. 1. Clupeid larva; *y.*, yolk, *n.*, notochord. 2–5. Upbending of notochord and urostyle; *m.*, muscle fibres; *h.*, hypurals; *u.*, uroneurals or parts of urostyle; *ep.*, epurals; 1–8, hypurals.

influence of what the head does and often remains more or less straight, in which case it is called diphycercal. The embryonic fin is a fold of skin that surrounds the body above and below and has almost certainly arisen simply from the beating of the body from side to side against the resisting medium. It is little or not at all developed in the forms like the Cyclostomes and Elasmobranchs, which pass through the larval and postlarval stages whilst still in the egg-capsule. Nor is it well-developed in the larval forms which are rounded in section and display a wriggling form of motion, like the Pipe-fishes and some of the elongated Blennies.

Perhaps the most interesting and instructive way of forming a tail is that shown by the Eels. Here the body is long, very muscular and strongly flattened from side to side. On examining a series of Leptocephali from about 18 mm. and onwards we notice that a dark line is gradually forming out in the embryonic fin close to the margin, and under the microscope it is not difficult to see that this line is formed from the collection there of numerous muscle cells. These can be traced back to the muscles of the body near the tail, which gradually become thinner and more transparent than the muscles further forward on the body. A study of Plate IX (1 and 2) will show how the end of the notochord becomes deprived of the muscle fibres at the sides, just owing to the streaming of the muscle cells towards the tip and out on to the fin.

It is evident that the violent movements of the end of the tail cause the streaming of the muscle cells outwards, and they come to rest mainly along a fold in the embryonic fin. There they proceed to form rods of cartilage, two rods to a segment, right round the end of the notochord. The latter is but little bent upwards, but is pinched off as usual and the rods grow backwards towards the body (Pl. IX, 3, 4) to meet the end of the notochord and the developing neural and hæmal spines. The hypural cartilages are seen to be the same as (homologous with) the basal supports of the other fins, dorsal and anal, and have nothing to do with parts of the vertebræ. It has been shown above also, that the so-called uroneurals have nothing to do with neural arches.

We see, therefore, that movements and muscle cells form one of the most important structures of the fish. The other fins arise in precisely the same way. The loose muscle cells or buds are present naturally all round the body, even along the abdominal region. The concentration of the muscle buds into local areas in the Dog-fish has been described by Goodrich (1906), who remarks that every

PLATE
IX

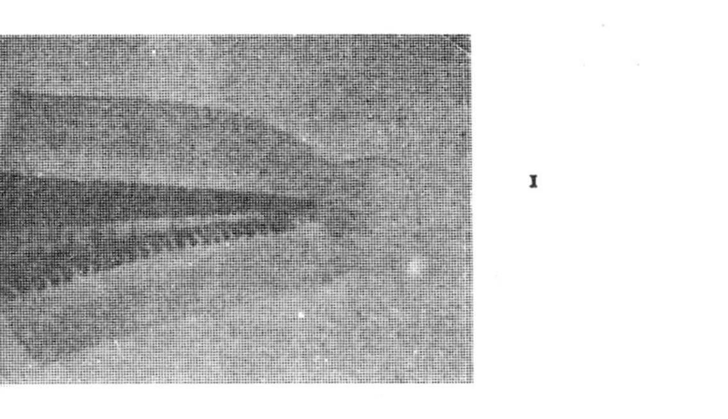

1

3

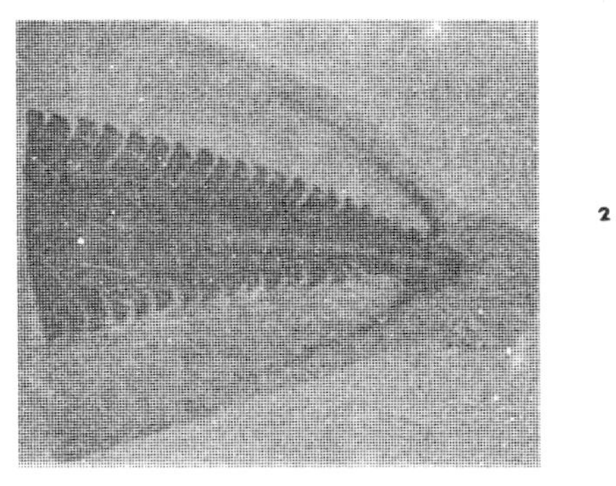

2

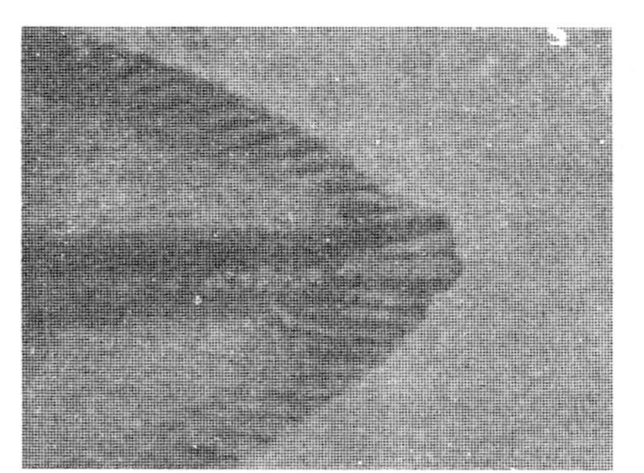

DEVELOPMENT OF THE TAIL IN THE EEL.

Figs. 1 and 2 from drawings *ad nat.* by Helmuth Lissner, 3 and 4 from Johs. Schmidt.

segment has these buds. The reason why they do not all develop is, that the movements of the fish do not affect all parts of the body equally, that is, if the movements are at all rapid or violent. As the body bends and twists, some places are more at rest than others, and it is just in these that the muscle buds are able to develop. The forms displaying greatest movement, in the young stages, have the fins

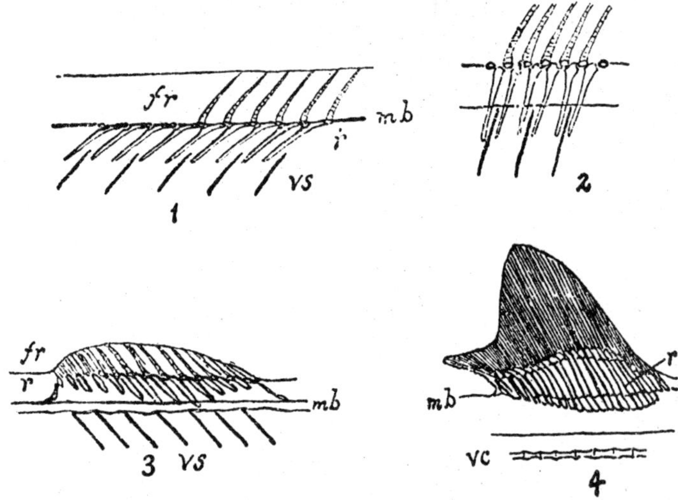

FIG. 19.—Formation of the fins. 1 Slow movement (Ammodytes); radials (*r*) within margin of body (*mb*). 2. Stronger movement (Flat-fish); radials beyond margin of body. 3. Rapid movement (Clupeids and Argentina); radials pressed out to margin of embryonic fin; *vs*., vertebral spines; *fr*., fin-rays. 4. Violent movement (Dog-fish); radials well beyond margin of body and tri-segmented, no true fin-rays; *vc*., vertebral column. (4, from Bridge, after Mivart.)

concentrated at definite spots, the zones of comparative rest, as in the Sharks and Dog-fishes, whilst in the slow-moving or wriggling forms the fins are more evenly distributed along the margins of the body.

The accompanying figures show how the structure of the fins varies according to the movements of the body. Where these are rapid, as in the Sharks, the radials or supporting rods of the fin are pressed well out into the embryonic fold (Fig. 19, 4), just as in the case of the Eel.

The fin-rays are also affected; they do not get a chance to develop, one may say, in the Sharks. In the Eels and Eel-like fishes they are very short, or also absent. Whilst in the Salmon and Herring tribes they are only of moderate length as a rule, and the radials are formed well out in the embryonic fin (Fig. 19, 3). Broader fins are met with in the more slowly moving larvæ, Carangoids, Scombroids, and Perches,

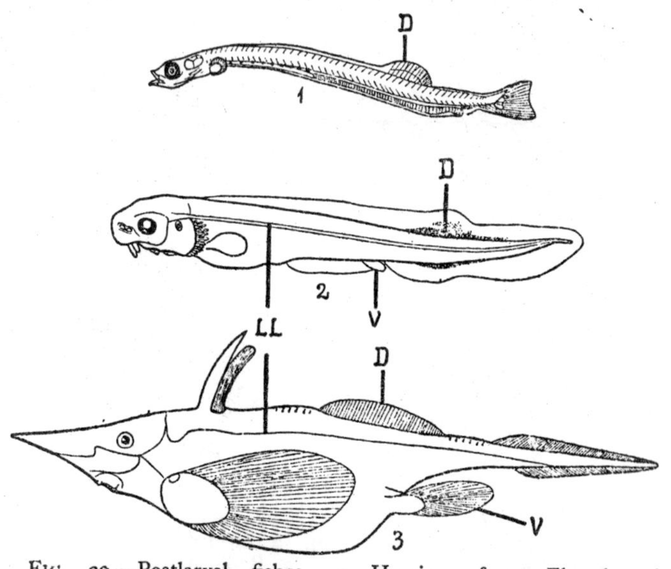

FIG. 20.—Postlarval fishes. 1. Herring from Ehrenbaum); 2. Sturgeon (from Ehrenbaum); 3. Harriotta, a deep-sea Chimærid (from Goode and Bean); D, dorsal fin; LL, lateral line; V, ventral fin.

and the fin-rays become stronger and ossified as spines, in the parts of little movement, as in the first dorsal fin above the abdominal cavity.

The movements of the body thus determine the structure and position of the fins, and as the movements change during growth, the fins may alter position and character. As Goodrich has shown in the case of the Dog-fish, the posterior part of a fin may cease to grow whilst new radiais may develop in front. The young Mackerels have uniform, continuous second dorsal and anal fins, but as the tail increases in length and strength, the continuity is broken and only isolated fin-lets are able to persist. In the Salmon tribe a developing dorsal fin has its growth checked and remains as an adipose fin.

The changing modes of movement are expressed in the order in which the fins appear. The caudal fin is the first to develop, then among the Teleosts the second dorsal and anal appear together. But the common Clupeids are unique in that the first dorsal develops before the anal and whilst the caudal is still heterocercal (Fig. 20, 1). This is the condition seen in the earliest fossils of fishes, and it is also present to some extent in the Sturgeon (Fig. 20, 2). But in the latter, and more especially in the Elasmobranchs (Fig. 20, 3), we see that the body tends to broaden out, owing to the fact that it is not actively engaged in swimming, and the ventral fins (V) thus appear early.

Where the movements are comparatively slow and sinuous, the dorsal and anal are long and develop together, as in the Flat-fishes and Blennies. If more rapid, these long fins break up into parts which develop separately at a number of nodes and then gradually extend their bases until they meet. The Chimæroids and Dipnoi have also retained this sinuous form of movement, and if we draw a line from the mouth along the middle of the body to the centre of the tail, we can see the condition that determines it. The parts above and below this line are equal or approximately so, in other words, the centre of gravity lies in the middle line (cf. Fig. 2).

If now we extend the series of diverse forms, we can see how the shape of the fish and its structures change according as the centre of gravity lies below or above this line. In the Clupeid forms the centre of gravity is below and the anal fin develops along the ventral margin, whilst the dorsal fin remains small about the middle of the body. In the Physoclistous forms (Carangoids, Scombroids, Perches, etc.) the centre of gravity lies mostly well above the middle line and the anal remains comparatively short, whilst the dorsal develops along the greater portion of the back. We can understand this phenomenon on comparing it with the rocking movements of a vessel. When the centre of gravity is well below that of buoyancy, the upper part has a short but rapid swing and the dorsal rays have difficulty in developing ; where the centre of gravity is above

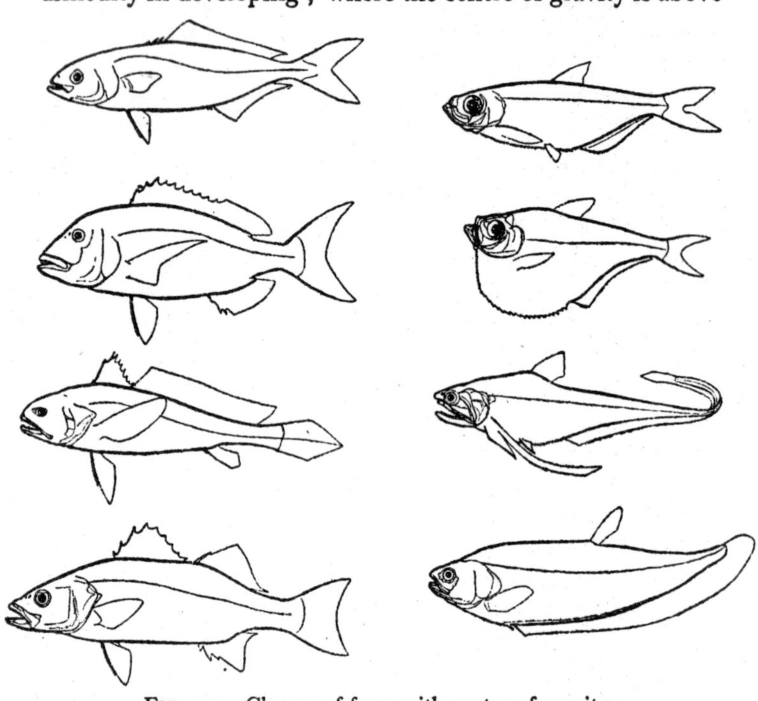

FIG. 21.—Change of form with centre of gravity.

Left : Naucrates, Dentex, Sciaena, Labrax (Morone).
Right : Pellona, Pristigaster, Coilia, Notopterus.

that of buoyancy, the upper part has a slow and long swing, and the dorsal fin is the longer. A metronome might also be used to express the same differences. Hence the lateral swing movements as well as the longitudinal have a great influence on the formation of the fins.

We thus come to the conception of balance in fishes, a matter that has been all important in their differentiation.

It is not easy to follow the intricacies of the many movements in a fish's body, but a simple and natural picture happens to be present in the colour markings of various forms. The distribution of pigment in a fish is not altogether due to the movements of the body, as other factors, the condition of the tissues for example, have also to be taken into account, but in many cases the influence of the movements can be clearly seen. In the Chætodont Holocanthus (Pl. X) the centre of activity obviously lies in the caudal region, from which the lines of force, as represented by the markings, spread out forwards, backwards, and above and below in very symmetrical lines and curves. The line from the mouth to the middle of the tail divides the body into equal halves above and below. From the pectoral region forwards and on the tail the curves are replaced by almost straight, vertical lines, showing that the movements here are from side to side. The vibrating movements of the dorsal and anal are also represented in the crossing of the curves at their bases.

These markings, we may believe, are formed in the same way as the scales, by the wrinkling of the skin caused by the action of the muscles underneath. The darker lines probably represent the zones of comparative rest, the light lines where the skin is wrinkled. Holocanthus is a slow-moving fish which probably rocks above and below as it moves, with the forward movement approximately equal to this rocking movement. If the rocking movement is the greater, however, the curves would break up into spots of various kinds, as is seen in the Argus-fish, Scatophagus, whilst greater forward movement leads to the development of transverse bars, as in the young Salmon, etc.

Before proceeding with a discussion of the phenomena connected with the balance, we have to consider the origin of the paired fins, pectorals and ventrals. In the true fishes these are formed in precisely the same manner as the vertical fins, that is to say, from muscle cells or buds pressed out along the ventral margin. But while the vertical fins are formed of two halves, each belonging to one side of the body which fuse together—except in the case of abnormalities where the double nature of the fins can be seen ; in the paired fins the buds are most usually separated from each other and pressed out to the side. In some cases the marginal skin-fold is also continued forward under the abdominal cavity, but this seems to be a secondary condition, due to the weak development of the ribs. As the latter grow they usually reach down to near the ventral margin, and the space for the paired fins is thus restricted.

The possibility of fins developing anywhere along the abdominal margin is shown, however, by what we find in a number of cases. It has been mentioned, that the muscle buds when they get the chance form cartilage, and in many of the Teleosts we find cartilaginous bars along the ventral margin of the abdomen. These are most developed in some Flat-fishes where they extend along the whole distance from the shoulder-girdle to the anus. Then in forms like Merluccius, Atherina, and the Mugils the middle portion is well developed. In Gastrosteus, the Plectognaths, and the ancient Arthrodirans (Fig. 48), the bar extends from the shoulder-girdle a long way back under the abdominal cavity. In Balistes there is a noteworthy marginal fin (or fused ventral fins) just behind the massive, fused bars.

These variations can be understood from the following considerations. The muscle buds which go to form the cartilaginous bases of the paired fins are pressed out into the skin or membrane along the margin of the body. Under the abdominal cavity there is no space for them and they come to rest up on the sides, between the abdominal muscles and ribs and the skin. Consequently, they have a better chance of developing when these muscles are but feeble, as in the Flat-fishes and the other forms mentioned. But where the muscles come into use early, the ventral fins may not be able to develop, as in the Eels. In the Cyclostomes the pectorals have suffered the same fate. But in some of the Eels and Flat-fishes the pectorals may be discarded even after they have begun to develop.

These cartilaginous bars along the ventral margin

PLATE X

(*a*) Movements and Colour-markings.

In Holocanthus (left) and Scatophagus (right). (From Brehm's "Tierleben," 4th ed.)

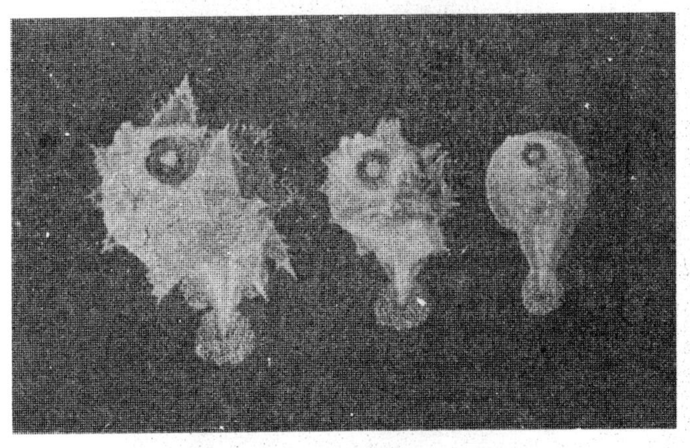

(*b*) Metamorphosis of Sun-fish (*Ranzania*).

(From Johs. Schmidt.)

should be close together to begin with, and in the Clupeids we find that the anterior portions are fused across the middle line (Goodrich, 1920), forming a coracoid plate. In the Plectognaths, as mentioned, the two bars are fused along their whole length and in some other forms, like Gastrosteus and Monocentris, they are closely apposed to one another. This may be considered the primitive condition, or a return to the primitive condition. In the great majority of cases, the bars and coracoid plates are either worked up the sides or disappear. Possibly the pelvic bone in the more specialised Teleosts is the real homologue of the primitive coracoid plate ; in which case the cartilage bones forming the base of the pectoral fin should really be called scapulars and not coracoids, as is usually done.

The removal of the pectoral fins up the sides may be referred to the tendency of the body to swing from side to side just in this region, and is more strongly marked in those forms which have an anterior air-bladder or none at all. In the Elasmobranchs the bases of the fins are spread well out to the sides and upwards.

Whilst the pectorals of fishes have thus undoubtedly arisen from the ventral line, it is possible that they are in no way homologous with the fore-limb of Higher Vertebrates. In the Dipnoi, for example, the long pectoral seems to have arisen from an interlamellar gill-bar (Graham Kerr), as is indicated both by its position and structure (Fig. 26). The peculiar mode of development of the fore-limb in the frogs also points in the same direction.

There is also evidence that the position of the ventral fins is determined by the balance and movements of the body. Where the ventral fins are abdominal, they develop just under the expanded portion of the air-bladder, thus in a position of comparative rest. In other forms they move forwards with the air-bladder and the anal fin (Fig. 21) or disappear.

The structure of the paired fins follows in the main the same rules as the vertical fins. Under strong movements of the fins the cartilaginous bases or radialia grow well out into the fin-membrane and form various kinds of limb-like structures whilst the rays are little developed, as in the Elasmobranchs and Dipnoi. Or the bases remain small and the fin-rays are well developed under less movement, as in most of the Teleosts. In general, the stronger the movements of the fin are, the longer it grows and the narrower, that is, the central element or series of elements becomes more prominent than the lateral ones, and then the median axis may become segmented in the same way as the vertebral column (Fig. 26). With less movement the breadth of the fin and number of rays increase, as in the Teleosts.

2. Causes of Change in the Balance

It is perhaps difficult to realise that fishes are not well-balanced. The Pipe-fishes and some others swim about in an almost vertical position, head upwards, Centriscus and some of the abnormal Goldfishes with their head downwards. The Flat-fishes, the Ribbon-fishes, and even some of the Labroids lie on the one side. The lower half of the Clupeids is heavier than the upper, whilst the upper part is heavier in the specialised Teleosts and the Sharks. The balance of the fish may thus be imperfect in all three dimensions.

If we examine more closely into the structure also, we find that few fishes are really symmetrical. The John Dory has more heavy plates on the one side than the other (Byrne). In the Serranoids and particularly the Sparoids the vertebral column and the pectoral arch are frequently so twisted to the one side, that one wonders how they have managed to escape becoming Flat-fishes. About 70 per cent. of the common Haddock have the snout bent to the one side or the other, slightly, it is true, but sufficiently to indicate the presence of some disturbing factor. As the Cod grows old, the bones in the front of the mouth tend to droop downwards and to the one side. These may seem to be exceptional cases in the apparently well-formed world of fishes, but if we also take the variations of normal forms into account, we see that this matter of balance has probably been the determining influence in the differentiation of the forms from one another.

The balance of a fish may be expressed simply as the disposition of the body substance. The young Clupeids have a small head and a moderately long body, the Eels a still longer body. The Perches and the Mackerels in the early stages have enormous heads in comparison with the body. These are able to maintain the head in a horizontal position in front of the vertebral column, but in a large number of cases (Carangoids, Figs. 6 and 22) the head is strongly bent down on the vertebral column. If we remember, that these fishes have almost certainly come from the Clupeids, we are induced to ask what can have been the factors which led to these changes.

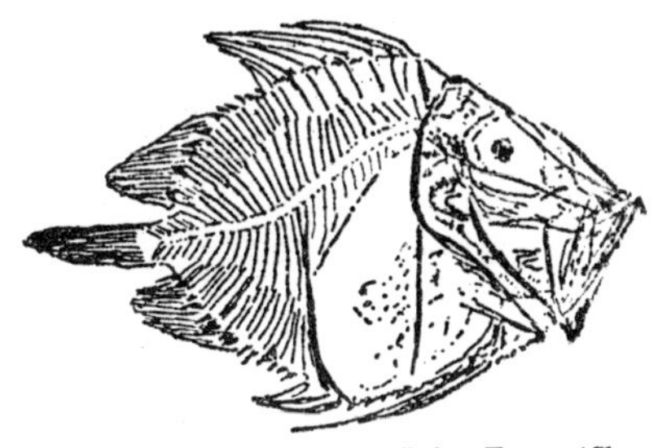

FIG. 22.—Radiogram of John Dory (*Zeus faber*), showing how skull is bent down on vertebral column.

A change in the disposition of the body substance can readily be pictured, when we compare the notochord of the early stages to an elastic string, which becomes narrower when extended, broader and more concentrated when contracted. Such a string would vibrate a good deal more in the former case than in the latter. Hence, if we start from a moderately long form like the Clupeid larva, we have to consider the influences affecting the length of the notochord.

A characteristic feature in the early development of the Clupeids is the way in which the gut seems to move forward to make room for the anal fin. In the beginning the gut is quite straight and reaches to the tail (Fig. 23). But when the larva has been moving about for some time, either the gut contracts or the notochord behind extends under the influence of the movements of the tail. Both of these processes may be going on at the same time, for the gut in these forms has an elastic or muscular coat. In most forms, however, this coat does not appear to be present, and the moving forward of the gut is due apparently to the movements of the caudal region. In Atherina (Fig. 23, 7) and the Mugils the gut moves forward to begin with, and then later extends itself again, and some other forms show even

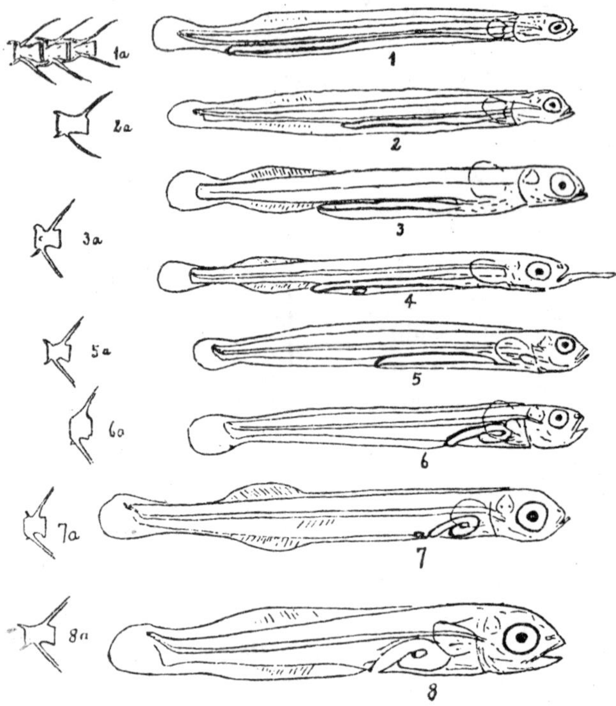

FIG. 23.—Change of form and structure with change in gut. 1. Herring; 2. Sand-eel; 3, 4. Rhamphistoma belone; 5. Gunnel; 6. Blenny; 7. Atherina; 8. Mackerel.

greater changes. In many cases also the gut forms a coil, which remains anteriorly, but the whole may grow backwards again later, as in the Trichiuroids.

The changes in the position of the gut and coil are exemplified in the accompanying figures. It will be seen that as the tail-end becomes larger and the gut moves forward, the whole body becomes deeper and the head larger. We may say, that the centre of balance or gravity is moved forward, with a greater concentration of the body substance anteriorly, and this again leads to a change in the mode of movement.

Whilst the changes in the gut have probably been the principal factor in altering the disposition of the centre of gravity, two other factors have to be noted. The air-bladder in the Teleosts has begun in a posterior position and has been obliged to move forward with the gut, thus influencing the movements and balance of the structures. Its influence in a special case will be noted later, but it may be noted that in Belone (Fig. 23, 4), where it remains posteriorly, the ventral fins develop underneath, thus abdominally, whereas in the other specialised forms figured they develop anteriorly (Fig. 23, 7). In the Dipnoi and some Teleosts the air-bladder is very long, reaching from the gullet back towards the tail, and the body also remains elongated.

The third factor affecting the movements and shape of the body has been the formation of the mouth. This has been a difficult matter for the Teleosts, as will be described below. In many cases, it appears, the young larvæ have to work hard for a living whilst the mouth is being formed.

The movements of the body in affecting the relative proportions of the different regions and in moving forward the gut and air-bladder have thus led to the change of form and structure among the fishes. It is probably impossible for the larva to perform these movements symmetrically, and this seems to have been the origin of the trouble. However symmetrical a body may be, as soon as it begins to move it swings and rolls from side to side owing to the unequal resistances of the water on its different parts, and the faster it tries to move the worse these involuntary motions become.

It might be thought that the fishes had long ago overcome these difficulties. And each species in its own particular way has done so and may now be said to have its own specific balance of structures, which is inherited. But the early larvæ have still to go through the same difficulties, and an interesting suggestion has been put forward by Bethe (1894), which goes far to explain why the development of the Teleost is such a complicated matter. He points out that the labyrinth of the statocyst or "hearing" organ is very slow in forming, and until that occurs the larva has, so to speak, but little control over its movements. The time of development of this organ may possibly serve as an index of the progress and differentiation of fishes. The Herring, for example, has no lateral line system on the body and the labyrinth is not linked up with the air-bladder until some eight to ten weeks after the larva is hatched. The young from demersal eggs, on the other hand, seem to have a well-developed labyrinth on hatching.

By constructing a series of models in the shape of a fish, Houssay (1912) has studied the various ways in which a body moves through the water. He found it by no means an easy matter to get any of them to travel without wobbling or swinging from side to side, even though made quite symmetrical. After many tests and failures he succeeded in making several which he could pull through the water at varying speeds in a fairly straight line. From these experiments he came to the conclusion, that the pressure of the water against a moving flexible body could account for all the morphological characters of the fishes.

Pressure is the main external factor in moulding the shape and structure and Houssay's work is of great value in helping us to understand the mechanical difficulties the fish has encountered, and perhaps more particularly, how the different rates of movement make a profound difference in the shape of the moving body. Yet, after all, the fish itself has been the experimenter and has been obliged to regulate itself. Had it been content, or been able, to remain a slowly moving animal, the waters would contain few real fishes.

A biological summary of the development of form and structure would appear as follows :—

1. First form of movement : very slow in regard to

distance covered, worm-like, sinuous; all parts of body approximately equal and flattened from side to side; no definite balance; tail straight or diphycercal; gut reaching to tail. Larva of Clupeids, Dipnoi.

2. Second form of movement: slow, sinuous in front, twisting or rocking behind; head becoming heavier, tail bending upwards; dorsal fin arising posteriorly; balance indefinite; gut being moved forward away from tail. Early postlarvæ of Clupeids, Eels, and Sturgeon.

3. Third form of movement: very rapid movements, whole body swinging and rocking from the heavy head backwards; tail strongly heterocercal; end of gut removed from tail; ventrals appearing; balance indefinite, but centre of gravity near the head. Sharks.

4. Moderate movements, wave-like, about a definite centre in middle of body; air-bladder developing; caudal fin becoming homocercal; anal fin appearing; balance more definite, horizontal. Later postlarvæ of Clupeids.

5. More rapid movements; front part of body moving as one piece from side to side, posterior half swinging, rocking or vibrating; caudal region well developed; gut and air-bladder moving forward; dorsal and anal developing at the same time; tail varying from homocercal back to diphycercal form; balance horizontal but centre of gravity indefinite. Eels, Blennies, Gadoids.

These will suffice to indicate how the various forms may have arisen and how they develop. Many kinds of intermediate forms of movement can be distinguished and slow forms appear at every stage. The body then becomes less compact, ill-balanced, even shapeless, as in the early Ganoids, Plectognaths, Pediculates, and Ribbon-fishes. Or it may become long and Eel-like, or even worm-like, as in the Dipnoi and some of the Blennies. Or the rate of movement may be speeded up to an intense degree with the whole caudal region developed into a remarkable propelling organ, as in the Mackerels and Tunnies.

3. Formation of the Head

In Houssay's experiments the head of a yielding body was flattened above and compressed from side to side on being drawn through the water. This we can readily understand, and it indicates one of the main factors in the formation of the head, namely, the pressure of the surrounding medium. If we compare the Herring or Cod (Figs. 1 and 2) with a form like the Lumpsucker, Cyclopterus (Pl. XIV), which from the beginning moves but little through the water and has a shapeless mass for a head, we see at once what a difference movement makes.

The head of a Clupeid larva just after it has hatched (Figs. 18 (1) and 27 (1)) corresponds very closely to the plastic body of Houssay's experiments. It is rounded, almost spherical, with the brain curved round in front and no mouth below. Internally it has practically no structure; there is a slender double rod of cartilage between the eyes and backwards towards the body, but the roof is entirely membranous and supported only by the brain, a condition that may certainly be regarded as very primitive (Wells). The "hearing" organ or statocyst is represented only by a cartilaginous nucleus, so that even the cranium proper, the part behind the eyes, has still nothing solid about it.

During the larval and postlarval periods this head gradually becomes narrow and pointed. The brain is forced back between and behind the eyes into the cranium, and a complicated mouth with projecting jaws develops. As a rule, these phenomena are simply referred to as growth changes, but there is good reason to believe with Houssay, that the definite shape attained is due to the pressure of the water as the larva swims through it. Growth if unhindered tends to be spherical in three dimensions, as in the case of the Lumpsucker; hence the actual form and therewith the underlying structure are determined by the movements and the surrounding pressure.

In contrast with this we have the Elasmobranch embryo which develops under sheltered conditions within an egg-capsule. The head of these forms might be called shapeless (Fig. 14), simply because they are not exposed to the pressure of the water ; nevertheless, one can see that they seem to be flattened on the top in front. But we have to take this in connection with the fact, that these embryos display a very considerable amount of movement from side to side. Thus, the pliable substance of which the head is composed is, so to speak, pressed out to each side by this movement, whilst

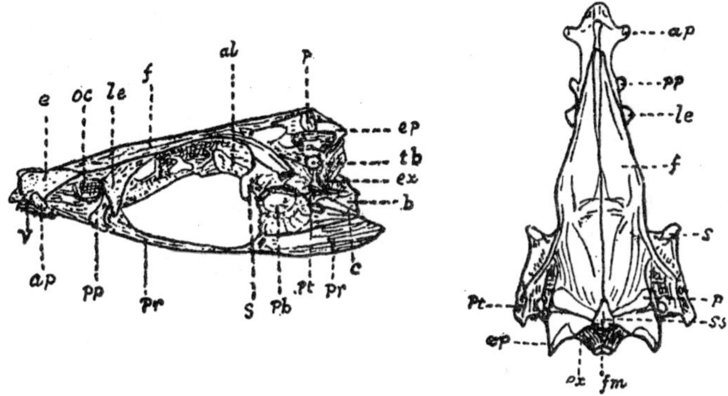

FIG. 24.—Skull of Herring from the side and above. *ap.*, anterior palatine process ; *pp.*, posterior palatine process ; *e.*, ethmoid ; *le.*, lateral ethmoid ; *v.*, vomer ; *oc.*, olfactory cavity ; *pr.*, parasphenoid ; *f.*, frontal ; *al.*, alisphenoid ; *s.*, sphenotic ; *pb.*, periotic bulla of lateral line system ; *tb.*, tympanic do. ; *pt.*, pterotic ; *c.*, occipital canal of lateral line system ; *b.*, basioccipital ; *ex.*, exoccipital ; *fm.*, foramen magnum for nerve cord ; *ep.*, epiotic ; *ss.*, supraoccipital ; *p.*, parietal.

the central cartilaginous base still continues to grow forward. Hence the skull comes to have a flattened appearance.

Another contrast is presented by certain deep-sea fishes like the Gastrostomids (Fig. 10), which have all the appearance of being rapid swimmers or of trying to be so. Here the head is almost completely flattened from side to side by the greater pressures of the deep sea.

If we examine now the skull of an adult Herring (Fig. 24) we see how, in agreement with this view, it is shaped like a triangle, viewed from the side or above, with the apex in front. The posterior part is high and broadened out where it is pressed against the trunk. But the most interesting thing is the position and nature of the bones. The upper part or roof of the skull is formed by the long membranous frontals (*f*) with very small parietals (*p*) and supraoccipital posteriorly (*ss*). Only the last is a cartilage bone.

The distinction between these two different ways of forming osseous tissue is a very important one. It has been assumed by many that cartilage preceded bone, and consequently, if bone forms in membrane without any appearance of cartilage, the latter must somehow have disappeared. But as a rule the possibility of bone forming directly in membrane without the aid of cartilage is now generally recognised, and we see here in the Clupeids that the roof of the skull is formed entirely of thin membrane bones. The older idea was based to a large extent on the belief that the cartilaginous fishes, Elasmobranchs, preceded the bony fishes, hence the term "Teleosts." But as a matter of fact, the earliest fishes had bony plates on the head of a membranous or dermal nature just like the modern Teleosts, and what have been taken to be the earliest Elasmobranchs were also provided with similar ossifications. It is unnecessary to believe, therefore, that cartilage preceded bone. It seems, indeed, that cartilage develops mainly if not entirely in places not exposed to great pressures, and where muscular tissue (mesoderm) is present. This cartilage may or may not become ossified by the in-wandering of cells from the surrounding membrane or by the deposition of calcium salts within, but it seems in all cases necessary that the fish or larva must be free-swimming before this occurs.

The two forms of ossification show a remarkable distribution in some cases. Thus, the neural and hæmal spines are mostly formed from bone laid down in the membrane between the muscles, but the radialia and the ribs are first of all formed of cartilage, derived probably from the muscle cells. Similarly, in the head region the parasphenoid (*pr*), which forms the strong supporting axis of the base of the skull, is really a membrane bone though it is formed underneath and close to the primordial cartilage. The palatine is usually formed from cartilage, but it may have an entirely membranous origin.

The formation of bone in the membrane above the head has come as a direct response to the pressure of the surrounding medium. Originally, as explained in a previous chapter, the bone was probably laid down in the head and body simply as an excretion, and its arrangement later into definite forms and positions has come just from the outer pressure. Then it came to serve as a protection to the brain.

Another factor of importance in determining the shape of the head and its structure is the inherited specific balance. The normal tendency is for the head to grow forward horizontally from the trunk, but very frequently it is bent downwards on the vertebral axis, which means a restriction of the space for the organs and structures under the head, and thus a special rearrangement of the latter. This depression of the skull is mostly seen in the advanced or specialised groups of Teleosts, which have lost the open communication of the air-bladder to the gullet and in which the head is massive in comparison with the caudal region (Carangoids). It is clearly due to the balance being imperfect, that is, to a surplus concentration of the substance anteriorly.

Perhaps the most important factor in determining the shape of the head, and even of the body in many cases, has been the development of the various structures connected with the mouth. From the great variety of these structures we can see that this has been a very difficult matter for the fishes. There are three different kinds of mouths and each naturally has led to differences in the skull above and the arches behind. The relations of the various parts of this complicated region are shown in the accompanying diagram of the structures in the adult Herring (Fig. 25). In some ways this is the most advanced type of mouth, but it lends itself also to an explanation of the simpler, though not necessarily more primitive types.

It will be seen that the tip or symphysis of the lower jaw is connected to the foot of the pectoral arch (thus to the body) by way of the hyoid arch (broken line). This is the case in all Vertebrates and usually the mouth is placed in a

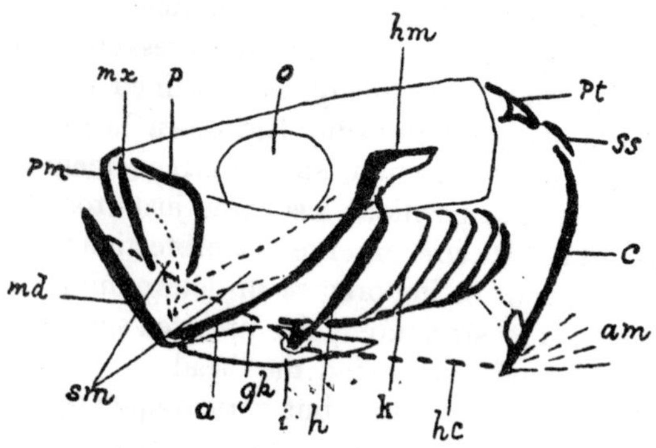

FIG. 25.—Head arches of the Herring (quadrate and gill-cover omitted). *pm.*, premaxillary; *mx.*, maxillary; *p.*, palatopterygoid; *o.*, orbit; *hm.*, hyomandibular; *pt.*, posttemporal; *ss.*, suprascapular; *c.*, clavicle; *am.*, abdominal muscle; *hc.*, hyoclavicular connection; *k.*, gill-arches; *h.*, ceratohyal; *i.*, interoperculum; *gh.*, geniohyoid muscle to tip of lower jaw; *a.*, suspensorium; *sm.*, cheek-muscles to close mouth; *md.*, mandibular.

ventral position, so that the connection serves as the opening apparatus. But the compression of the head laterally in the Teleosts has led to the point of the lower jaw being produced forward, and this has been the origin of the difficulties.

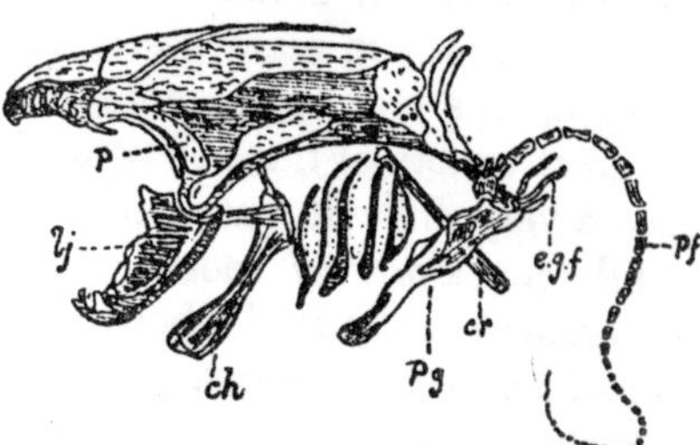

FIG. 26.—Skull of Protopterus. (From Wiedersheim.) *p.*, palatine; *lj.*, lower jaw; *ch.*, ceratohyal; *pg.*, pectoral arch; *cr.*, cranial rib; *e.g.f.*, external gills; *pf.*, pectoral fin.

The pectoral arch is represented here as consisting of three parts, the uppermost connected with the skull, but in the young stages as in all Higher Vertebrates the arch is quite free. Sometimes the lowermost bone (clavicle) or the second (suprascapular) gains a direct connection with the side of the skull, as in the Plectognaths and the ancient Arthrodirans, or it may be connected by ligament to the base of the skull, as in Protopterus (Fig. 26), but as a rule it is quite free and is able to swing backwards and forwards. The muscles working on its base are indicated in the figure. On the one side are the oblique abdominal muscles (*am*), on the other the muscles to the branchial arches, the hyoid arch (*hc*) and thence to the lower jaw. For the mouth to open properly it is clear that these arches must be nicely adjusted to one another. This is seldom the case among the Teleosts, and the various ways in which the difficulties have been overcome have led to the differentiation of so many groups of fishes. Sometimes the mouth remains large and horizontal, more often it is inclined upwards, becomes smaller and can hardly be opened. As the jaws are working on the front part of the skull we thus find a variety of changes in the shape of the head.

The chief weakness has lain apparently in the lower jaw. To begin with, this consists of a rod of cartilage ; it may have been part of some muscular tissue before that, but it is generally known as Meckel's cartilage and is supposed to represent the first of the gill-arches. If this had remained in connection with the skull, as in Protopterus (Fig. 26) and higher Vertebrates, or had the skull not been pressed upwards away from it, there would be little difficulty. But in the Teleosts it is long and thin and soon breaks up into a number of separate pieces. In other words, it crumples up and is unable to serve as a jaw. That this is due to the opening apparatus will be exemplified later in describing the transformation of the Pipe-fishes.

Some new arrangements now appear, which go to form a stronger lower jaw. The chief of these is the descent of the palato-quadrate arch. This comes to support the middle apparently of Meckel's cartilage, where the lower jaw is jointed on to the posterior part. Thus a new mouth is formed and this persists as the mouth of the Sharks and Rays. But in the Teleosts it rather adds to than diminishes the difficulties. It keeps the suspensorium (*a*, hyomandibular, symplectic, quadrate) away from the skull and thus forms a wide mouth ; but when the lower jaw closes above the level of the hyoid arch, it cannot be opened by the mechanism from that arch. The difficulty is increased at this time by two things : the lower jaw ossifies and a long rod of cartilage, the preoperculum, forms along the posterior margin of the suspensorium. The latter is then unable to bend. It may be drawn far backwards and thus allows an enormous mouth to be formed, as in some deep-sea Teleosts, but in many others it remains forward and then for a time the mouth cannot be opened, and the postlarvæ have difficulty in obtaining food.

This remarkable condition has now been revealed in a large number of Teleosts. What it leads to in the case of the Flat-fishes and the Pipe-fishes will be described later. But it seems clear that the little fishes make violent efforts to overcome the difficulty. Some elongated forms work their muscles to such an extent that the digestive tract is forced forwards right under the head. In others the palatine arch gives way, as in the Pipe-fishes, and in a strange form, Stylophorus, recently described by Tate Regan (1924), the whole of the upper part of the mouth drops or is pulled away from the skull. How the mouth is able then to open or close is still a mystery. In some other forms, as in Regalecus (Fig. 9) and Zeus (Fig. 22) the whole skull is drawn downwards.

What saves the situation in most Teleosts is the development of the interoperculum (Fig. 25, *i*). This like the preoperculum has little or nothing to do with the real gill-cover. It forms along the outer side of the hyoid arch (ceratohyal) and its anterior end is connected with the outer face of the lower jaw, at first simply by skin, later by a ligament in which develops a small sesamoid or angular bone. By this means the lower jaw is rotated about its joint, when the fish begins to open its mouth, and then the original mechanism continues to function.

The changes in the upper jaw are equally interesting. The primary condition seems to be that shown by the Arthrodirans, Protopterus (Fig. 26) and the Holocephali (Chimæra, etc.), in which the palato-quadrate forms part of the skull. This was the case also in the ancient Cœlacanthids (Stensiö, 1921), and possibly Crossopterygii (Watson, 1912). In the larval Herring the palatine appears to separate from the side of the ethmoid cartilage in the early larva. Then, as mentioned above, it drops down to form the upper jaw of the Elasmobranchs and Ganoids. In the Teleosts, however, it remains in touch with the ethmoid and serves as a buttress for this above and the suspensorium below; in this position it has considerable influence on the shape of the skull. The membrane of the mouth is continued over this arch well in front and in it there forms the maxillary, an important membrane bone in the Teleosts with posterior ventral fins (Abdominales), and later, in front of the maxillary, the premaxillary. These bones naturally are subject to great variation, like the lower jaw.

That these changes in the mouth and head have come from the movements of the body and the changes in the alimentary canal, can hardly be doubted. The complications are so numerous, however, that any general description of the relations and events cannot be given. Each group of fishes has to be dealt with separately and the few examples that follow may serve perhaps as an indication of what has happened in other cases.

4. Transformations

In the course of the many changes gone through by the bony fishes certain critical periods, more critical than others, have arisen in which it would seem impossible for the individuals to survive. These are indicated by various signs, jaws working on the top of the skull, skull bent downwards with mouth and teeth reduced to tiny proportions, huge spinous processes on the head and body, fins enormously developed or discarded—indicating the severity of the struggle the organism has experienced in co-ordinating the growing structures. In some cases we can see in detail how the form has changed.

It has been indicated above, that the balance of fishes has been quite indefinite in the beginning. In making their way through the water they might be horizontal at a given speed, if moving faster they would tend to rise, if slower the head or the tail might sag. The acquirement of an air-bladder might seem to be a remedy for this state of matters, but a great deal depends on its position and how it grows. The Clupeids are able to do without an air-bladder for some time, and when it appears, it pushes rapidly along the whole length of the abdominal cavity, so that the greater part of the body is supported by its means. This arrangement also has its disadvantages, however, and the Clupeids had to make various other regulations, but for the moment we are more concerned with cases in which the air-bladder is cut short, as it were, in its growth and does not extend forwards or backwards. The little fish, supple and flexible, has then to balance itself on a knife-edge.

Pipe-fishes.—Such a condition is seen in the larva of Fistularia recently described by Delsman (1921). It is in every way a Clupeid larva, but within a short time its appearance completely alters. The air-bladder remains about the middle of the body (Fig. 27, 2, 3) and we can follow the signs of an alternating see-saw movement of the head and tail. At first the tail end preponderates. The head is being tilted upwards, not merely externally, for the ethmoid plate (or parasphenoid axis) is bent upwards relative to the mid-line of the body (Fig. 27, 3) and the forebrain is being pushed backward between the eyes. A permanent sign of this early compression is in some cases found in the fusion of the anterior vertebræ.

The ventral apparatus closing the mouth, which has been shown above to be all-important in the formation of the mouth, is endeavouring to counteract this upward tendency. The pectoral arch is retracted from the head by the abdominal muscles; the hyoid arch is strongly retracted also and thus pulls on the lower jaw, upwards and backwards, as well as on the gill-arches.

The lower jaw then closes and the critical stage is reached. The long and attenuated appearance of the body and the narrow gut indicate that the little fish travels far and fast in its efforts to obtain food, and in the meantime the mouth structures are being reorganised. The mouth apparatus, palato-quadrate arch, has completely broken down; the

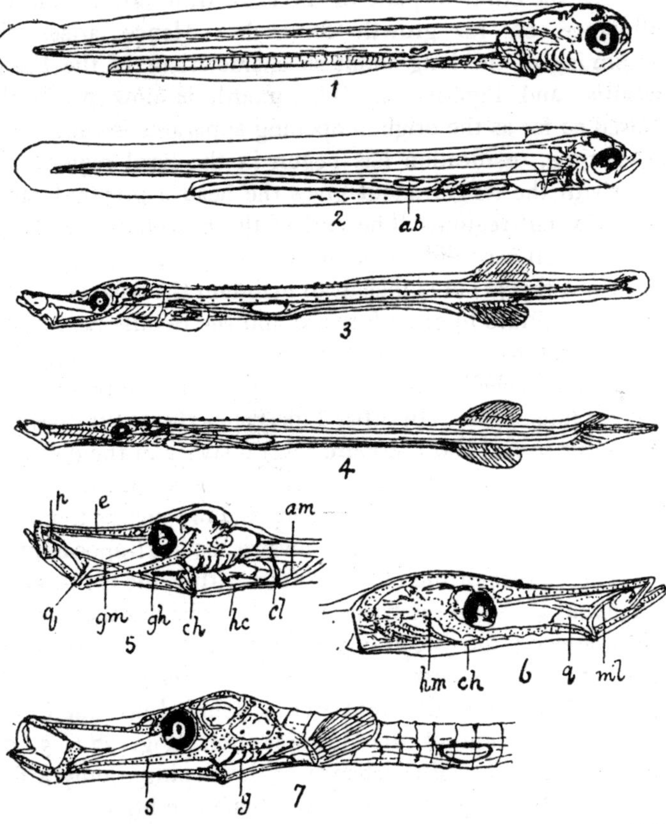

FIG. 27.—Transformation of a Clupeid into a Pipe-fish (Syngnathid). 1. Larva of Clupea finta. 2, 3, 4, 6. Stages in development of Flute-mouth (Fistularia serrata). 5. Nerophis aequoreus. 7. Siphostoma typhle. *ab.*, air-bladder; *am.*, abdominal muscles; *ch.*, ceratohyal; *cl.*, clavicle; *e.*, ethmoid; *g.*, gills; *gh.*, glossohyal; *gm.*, geniohyoid muscle; *hc.*, hyoclavicular muscle; *hm.*, hyomandibular; *ml.*, maxillary ligament; *p.*, palatine; *q.*, quadrate; *s.*, suspensorium.

suspensorium is separated from the quadrate, which is rotated upwards, and the latter is again separated from the palatine. These features are well seen in Ryder's figure of the larval Hippocampus (1882), and in the Pipe-fishes represented here (Fig. 27 : 5, 6, 7).

The effect of this breakdown of the palatine arch is to set free the ethmoid plate of the skull from any restraining influence on its growth forwards. Hence arises the beginning of the long snout so characteristic of the Flute-mouths and Pipe-fishes. The mouth is now practically closed so far as the original opening apparatus is concerned. Meantime, ossification has begun in the head region, and this with the longer snout gives the head a preponderance over the tail region. The end of the notochord bends up in the usual way and a caudal fin is formed. In Fistularia, however, this is not sufficient to maintain the balance. The snout continues to grow forward and the middle rays of the caudal fin grow backwards until they become a long append-age, almost one-fifth of the total length, characteristic of the adult Flutemouth. In other Pipe-fishes the snout does not grow out so far, and there are several stages in the develop-ment of the caudal region ; there may be a tail fin, but the preponderance of the tail becomes more marked, no fin develops and the tail curls downwards, as in Hippocampus.

The closure of the mouth brings with it the elevation of the hyoid arch and the basibranchials ; this has the effect of disorganising the breathing apparatus. In Fistularia the gills are able to develop, but are compressed and small. In the Pipe-fishes the hyoid arch is doubled up, due to the greater retraction of the pectoral arch (Fig. 27 : 5, 7), and the gills have little chance of developing ; only a few tufts are able to grow (Lophobranchiate condition).

The new mouth develops just as in the Clupeids. The maxillary is attached at its head to the palatine, carried forward by the ethmoid plate, and has the expanded jugal portion characteristic of the Clupeids. Normally this jugal bone is the base for the insertion of the closing muscles of the mouth, which arise from the suspensorium and skull behind the eye-ball. By the closing of the mouth these muscles are put out of action and other muscle fibres appear in front of the eye-ball which come to operate on the jugal.

When the mouth is firmly closed (Fig. 27, 4) this jugal projects above the snout, but when the new muscle works on it the jugal rotates downwards and the maxillaries in front open upwards like a lid. The closing muscle has become the opening muscle.

The opening muscle, as mentioned, became the closing muscle. The interopercular apparatus, however, tries to develop and in some forms like Syngnathus, it comes to help in opening the mouth by working round the angle of the lower jaw, but it develops too late to prevent the main changes in the mouth.

These changes can be followed in the developing young fishes. And if we compare a series of forms, from the more nearly normal Fistularia through Syngnathus and Nerophis to Hippocampus, we find that the changes are intensified and occur at an increasingly early stage. Further, through the disturbance of the gills and circulation the old tendency to deposit chalk in the skin revives and these fishes become encased in Ganoid-like plates, quite different in growth and appearance from the ordinary Teleostean scales. To counterbalance this retrograde step externally these fishes, of which about two hundred species are known, have apparently increased in intelligence. Not only have they adapted themselves with scrupulous nicety to particular conditions, they have even acquired the habit of nursing their eggs.

Flat-fishes.—The Pipe-fishes exemplify an imperfection in the balance fore and aft ; the Flat-fishes are ill-balanced from side to side. Some five hundred species are known, and these can be divided into two main groups according to the way this lack of balance affects them. In the one group the structures of the body grow asymmetrically, in the other the asymmetry that develops can be referred more directly to the irregular movements of the body.

Like the Pipe-fishes these forms do not show any asymmetry to begin with ; they are quite normal in the larval stage, but as soon as the coil forms in the alimentary canal trouble begins in the mouth and skull. The mouth goes through the same preliminary changes, bending upwards and closing, as those described in the Pipe-fishes. But instead of the palato-quadrate arch giving way, it begins to grow asymmetrically and other structures give way.

Owing to the imperfect balance in the body, the fish when it moves bends its head to the one side and the growing structures of the mouth are pulled to that side. The base of the skull is also pulled in the same direction and the buttressing palatine presses upwards on the ethmoids, until they also grow asymmetrically and thus affect the growing frontals. In this way the upper part of the skull gets pushed over to the other side. These changes occur when the little fishes are swimming freely in the water and thus have nothing to do with the later bottom-life.

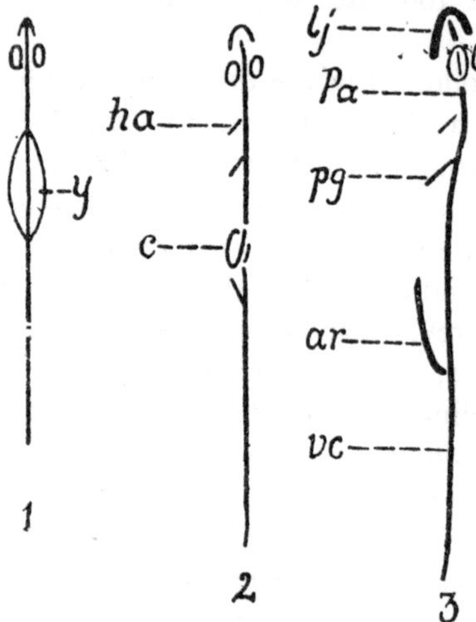

FIG. 28.—Loss of bilateral symmetry in Pleuronectes.

The principal course of events can be understood perhaps from the accompanying diagrams (Fig. 28), which show the changes as viewed from above. The larva with yolk-sac still attached (*y*) is quite symmetrical (1). Then a coil forms (*c*) and the various arches (pectoral girdle *pg*, hyoid arch *ha*, lower jaw *lj*) incline and grow to the side (2). The abdominal rod (*ar*) also grows to the same side and the asymmetry affects the vertebral column (*vc*) and its anterior continuation, the parasphenoid or ventral axis (*pa*) of the skull (3). The lower jaw may regain its symmetry to a great extent but it lies to the left of the median, longitudinal line (Pleuronectes) and the eyes are thus forced over to the right side.

If no air-bladder develops, the deflection of the mouth

is always to the left side, that on which the coil of the gut is formed. If an air-bladder develops, it goes naturally with the coil to the left side (with a notable exception, to be mentioned below) and thus serves to counterbalance the coil in the first stage. Later, as the abdominal organs settle down to a better balance among themselves, the air-bladder becomes too great a counterbalance and the post-larva is bent over to the right side. The asymmetry of the body then passes along this side to the mouth and skull in the same way as before, and the *upper* part of the skull is deflected over to the left side. In this way we obtain flat-fishes with their eyes on the right and left side respectively.

A curious thing is, that a cross between these two kinds may occur. Lepidorhombus and a few more have no air-bladder, and they begin with a deflection to the left side; their mouth parts retain the imprint of this early deflection. Then the body swings over to the right side and the upper part of the skull is deformed, as in the turbot (Rhombus). Again, the common Flounder usually inclines over to the left side, but about 30 per cent. in some places bend over to the right side. A perfectly symmetrical Flounder has even been found, and many examples of imperfect asymmetry, abnormalities, have been recorded. These illustrate how unstable the balance must be in these forms, and we find also that the abdominal organs are arranged in various ways, apparently to counteract the growing asymmetry.

The determining influence on the course of the growing asymmetry in these forms is the large abdominal rod, which grows downwards behind the terminal portion of the gut. So long as this remains vertical and leaves the organs free to swing about, the final deformation of the skull does not take place, even though the initial stages have begun. And many normal fishes among the Carangoids show this stage (Fig. 22). But when the rod grows forward below to enclose the abdominal organs, a characteristic of these Flat-fishes, the transit of the eye is rapidly accomplished. This rod grows to the one side and in the abnormal forms mentioned above (Lepidorhombus) it has a double curvature, first to the left, then to the right.

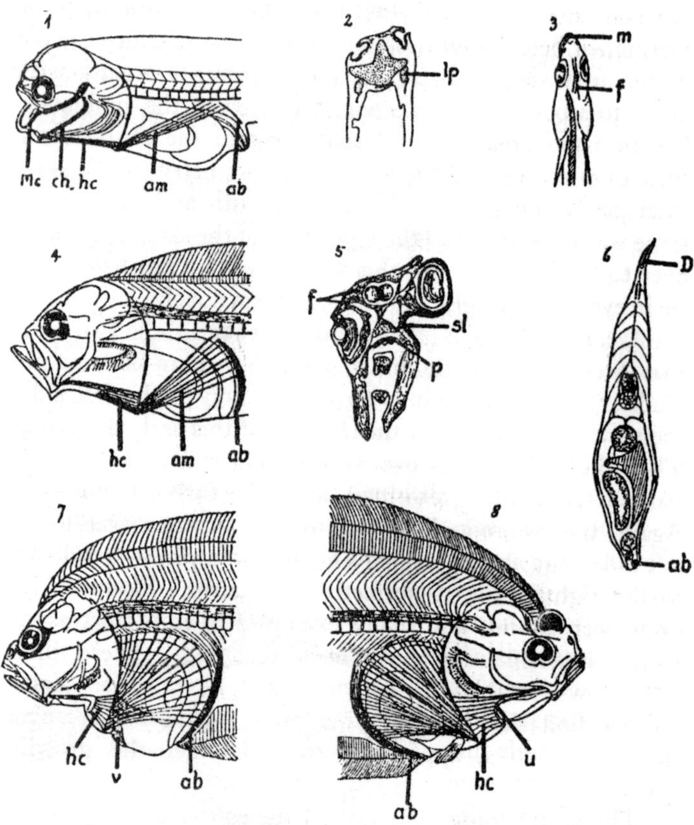

FIG. 29.—Metamorphosis of a Flat-fish (Pleuronectes). 1, 4, 7, 8, Distorsion of mouth to left side and rupture of opening muscle (*hc*) of mouth; *ab.*, abdominal rod growing to one side; *u.*, urohyal; *v.*, ventral fins; *am.*, abdominal muscles operating on pectoral arch. 2. Section through olfactory region, left palatine (*lp*) pushing up ethmoid on one side. 3. Mouth (*m*) and frontals (*f*) asymmetrical in opposite directions. 5. Raising of migrating eye in a sling (*sl.*); *p.*, parasphenoid; *f*, frontals. 6. Section through asymmetrical body of post larva; *ab.*, abdominal rod; D., dorsal fin.

The actual transit of the migrating eye is caused by the development of a ligament (*sl*, Fig. 29, 5), which arises from the compression of the tissues below the eye on the side towards which the mouth is originally bent. It is due to the compression and thickening of the connective tissue between the lateral ethmoid in front and the sphenotic behind. This thickening gradually increases and the eye-ball is lifted as in a sling and pressed over to the other side. In many cases the frontals have by this time been deflected and the one nearest the eye-ball may develop only partially, but in Bothus the frontals are strongly ossified long before the eye is compelled to move. And so powerful is the force used in moving the eye, that the frontal in this form is completely smashed and disintegrates. The other frontal, however, maintains its position, and the result is, that the migrating eye can find no permanent socket and continues to wander backwards above the skull throughout the life of the fish.

The close connection between the growing asymmetry of the body and the deformation of the skull can be judged from what happens to the opening muscles of the mouth. In the first stage, before the lower jaw is bent upwards, these muscles are quite normal. When the lower jaw ossifies and closes, the muscles from the pectoral arch cannot open it. Something must give way, just as in the Pipe-fishes ; here the lower part of the opening muscles breaks away from the anterior insertion and gradually forms a curve downwards (Fig. 29 : *hc*, 1, 4, 7). Along this curve the torn muscle cells are replaced by cartilage and a new bone, found nowhere else but in Flat-fishes, is formed. The result of this operation can be seen in the adults. When we pull on the lower end of the curved bone (urohyal) it rotates and helps the interoperculum to begin the opening of the mouth, then the upper part completes the process. Many other changes occur in the region of the head ; the arches are all inclined to the side, the muscles are partly torn away from their connections and the eye-muscles, symmetrical to begin with, find new attachments. The important thing is, that the organism has once more triumphed over difficulties that might seem insuperable.

These fishes are called flat, and no doubt the narrow-ness of the body helps to emphasise the imperfect balance, but many other fishes are just as "flat" without becoming asymmetrical in the head region. Some approach very near to this condition, but so long as the abdominal organs remain free, the asymmetry of the body does not make the skull asymmetrical.

In the other group of Flat-fishes (Symphurus) the body is not flat; it is rather rounded to begin with, such as we find in the Blennies and elongated Gadoids. The caudal region is comparatively long and there is a large air-bladder, as is so often the case, on the left side. The asymmetry is thus very marked in sections (Fig. 30, 2) and one can see that such a body must roll and twist when moving through the water. And this rolling has a bias to the left side affecting the tissues to such an extent, that the left ventral fin is unable to develop, whilst the right ventral is twisted down to the middle line. The pectoral fins are large and heavy during the postlarval stages. In these forms there is always a large amount of lymphoid tissue, under the skin and inside and outside the skull, showing the presence of friction between the different parts.

Here the asymmetry develops mainly from the violent swaying of the head from side to side. With the bias of the body the mouth parts are drawn to the left; the tissues of the right side of the mouth are ruptured and new connections are formed between the premaxillary above and dentary below.

When the front part of the anal fin with its supports develops forward, asymmetrically, just as in the forms with an abdominal rod, it compresses both the abdominal organs and the air-bladder. The following changes are rapidly gone through (Fig. 30, 3). The air-bladder suddenly disappears; the shoulder-girdle is pulled violently backwards; the mouth is compressed upwards; the front end of the snout (ethmoid) is also bent up and to the right. The third figure shows a little fish undergoing these changes. Behind, the pectoral fins are abruptly thrown off; in front, the right eye disappears from the right side and is squeezed through the head to the other side, the bending of the

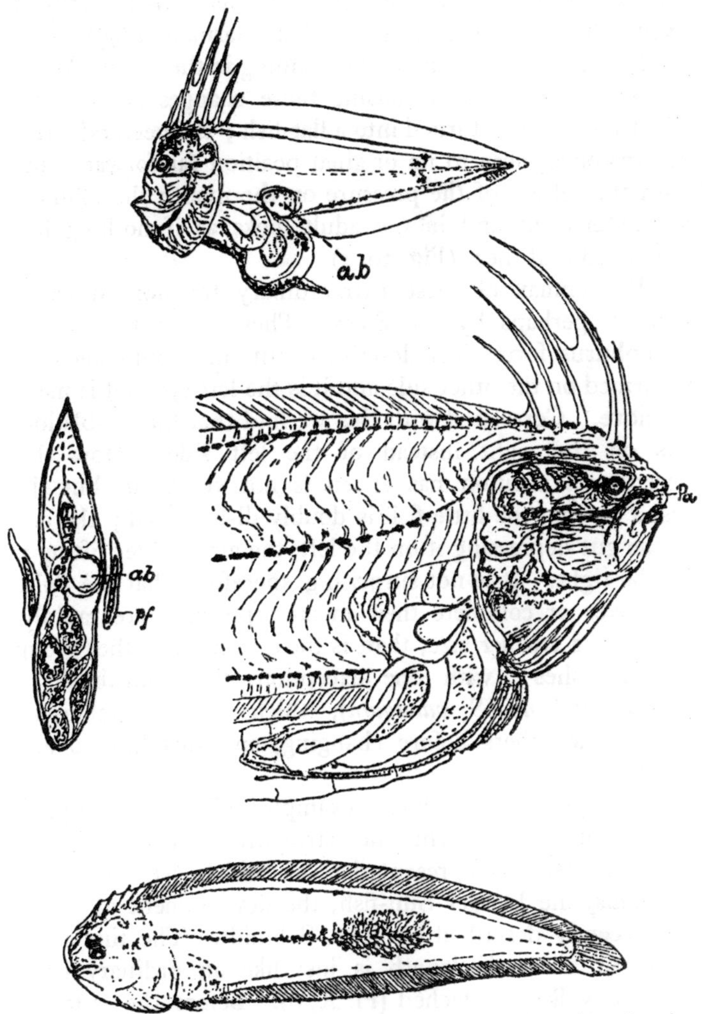

FIG. 30.—Metamorphosis of a Flat-fish (Symphurus). *ab.*, air-bladder ; *pf.*, pectoral fin ; *pa.*, parasphenoid axis.

Above, (1) young postlarva, with mouth distorted to left. To the left, (2) asymmetrical air-bladder seen in section. To the right, (3) right eye disappearing. Below, (4) metamorphosed specimen.

ethmoid by the jaws being the immediate cause of this transit. As in Bothus the right eye often fails to obtain a fixed orbit, and one case is recorded (Garman, 1899) of the eye failing to get through and remaining in the head. When the violent convulsion causing these changes is over, the Blenny-like fish is turned into a flat-fish ; the pectoral girdle returns once more to its original position far forward, and with the release of the pressure on the mouth the ethmoid straightens out and in the adult looks as if nothing had happened to it there (Fig. 30, 4).

The climax of these extraordinary transformations is reached perhaps by the Soles. These are very like the Symphurus forms just described, but the mouth becomes deformed on the other side and it is the left eye that is made to move. Correlated with this we find that the air-bladder lies on the right instead of the left side. How this exceptional position has arisen is not yet clear, but the result is, that the little fish is doubly ill-balanced ; the coil of the gut depressing it on the left, the air-bladder lifting it on the right. It is not surprising, therefore, that the Soles turn over or complete their metamorphosis more rapidly, that is at a smaller size, than is the case in any other group of Flat-fishes ; with one exception, the Thickback or Variegated Sole (*S. variegata*), which reaches double the size of the others before changing, and this has no air-bladder.

Plectognaths.—A third example of the complete transformation of form and structure is given by the Plectognaths, as representative of which we may take Ranzania, the Longer Sun-fish, the development of which was recently described by Johs. Schmidt. The tiny larva of about 2 mm. in length is just like an ordinary larva with the yolk-sac attached (Pl. X, *b*). Before the latter is absorbed, however, apparently when the mouth forms, water penetrates from the latter into the yolk, which has a much less pressure within it. The result is, that the whole ventral half of the larva swells up and the ordinary movements are prevented. The larva becomes a passive body rolling about in the water and the skin is thrown out into spinous processes which give it a curious appearance (Pl. X*b*, 2, 3). The tail consequently develops but little or not at all. But the little fish makes use of the dorsal and anal fins, which are not abnormal to begin with, to such an extent that they become the principal organs of propulsion, and the muscles operating them develop right back to the vertebral column.

These examples indicate that the regulation of the structures in fishes has been, is still, and always will be, imperfect. The little fish, if it is to swim freely, has to organise and co-ordinate many dissimilar parts and processes, which persist in nearly the same condition as when they first originated. Any change, any improvement or differentiation, can only come from less or greater movement, and then the environment plays its part. And at various periods the organism has given up the attempt to remain a free-swimming animal and has regulated its structures in relation to life on the bottom or on land.

CHAPTER VI

ECONOMY OF THE BODY

I. Production and Transport of Energy

Fishes must eat to live and the food must undergo many changes before it becomes stored in a condition that may be called potential energy ready to express itself in a kinetic form as movement. The processes of assimilation and excretion are very much the same in all animals, but the apparatus shows many degrees of specialisation ; we can hardly say perfection, since we do not know what would be the best. As with the skeleton the various organs of fishes connected with the digestion and transport of food-materials have not arisen owing to their utility, and one of the chief points to be noted in what follows will be the variation in directions away from what we should regard as improvements or progress.

1. Digestive System

As usual the food passes from the mouth through the pharynx into the gullet or œsophagus and thence into some kind of stomach and intestine. These different regions are not so distinctly separated from one another as in Higher Vertebrates.

Fishes have many sets of teeth ; in fact, there is hardly a bone in the mouth that does not bear teeth in one form or another. These teeth or denticles, which are of the same nature as the denticles or tubercles of the skin, are not specialised to any great extent, though in the Sharks and Rays the jaws may contain several kinds partly owing to varying degrees of use and renewal and partly to the fusion of some of the roots together. As they are seldom planted in sockets, but merely apposed or ankylosed to the jaw bones, they are readily lost and as readily replaced. Frequently, owing to some disturbance of the tissues in the young, several rudiments may fuse together to form composite teeth, and they are often absent altogether.

In addition to the " front " teeth, which may be preceded in deep-sea forms by " milk " teeth, there are pharyngeal or gill teeth on the last branchial arch where the oral cavity narrows towards the gullet. These are of a similar nature but are usually of a stronger build than the jaw teeth. In the majority of species there is no chewing of the materials, the fish simply bolt the food as fast as they can and they have no salivary glands. But in the Lung-fishes and perhaps the Mullets some kind of mastication seems to take place.

The distinction between gullet or œsophagus and stomach is not well-marked in fishes. In the Clupeids, indeed, the stomach is simply the swollen, hinder part of the gullet and forms a blind cul-de-sac, at the end of which we find the duct leading into the air-bladder (Fig. 31, *d*). Hence the statement in earlier literature that the air-bladder communicates with the stomach in these forms, is quite correct, judging from the adult structure. The development shows, however, that there is no stomach to begin with. Sometimes the cul-de-sac is of large dimensions and very distensible, as in Ammodytes and the Trachypterids. In the latter it extends backwards almost to the tail, and in a neighbouring form, Lophotes, it seems to have obtained a direct communication with the exterior.

In the beginning, therefore, the stomach has been more a retaining organ or crop than a digesting organ. In some fishes the muscular coat is so strongly developed, that it forms a regular gizzard. This adaptation is found in the Gillaroo trout and in the Mullets ; in some of the Wrasses we find the opposite extreme, a simple, straight tube gradually narrowing backwards.

These examples illustrate the low stage of development

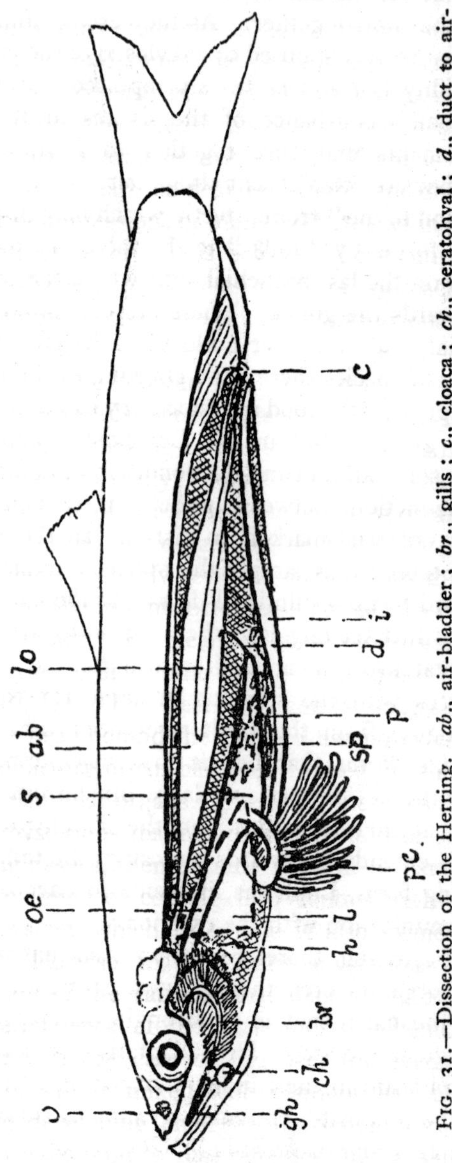

FIG. 31.—Dissection of the Herring. *ab.*, air-bladder; *br.*, gills; *c.*, cloaca; *ch.*, ceratohyal; *d.*, duct to air-bladder; *gh.*, glossohyal; *h.*, heart; *i.*, intestine; *l.*, liver; *lo.*, left ovary; *o.*, olfactory organ; *oe.*, œsophagus; *p.*, pancreas; *pc.*, pyloric cæca; *s.*, stomach; *sp.*, spleen. (Liver removed, but position indicated by broken line.)

still persistent among the fishes. In many cases also the epithelium is ciliated to help the food along. This reminiscence of a primitive character is common in the Dipnoi (Protopterus) and the Ganoids, and is possibly connected with a sluggish temperament, but more probably they cannot help themselves and their sluggishness comes from the retention of this and other primitive characters. We can imagine, at any rate, that a greater development of muscle would react on the digestive organs and lead to increasing metabolism, thence to greater differentiation of the various parts.

It is worth noting in this connection that the Clupeids, which begin as it were at the very beginning of the fish series, take their place in the adult condition among the most highly organised. They are active fishes everywhere and show a variety of innovations, which other fishes and higher animals have turned to special account. Not content with turning part of the gullet into a stomach (*s*), they have expanded the latter into a sigmoid flexure, so that the food travelling round the stomach returns almost to where it started from. Though many fishes have given up this peculiarity, we still retain it.

There is a constriction, but no valve, where the stomach leads into a narrow but well-made duodenum. This coils on itself in the Clupeids and the food must of necessity pass through slowly. It is just at this place that various important glands are developed and pour out their ferments for the digestion of the food. Then follows a long thin intestine (*i*) with numerous folds in the mucous membrane for the retention and absorption of the digested "brew." There is no large intestine or colon, the thin-walled gut passes into a short and muscular rectum, which opens out to the exterior. In other Teleosts the rectal opening is distinct from the urinogenital opening; in some Clupeids, Herring, Sprat, etc., there are four openings all together in a single depression which may well be called the beginnings of a cloaca (*c*).

Some interesting variations in the above apparatus may just be mentioned. Bennett (1879) found that the four external openings in the Herring could be reduced to three, and this in different ways. In one case the air-bladder and the genital ducts united at some distance within the body, in another there was a common urinogenital opening. Ridewood (1898) found that the air-bladder did not open to the exterior in the Twaite Shad (*Clupea finta*).

These and other variations are symptomatic of the position of the Clupeids. The withdrawal of the rectum within the body, which we can see beginning in these forms, has led to the formation of a definite cloaca in the Sharks, Rays, and Dipnoi. The persistence of a separate rectal opening, but fusion of the kidney and genital ducts into a common urinogenital opening, is a characteristic of the Teleosts.

Internally the long thin intestine frequently folds and twists on itself—simply as a matter of growth. The folds may remain separate, as in the Perch and many others, but their fusion has apparently led to the formation of spiral valves of various kinds, as in the Clupeid Chirocentrus, Ganoids, Dipnoi, and the Elasmobranchs. In many Teleosts, on the other hand, the twisting and folding begin further forward in the pyloric region, and coils or windings are formed, as in some Clupeids, the Gadoids, Flat-fishes, and Mugils. In the Loricariidæ and particularly Plecostomus (Bridge) the intestine is disposed in numerous spiral coils like a watch-spring, just as in the frog.

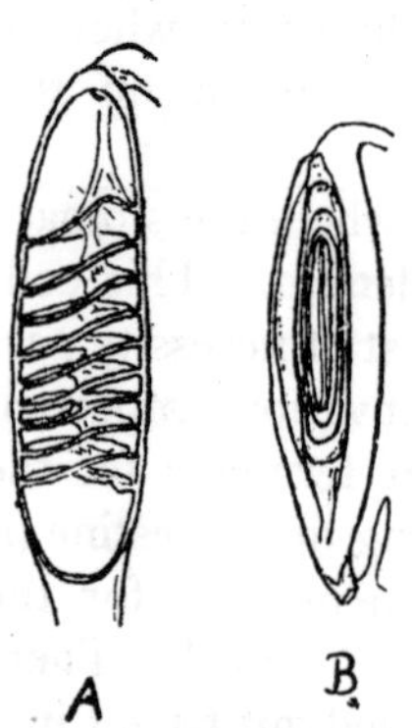

FIG. 32.—Spiral valves of the Elasmobranchs. A., Ray; B., Hammer-headed Shark. (From Bridge, after Jeffery Parker.)

In the fishes, at any rate, we cannot speak of these varying phenomena as adaptations, since they are not connected with any particular habits. The spiral valve (Fig. 32), some have believed, is an adaptation in that it increases the absorptive area of the intestine, and if the fishes possessing it were vegetable feeders, there would be some meaning in such a view. But the majority are carnivorous, the Elasmobranchs exclusively so, and one may think with some reason, that the valve is a nuisance and a source of danger to the latter. In any case, it is not a progressive character. If the frogs came from the Lung-fishes (Dipnoi), they showed their greater wisdom by at once getting rid of this undesirable character.

The process of digestion in fishes is very similar to that in higher animals and is carried on by the same means. In general, that is, but in particular there is a great amount of variation in the distribution of the secreting glands and ferments. Mucus glands in the form of goblet cells are abundant everywhere from the mouth to the rectum. The stomach, as mentioned, is chiefly a retaining organ and gastric glands are occasionally absent and often restricted to the first part. On the other hand, the digestive ferment pepsin is found in some cases (Eels, Carp) not only in the stomach but also in the intestine (Krukenberg, 1877). Acid-secreting cells are also present in the stomach (Stirling, 1885) but are not specially differentiated as in higher animals.

The chief seat of digestion is the duodenum and it is here that we find the greatest variety in fishes. In addition to the two ordinary digestive glands, there is a mysterious kind peculiar to fishes, the pyloric cæca (Fig. 31, *pc*). The liver is a large organ, especially in the Sharks, and lies just behind the shoulder-girdle close to the heart (*h*). A gall-bladder is usually but not always present, and there may be as many as three separate hepatic ducts. The pancreas is well-developed in the Elasmobranchs, Acipenser, and in the larval Teleosts, but in the adults of the latter it becomes diffuse, isolated lobules occurring in the mesentery between the stomach and intestine and even in the substance of the liver. The pyloric cæca open from or into the latter part of the duodenum and are extremely irregular in their appearance. They are quite wanting in the Dipnoi and Elasmobranchs, with exception perhaps of the Greenland shark Laemargus (Turner, 1873) and some Skates (Gegenbaur, 1892). Among the Teleosts they are also absent in many groups, some species have only one or two, whilst in the Gadoids and Scombroids there may be over a hundred.

With regard to the function of these organs, the liver as usual is mainly a storehouse of sugar and fats. The bile is not an active agent in digestion in itself, but it appears to stimulate the other glands. Recently it has been discovered, that the strengthening powers of the oil pressed from the liver of the Cod depends in some way on the presence of the accessory vitamines, which though unseen are able to influence digestion. The pancreas secretes certain ferments (amylopsin and lipase) which are able to turn starch into sugar and fats into fatty acids. By itself the pancreatic juice has little power to digest proteid substances (Sellier, 1902), but requires the assistance of the other juices, particularly from the intestine (spiral valve in Elasmobranchs, perhaps the pyloric cæca in Teleosts) and even the spleen. The bile sometimes increases, sometimes diminishes the activity of the pancreas (Sullivan, 1907).

The pyloric appendages may be simply absorbing organs in some cases (Krukenberg, 1877), but in others all the ferments of the pancreas seem to be present (Stirling). Possibly the difficulty of separating the scattered elements of the pancreas from the other organs may have affected the last-mentioned result. The thin intestine is mainly an absorbing area and for this purpose the mucous membrane is developed into folds or ridges and even crypts. In some cases, however, the middle intestine is the principal seat of active digestion (Krukenberg). From his extensive studies this author came to the conclusion, that fishes in their digestive functions show the various stages of evolution from the Invertebrates to the Higher Vertebrates.

As a sign of the range and diversity of the digestive system in fishes it may be mentioned, that attempts to correlate the differences with one another and with the nature of the food or mode of life have so far been quite unconvincing. The pyloric cæca of the Teleosts have been compared to the spiral valve of the Elasmobranchs and both certainly mean an increase of the absorptive area, but the

Ganoids and Dipnoi have both and many Teleosts have neither. The fact is, that we know too little about the origin of these organs. It is said, for example, that the liver and pancreas are outgrowths from the mesenteron (primitive gut) and that the liver is the homologue of the simple sac found in Amphioxus. So far as the Teleosts are concerned, one may doubt this derivation; in such a case there would be no need for several ducts. It seems more probable, that the liver and pancreas arise from the hypoblast cells of the embryo which are actively engaged in converting the yolk-food before the gut is formed. The pyloric cæca, on the other hand, do not develop till much later, and it is possible that they arise in the same manner as that suggested for the air-bladder, from the accumulation of small globules of air or gases at the spot where the digestive tract alters its straight course and makes a bend; later, the duodenum is formed at this bend. The pyloric cæca may be correlated with the pancreas, as some have thought, in so far as their growth interferes with the growth of the latter and perhaps leads to its dispersal.

2. Circulation and Respiration

The varying activities of digestion and secretion within the organism would be of little avail were no mechanism present to convey the products from one place to another. On the one hand, the alimentary canal is like a retort in which the most intricate chemical processes we know of, are constantly reducing the food to a brew suitable for transport through the membranes. On the other hand, the living tissues are something more than a sieve permitting the passage of the dissolved constituents; active cells are present selecting the different elements and conveying them into the blood or lymph (leucocytes), whilst others are busy removing the harmful ingredients (phagocytes). There are two channels for the circulation of the absorbed elements, the blood and the lymph.

The circulation of the blood in fishes (Fig. 33) follows

in the main the same courses as in Higher Vertebrates—forwards ventrally, backwards dorsally above the abdominal cavity, with several ramifications between. These ramifications have arisen through the increasing size and importance of the various organs. For example, the venous blood with its stores of food collected from the intestine makes its way forward to the liver, where some of its constituents are removed by the cells of the latter, then assembles again into a large vein and passes forward to the heart. Similarly, the venous blood from the caudal region frequently passes

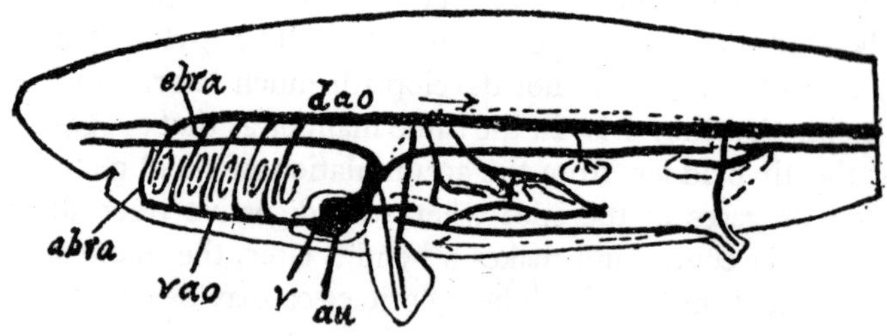

FIG. 33.—Circulation of the Dog-fish (*Mustelus*). The venous blood flows forward ventrally to heart (*au* and *v*) thence to ventral aorta (*v.ao*) and gills (*a.br.a* and *e.br.a*) and dorsal aorta (*d.ao*). (Simplified from Wiedersheim after T. J. Parker.)

through the kidneys, but there are many variations in the details.

The Teleosts show the more primitive conditions; the caudal veins may pass forward to the heart without touching the kidneys or liver (Cyclopterus), or one may go through the liver and the other to the kidneys (Cod and Perch), or both may go through the liver and kidneys (Eel). What differences these diverse arrangements make in the metabolism and mode of life of the fish, we can hardly say; we know only that the Eel represents the superior condition, since this becomes permanent in the Elasmobranchs, Dipnoi, and Higher Vertebrates (Amphibians and Reptiles).

With the venous blood from the head the veins of the body pour their contents into a large, thin-walled chamber of the heart, the sinus venosus. In the Teleosts this is continuous with the auricle and the blood then passes into the small, but muscular, ventricle (Fig. 34).

The ventricle is the first portion of the heart (indeed of the whole vascular system) to be formed, and it forms just where there is resistance to the circulation of the blood or lymph on entering the tissues of the embryo. It is at first a simple tube through which the materials of the food-yolk might pass in either direction. In the Clupeids and Teleosts generally the yolk lies well forward under the trunk

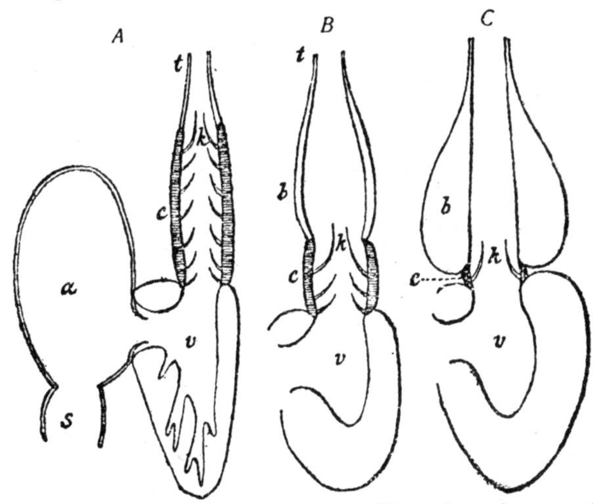

FIG. 34.—Structure of heart in: A., Elasmobranch; B., Amia; C., Teleost; *a.*, auricle; *b.*, bulbus aortæ; *c.*, conus arteriosus; *s.*, sinus venosus; *k.*, semilunar valves; *v.*, ventricle; *t.*, ventral aorta. (From Wiedersheim, after Boas.)

and head, in some even in front of the latter, so that the heart develops beneath or even in front of the head. In the Elasmobranchs the yolk-sac lies more posteriorly, in Bdellostoma still further back, and the heart and gills thus come to lie behind the head.

As the yolk diminishes, the heart or ventricle twists on itself and forms a coil to reach its final position in the Teleosts. The heart of the Elasmobranchs possibly, that of the Dipnoi certainly (Boas), undergoes the same twisting to a greater extent.

Though variations in the shape and structure of the heart arise from these early differences in its position, yet the principal variations are evidently correlated more with the rapidity of or resistance to the circulation in the young stages. To prevent the return of the blood into the venous system, a series of regulating valves has developed at various places, probably as a consequence of frequent regurgitation, like the "breathing valves."

The simplest mechanism is found among the Teleosts. There is frequently no definite valve between the auricle and ventricle, the contraction of the latter being sufficient to close the entrance, partly because the ventricle is more muscular in the Teleosts but mainly because the flow or pressure of the blood is not so great. Again, the exit of the ventricle in front has only a pair of valves, though in Albula, a near relative of the Clupeids, there are two rows of valves. In front of the valves the walls of the ventral aorta are greatly dilated, as if the flow of blood were here delayed, but this "bulbus aortæ" is not muscular. In the Elasmobranchs there is a pair of valves between the auricle and ventricle and several rows of valves on a muscular cone (*c*, Fig. 34) at the beginning of the ventral aorta. This arrangement not only prevents the return of the blood into the ventricle but helps it along to the branchial arches.

The greatest development is seen in the Dipnoi. The auricle is divided by a septum into two compartments. The venous blood from the liver and body flows through the right compartment, whilst the aerated blood from the air-bladder enters the left. Further, the ventricle itself has also an incomplete septum and the muscular cone in front is twisted into a spiral and several of its valves have coalesced into a longitudinal fold. The result is, that the venous blood is more or less completely separated from the aerated. The latter proceeds to the first two gill-arches, the former to the last two. There is a great similarity here to the structure and sequence seen in the frogs.

From the heart the blood proceeds along the ventral aorta to the gill-arches. After passing upwards along these and becoming aerated, receiving oxygen and giving up carbonic acid in the gill filaments, it is collected again into a single dorsal aorta. From this branches are given off to the head, brain, and shoulder-girdle, and then it passes backwards beneath the vertebral column to supply the body.

The secret of the differences between the various fishes and between these and the Higher Vertebrates is to be sought in the blood and lymph. All the tissues of the body, from the largest organ to the smallest cell, are bathed in fluids which pass in and out, leaving behind the materials selected by the cells and carrying away the products the cells want to get rid of. For even within one and the same body the cells are very different; what is food to one is poison to others. The cells therefore have the power of selecting and separating the materials. When this power ceases the cells perish and have to be removed; otherwise the body will die.

When people say, that "the blood is the life" they usually mean the red blood, but this is an erroneous notion. Judging from its manifold uses the lymph is the principal as it was the original blood. The fishes provide us with important lessons in this respect. Amphioxus has no red blood; the young Eels (Leptocephali) live for three years in the ocean without it; the embryos of many Teleosts have a well-developed circulation of colourless blood and some larval Clupeids do not obtain red blood until some days after they are hatched. The other groups of fishes, however, are well supplied with red blood from an early embryonic stage.

The lymph is a saline, aqueous solution containing albuminoids and other substances. Some of the albuminoids (chromoproteids) contain a colouring matter, which is able to take in a certain amount of oxygen and thus serves the purpose of respiration. But this colouring matter is not a necessity. Insects are able to do without it, simply because the branching tracheæ convey the oxygen directly to the cells. Many Teleosts make use of this primitive mode of respiration when they take air, or the gases dissolved in water, into their digestive tract or air-sacs, and it was just this mode of respiration that led to the appearance of the air-bladder.

The Teleosts thus show two primitive modes of respiration and this fact, taken with their feebler circulation, helps us to understand the differences in their organs and structures from those of the other groups.

Hæmoglobin, a complex chromoproteid, is always present in the blood corpuscles of the adult fishes and higher animals. How it is formed, we do not precisely know, but the head-kidney has something to do with its formation (Reuter). These corpuscles are oval in shape, biconvex and nucleated (except in Cyclostomes). The hæmoglobin within the corpuscles consists of carbon, oxygen, nitrogen, and hydrogen with a small percentage of sulphur and iron. Its peculiar quality is, that it has an attraction for oxygen, but this attraction is only relative. It takes oxygen from neighbouring media, which have a greater partial pressure of oxygen, and gives it up to cells with less than itself. The unfortunate thing is, that other gases have also an affinity for hæmoglobin, more particularly carbon monoxide with an attraction 140 times greater than that of oxygen. These poisonous gases, therefore, readily displace the oxygen, and very few cells or organisms can live without oxygen.

With the appearance of hæmoglobin and the formation of definite blood-vessels, a division of labour takes place between the blood and the lymph. The red blood corpuscles keep to the vessels and serve the purpose of respiration, whilst the lymph only holds a watching brief in this respect and concentrates on the feeding and cleansing of the cells.

The leucocytes are far outnumbered in the blood by the red blood corpuscles, but they are the main bodies in the lymph and lymphoid organs. They are mobile like the Amœba and have many diverse functions. For example, it is only by their aid that fats are transferred from the alimentary canal into the lymph, and for this purpose they push out between the other cells, which are busy absorbing the more crystalloid substances. Other uses will be apparent from the context.

In fishes the blood only amounts to about 2 per cent. of the weight of the body, but that does not include the lymph. In addition to the red corpuscles and leucocytes, the blood contains small blood plates and a fluid or plasma. The latter is of very great importance also, for it carries a quantity of albuminoids and various inorganic salts, of calcium, etc., as well as small quantities of other substances. The different albuminoids are special or specific, so that the blood of the Eel, for example, is a poison to the rat and many other animals. In this way blood relationships have been experimentally determined.

In the larval stages of the Teleosts, probably also in the embryonic Elasmobranchs, the lymph is of very great importance, not only in the respiration, but also in the formation of the skeleton. Above the head there is often a large lymph-sac, indicating that the bones of the skull are not being formed without some friction and difficulty. Sometimes this sac extends well back over the pectoral region. From its position and the function of the lymph, we may conclude that the armoured head and trunk of the ancient Teleosts (Arthrodira, etc.) were formed by means of such a lymph organ. In many forms also the skin seems to break away at many places from the underlying muscles, and lymphoid tissue comes to fill the spaces, as in the Soles, Toad-fishes, Pediculates, and Eels. The Eels especially have an abundance of lymphoid tissue and even a lymph heart in the tail. As in the formation of the ventricle, it is probable that some resistance to the circulation here has led to the accumulation of muscle cells (mesoderm) which the leucocytes have arranged into the shape of a tube. When the lymph collects there from the free spaces of the body, the heart propels it into the caudal vein, and it thus returns to the blood circulation. Many fishes also have masses of lymphoid tissue in various parts of the body, for example, on the heart of the Sturgeon, in the digestive tract and round the kidneys of Protopterus, etc. We do not know whether these masses have any special significance, but they indicate the activity of the leucocytes in collecting waste or discarded materials and storing them in safe places, where they might be turned to some other account. The special glandular organs, thyroid, etc., have arisen in this way.

The lymph shows its activity in another important direction, in the formation and conveyance of the black pigment, apparently as a waste product. For example, the divisions between the centra, when the vertebral column is being segmented, are frequently indicated by pigment spots; similarly, the loose air-bladder assembles a mass of pigment on its surface, also the gut, the top of the head, the base of the tail when the hypurals are forming and so on. This black pigment seems everywhere a sign of the leucocytes at work repairing the disturbed tissues. A striking example of this occurs in the flat-fish Symphurus. As metamorphosis approaches and a violent transformation of the organs and tissues is in progress, black pigment assembles in masses along the spinal column and under the skin about the middle of the body on the one side (Fig. 30). It is a sudden and evanescent phenomenon; the pigment becomes dissipated through the body after metamorphosis.

3. Excretory System

In the excretory system or systems of fishes there are many difficult problems for the morphologist and embryologist, particularly if one endeavours to see in fishes an intermediate stage between the conditions in the Invertebrates and Vertebrates. The uses of the system are quite clear. The solid residue of the food, which cannot be absorbed into the organism, is simply ejected from the open intestinal tract, by way of the rectum. The carbohydrates of the food are used up in supplying heat and energy to the tissues and organs, and the carbon dioxide is conveyed by the blood serum to the gills, where it is excreted as a gas. But the liquid waste products, formed from salts and proteins or nitrogenous compounds, must also be got rid of, and for this purpose, in Higher Vertebrates, the kidneys are specially organised. As might be expected, the fishes show a very early stage in the development of this apparatus.

Excretion may be performed in two different ways; by means of "wandering cells" without any special organ, or by tubules massed together and closely connected with an assemblage of blood capillaries, and opening to the exterior by a special duct or by ducts. "Excretion by means of wandering cells is certainly the original form of the separation of metabolic end-products" (Stempell and Koch, 1916). These wandering cells, the phagocytes, take charge of other cells laden with excretion and endeavour to convey them out of the body, either by definite channels (nephridia), if such are present, or through the skin or into the body cavity—wherever they can be deposited without harming the internal economy of the organism. Some Invertebrates, *e.g.* the insects, show various accumulations of these wandering cells which have been unable to penetrate the hard cuticle (nephrocytes, pericardial cells).

The fishes, especially the Teleosts, show a very high development of this early form of excretion. Guanin, a nitrogenous body derived from the purin nucleus ($C_5H_4N_4$), is deposited in great quantities in the skin (argenteum) and in the peritoneum. In all probability the deposition in the skin of inorganic compounds (carbonates and phosphates) is of the same nature, leading to the formation of scales, denticles, and plates of various kinds. This process has become an integral part of the organisation of the fish, and several things are noteworthy. In the Teleosts the scales are laid down in the skin before the kidneys (mesonephros) are developed. Secondly, when the circulation of the blood is disturbed, as by the compression and reduction of the gills, ordinary scales are replaced by large plates, which grow in a similar manner to the old rhombic scales (as in Gastrosteus and the Pipe-fishes).

In the Tunny and some others only the head kidney is developed, but a remarkable plexus of capillary blood-vessels extends in between the muscles posteriorly, where the mesonephros is formed in other fishes (Kishinouye, 1923). In these as in the other Scombroids the scales are very small, and the same is the case when large quantities of lymph are present under the skin, as in the Eels, Angler, etc.

As mentioned above, the wandering cells may convey their burdens into the abdominal cavity or ventral cœlom, and this has apparently led to a second kind of excretory organ peculiar to fishes. In many forms (namely, Elasmobranchs with a few exceptions, Sturgeon, Protopterus, the Teleost Mormyridæ and Salmonidæ) abdominal pores are present opening to the exterior near the anal aperture, and these may also serve for the extrusion of the eggs.

We come now to the true kidneys which stand in close connection with the blood circulation. The first portion to be formed lies under the anterior vertebræ near the skull and consists of but a few tubules with a long duct leading backwards on each side to the hinder end of the abdominal cavity. In the Cyclostomes and Tunnies, as mentioned above, this " pronephros " persists throughout life, but in many fishes its character changes at an early period and there is still some obscurity with regard to its mode of origin and how far it serves as a functional kidney. In some forms at any rate the duct does not appear till late in the larval development. In the flat-fish Symphurus, and in the Pipe-fishes (Huot, 1902) the kidneys and duct are only developed on the right side, and this seems to indicate that they have been formed later than the air-bladder, which has been able to displace them to the one side (Fig. 30).

The functional kidney (mesonephros) of other fishes develops in the postlarval stages and lies posteriorly in the body, but it makes use of the original ducts coming from the head kidney. The blood capillaries which go to its formation come from the dorsal aorta ; that is to say, the blood has to pass through the gills before it reaches the kidneys. As already mentioned, the venous blood coming from the caudal region passes through the kidneys in the Elasmobranchs, Dipnoi, and a few Teleosts, but in other Teleosts the venous system has little, if anything, to do with the kidneys. Since the renal portal system has become an integral part of the organisation of the Higher Vertebrates, it is not possible to explain the condition in Teleosts otherwise than as a primitive feature. Adaptation to any particular mode of life it cannot be, since some have it and some have not, under the same conditions.

We can understand the change better in the following way. The first attempt to form a kidney, anteriorly, interfered with the numerous blood-vessels that assemble just under the shoulder-girdle and the skull posteriorly. At any rate, these get mixed up with the kidney tubules and in some cases dissolve the latter. A second set of tubules then develops further back about the middle of the body. The Teleosts represent various stages in the development of these kidneys. As the posterior kidney grows in size, backwards and downwards, it comes to envelop more of the caudal veins, also found among the Teleosts, until there is a complete fusion or interchange, as in the Elasmobranchs and Dipnoi. In Higher Vertebrates a third kidney is formed, still further back in the body, but this metanephros does not appear in any fishes.

Lastly, reference may be made to the reproductive organs. These are organs not only for the multiplication and equipment of germ-cells, but also for producing internal secretions or hormones. They have no organic connection with the kidneys, but in the course of evolution the ducts have become mixed up together. The Teleosts again represent the early condition. The organs are developed in pouches of the abdominal cavity, separated from the kidneys by the peritoneum and air-bladder. The ducts from each side unite to form a single external opening, quite distinct from the urinary duct, in the Clupeids (Fig. 31). In the Eels, Salmon, and many others the ducts do not develop, the genital products simply drop into the abdominal cavity and escape by the abdominal pores. As the Teleostean form changes, with greater restriction of the space within, the genital ducts unite near their termination with the urinary duct to form a single urinogenital opening, as in the Perches, Cod, etc.

The development in other fishes, Elasmobranchs and Dipnoi, is quite different. For one thing, the air-bladder does not come in the way and the genital organs develop in closer connection with the kidneys. As in some of the Salmonidæ (Osmerus) the organs are free in the abdominal cavity (gymnoarian) and the products might escape through the abdominal pores as in the latter forms. In the males, however, the ducts remain essentially as in the specialised Teleosts, whilst in the females the condition foreshadowed in Osmerus is further developed. The duct remains open anteriorly and receives the eggs when they fall into the cavity. This arrangement has probably developed as a result of the closer union of the sexes.

Reviewing now the main features in the internal arrangements of fishes, it will be evident that we are dealing with the hidden secrets and stages of progressive evolution. The formation of blood corpuscles and a definite vascular system, though foreshadowed in some of the Invertebrates, now becomes an accomplished fact. From this has resulted the division of labour, whereby the lymph system concentrates on the work of feeding, repairing, and cleansing the system. The blood glands are the natural consequence, and we can now trace back to the fishes the origin of those qualities which have made the higher forms of life and activity possible.

In this internal evolution within we can read no special adaptation to particular conditions. The leucocytes continue their world-old activities, whether the organism lives in the sea or on land, and the organs they build up are of general not particular use. The construction of organs has been the consequence of the division of labour and led to futher growth. The internal phenomena of life thereafter can be interpreted as expansion, integration and differentiation, according to the ability of the nervous centres and blood glands to control the whole system. The taking in and absorption of food are the modifying and disturbing influences.

One thing is very clear in the fishes. The excretory organs have not kept pace with the growth of the other organs and body. They have been unable to follow the previously acquired rhythmic or segmental plan, such as we see in the skeleton, and for a time the lymph has reassumed an old function on a large scale, the removal of salts not wanted by the cells and blood to other places in the body where they can do least harm. It is in this way that we can understand the origin of a skeleton, internally as vertebral column and skull externally as plates, scales, and fin-rays.

The internal skeleton of fishes is clearly an adaptation to movements in a medium completely surrounding the organism, but the external skeleton was not necessary. It was an accident of growth and many fishes manage without it. Higher animals have not escaped from similar accidents.

The Teleosts have made a special use of the excretory products in the skin and have improved upon the heavy armour of the earlier Ganoids, but except from an æsthetic point of view it is doubtful whether this has brought them any advantage. The loss of salts from the blood has probably led to the latter having a diminished pressure relative to the surrounding medium, with the result that nervous energy is used up in maintaining the impermeability of the membranes.

CHAPTER VII

ECONOMY OF THE BODY

II. Utilisation and Emission of Energy

It may seem a little strange to speak of solid food-materials and bodies as the possessors of energy, but unless we have some common measure it is impossible to realise how food is converted into work. The division of cells, the building of the body and movement of the muscles can all be seen by the eye, but if we try to picture the relationships in space and time—what converts one cell into two, what changes solid food into components of the blood and lymph, what makes a muscle contract—we have to fall back upon some conception that is not to be appreciated by the senses. This is the idea of energy, and bodies are storehouses of energy, and these can be measured more or less accurately by physical standards. We have now to examine the structures and organs of fishes that are employed in the utilisation of energy.

1. Regulating System

Investigations of recent years, especially from the physiological side, have revealed that the body of living animals contains more than was formerly dreamt of. We have been accustomed to speak of the directing and controlling power of the nerve-centres, the obedient service of the muscles under that power, and the roving commissions of the blood and lymph, but a more refined power has now been recognised—one is not sure whether it stands above or between the other agencies, but it is something that is present from the beginning, even before the embryo appears, and persists so long as life endures within the body. This power has a number of organs to itself and is thus recognised as a system, the regulatory or regulating system.

The principal function of this system is the regulation of the flow of materials. We may say it is energy in a concrete form, but it is difficult, if not impossible, to find physical comparisons, and we may be dealing with only one or several agencies with a still higher co-ordinating agency. One characteristic is, that these agencies seem able to work against the known physical and chemical laws of nature, which means, of course, that we lack comprehension of their true relationships. We cannot say that they constitute life, since life also requires grosser materials, but they are its essences without which growth or progress would be impossible.

These several powers or agencies, so far as they are represented by distinct organs, make their first appearance among fishes, and it would seem that a period spent in the water was necessary for their development. But one of them is common to all forms of life—represents indeed the beginning of that integration or separation of the quick from the dead, as we understand it. This is the living membrane. We are able to make soap bubbles and splashes of colloids in a crystalloid medium which assume the shapes and some of the characteristics of a living cell, simply because they have a limiting membrane or surface of tension; but the living organism has the power of itself to mark off the exterior from the interior.

So long as a fish is alive, its internal constitution and particularly the composition of its blood are quite different from those of the surrounding medium. It is self-contained and only takes in the materials it wants to take in. This is also the case with Higher Vertebrates. The pressure of the blood in man is about eight atmospheres, measured by the concentration of the contained salts. In other mammals and birds it is a little higher, and the curious thing is, that in the bony fishes or Teleosts it is not much higher, though the surrounding water has a pressure twice to three times as great. When the fish is out of condition, however, or dies, the membranes gradually cease to have this control and the tissues become permeable (Sumner, 1906). In this way we can understand why fishes do not taste salt although they live in salt water, a puzzle to many people. Nevertheless, they can take in this and other ingredients when preserved.

Freshwater fishes like the Carp have a much greater pressure within the blood than that externally; by increasing the amount of salt in the water to 10 grm. *pro mille* Portier and Duval were able to obtain equilibrium within and without; but with still further increase in the salinity the gill membrane seemed to be forced and the fish died. As the pressure of the blood also increased, though slightly, in these experiments, it would seem that a small amount of the external salts in solution is able to pass through the membrane. But this is by no means certain, and other observers think that only water is able to pass through (Sumner, Dakin). It is possible that the fish is able to withdraw water and salts from the tissues and thus regulate its exchanges. The Salmon, for example, does not change its osmotic pressure all at once on entering fresh water; and the Eel, when transferred from fresh to salt water, does not assume the osmotic pressure of the latter. On the whole, therefore, the condition of the blood is determined by the inner metabolism and regulating system.

Different species of Teleosts vary considerably, and even the same species may vary at different seasons, probably according to the temperature, food, and condition of the reproductive organs. As indicated in a previous chapter, it is probably the changes in the osmotic pressure of the blood, with the accompanying changes in the tissues, that lead the fishes to undertake the spawning migrations.

The Elasmobranchs, Sharks and Rays, are apparently different from the Teleosts in regard to their relations with the outer medium. Owing, perhaps, to a large amount of urates in the blood the osmotic pressure of the former is but little less than that of the salt water. Possibly the great difference in constitution and scaly covering, and the communication of the pericardial cavity with the abdominal cavity, and thence with the exterior, may contribute to the difference. The tissues are more permeated by the medium.

How far the tissues of the young larval and postlarval stages can be permeated directly by the water and soluble salts, is an interesting question. We may certainly believe that water can penetrate through the thin epidermis, and it is probable that mucous pores enable the water to come into closer contact with the tissues, thus causing these to swell as in the case of the Leptocephali ; and it is also probable that the formation of the scaly covering depends just upon the dilution of the colloid substances in the dermis by the water. But we know extremely little as yet regarding these interesting and important phenomena. If Pütter is correct in his views, that dissolved substances can be taken in through the skin as food, just as poisons can enter in the same way, it becomes a doubtful matter if the young fish is so independent of the external medium.

Whilst it is clear that the living membranes are regulating forces in the relation of the fish to its surroundings, the regulatory system proper consists of a number of internal glands, which have the peculiar property of being able to secrete various organic substances that pass into the blood or lymph, not into the alimentary canal, and are thus conveyed through the system to all parts of the organism. These substances have been called " hormones," the stirrers up, since they are specific in their action, affecting special organs and thus stimulating or retarding the activities and growth at distant places. The " vital humours " of the body is perhaps the most fitting expression for them in popular language.

The peculiarity of acting at a distance is, however, not restricted to these substances. It has long been known, for example, that the carbonic acid formed in muscle affects the breathing centre in the medulla and that albumen in the blood affects the liver, whilst the pancreas requires the stimulus of other secretions. The chief difference seems to be, that the influence of the hormones is spread over a longer period of time.

Very little is known regarding the origin and activities of these substances in fishes, but it may be assumed that their function is similar to that in higher animals. The spleen (*sp*, Fig. 31) lies alongside the stomach in close connection frequently with the pancreas. In addition to its influence on the latter, it is an important centre for the dissolution of old red blood corpuscles and formation of leucocytes, though this may be part of its business as secretion-former.

The pancreas is in reality a composite gland, one part pouring its secretions into the duodenum, the other part enclosed within itself and delivering its secretion only to the lymph and blood capillaries in its substance. The latter part consists of a number of detached groups of cells or "islets," the presence of which in fishes was determined by Rennie (1904). They have become more important within recent years by the discovery, that the substance obtained from them, insulin, is a palliative for diabetes. It regulates the amount of sugar formed in the liver under normal conditions and has the effect, when introduced into the blood of patients, of diminishing the excessive amounts of sugar which would otherwise overburden the kidneys.

The thyroid and thymus glands arise in connection with the gill-arches, the former below, the latter above. Originally, they have played some part on the surface, as mucus glands, in the conveyance of food and as protection for the adjacent tissues; but they become detached from the surface epithelium during development, sink into the dermis, and assume the structure of lymphoid organs. The thymus, like the spleen, has been supposed to be an organ for the destruction and re-formation of the red blood corpuscles, but it may be that this is part of the process of secretion. An interesting thing is, that these glands seem to counterbalance one another in some ways. Gudernatch (1911) found that the metabolism and metamorphosis of tadpoles were accelerated by thyroid extract, but prolonged by thymus extract. The effects of the thyroid are indeed very similar to those produced by increase of temperature. What part they play in the development and life of fishes is still quite unknown. The same may be said of the supra-renal bodies, which are well-developed in fishes, and the head kidney.

The latter is one of the most complicated organs in fishes. Even in the earliest stages, when perhaps still functioning as a kidney, it seems to take some part in the formation of the red blood corpuscles. What its function is later, is problematic; it is always richly supplied with blood. Other mysterious glands are found in the skull in connection with the brain, pituitary body below, pineal and parietal bodies above (Fig. 44), endolymphatic sinus of Dipnoi. The pineal body and parietal gland form a kind of eye in the Lamprey (Petromyzon), but in other fishes the "eye" does not develop and they seem to have been diverted, like the thyroid and thymus, to more general purposes.

Lastly, mention should be made of the hormones of the sexual organs. Like the pancreas these organs have at least two functions, and the hormones carried away from them by the blood and lymph to other parts of the body have the most profound influence on the general activities of the organism, as well as on the development of special parts such as the secondary sexual characters.

It has always been a mystery that the secondary sexual characters, such as different coloration, fins larger in one sex than in the other, should be present long before the respective fishes became mature. But there seems no mystery about the possibility of these hormones acting as soon as the sexual organs are present, and becoming more influential as the latter ripen. Hence we can understand why the pectorals of the male Cottus, for example, become serrated before the eggs are deposited, why the male Gastrosteus begins to construct a nest before he has chosen a partner, and why the male Carplings, remarkable for their brilliant coloration at all times, are always so urgent in their behaviour towards the females.

Our conception of what is meant by heredity is also

profoundly influenced by these new ideas regarding the hormones. Heredity can no longer be considered as the transference of morphological or structural characters only; the regulating influence is likewise handed down from one generation to another, and if we try to express what this means, we pass well beyond the mere physical phenomena. The hormones in fact lie on the borderland between the physiological and the psychological worlds.

2. Muscular System and Electric Organs

The conversion of potential energy within the body into external work becomes most obvious to us in movement, and this is associated with the muscular system. In fishes the muscles compose the greater part of the body, and they are concentrated more especially in the tail region, the propelling organ. Some fishes have no tails, it is true, others are almost all tail of a slender kind, but exceptions do not affect the general rule.

In simple forms like the Clupeids the muscles are arranged in parallel series from head to tail, the position and musculature of the fins having little influence on their primitive segmentation. Of more importance is the wide separation of the lower half of the muscles caused by the development of the abdominal cavity. This does not spoil the symmetry in the Clupeids (see Pl. VIII), but it has a greater effect in later forms. Theoretically, as in the beginning, the muscles of a segment are arranged symmetrically, in two then four sets, corresponding to the division of the body into two lateral halves and two dorso-ventral halves. As soon as the fish begins to move, however, whether in the egg or free in the water, this primitive arrangement alters. The parts nearest the vertebral column incline backwards, especially in the caudal region, whilst the parts outside retain their connection with the skin at the margins. Hence the muscles of any segment come to have the shape of a W (⋚) and a cross-section of the body cuts through a varying number of muscle segments. Further, the two lateral halves in such a section are usually not quite symmetrical or the same.

The segmental arrangement persists on the whole throughout the fishes, the principal changes depending simply on the relative development of the more central muscles, directed towards the tail, and the marginal muscles directed towards the head. In swiftly moving fishes like the Sharks and Mackerels, the latter become of minor importance except near the head, whereas in deeper forms, which move more slowly, the marginal portions may be the larger, as in the Flat-fishes. This has to do with the working and balancing of the body and fins, and in the Plectognaths, where an " accident " in the larval period deprives the fish of the use of its tail, these marginal muscles attain to an enormous development. In the Elasmobranchs the muscles tend to arrange themselves in groups along the vertebral column, as in the Higher Vertebrates.

Along the side of the body the meeting-place of the dorsal and ventral halves of the muscle segments is usually marked by a longitudinal furrow. This is sometimes filled with muscle fibres, forming it may be a separate muscle, but probably detached portions of the principal muscles, which are richly supplied with fat and blood capillaries. In some fishes the furrow thus obtains a reddish colour. It is quite wanting, however, in the Lamprey (Petromyzon), where the muscles are arranged in compartments which show a remarkable resemblance to the electric organs of other fishes (Stannius, 1854).

The distinction between smooth or non-striated and striated muscle fibres expresses the general difference between slow but powerful contractions and rapid but weaker contractions. Yet many intervening stages occur, and the one may change into the other or into a different kind of cell tissue. The alimentary canal of the Clupeids and several other forms has striated muscle in the larval and postlarval stages, and this persists in the adults of many forms (Oppel, 1896), though usually the smooth variety only is present. Hence, in some cases the contractions of the stomach and gut are not altogether "involuntary" but under the control of the fish. On the other hand, although the muscles opening and closing the mouth consist mostly of striated fibres, yet in forms with large and powerful mouths the basal part near the skull shows the smooth kind. Thus a stronger and longer contraction of the jaws is permitted, whilst the rapidity of opening and shutting of the mouth is not interfered with. This adaptation is seen very well in the Eel. Similarly, the muscle fibres of the fins are less striated as a rule than those of the body, showing that their movements are to some extent involuntary.

These differences are due apparently to the varying proportions of the ground-substance (sarcoplasm) and fibrillæ and probably represent a difference in the use of the various muscles, but very little is known as yet regarding the connection between structure and function of the different kinds of muscle cells.

The colour of the muscle in fishes is usually white or grayish, and this may be connected with the curious fact, that the consumption or oxidation of foodstuffs is very complete in fishes (Pütter). Corresponding to this the heat produced is usually small in amount, the temperature within the deep-seated muscles being only about ½ to 1° C. higher than that of the surrounding water (Simpson, 1908). The Mackerel family, especially the Tunnies, form a remarkable exception. The temperature of the deep-seated tissues may be 10° C. higher than that of the water (Kidder, 1880), and here the colour of the muscle is of a deep brown or red in parts. The heart of fishes is also brown, at least in dead fishes, and the darkest part of the muscles in the Tunnies is richly supplied with blood capillaries (Kishinouye, 1923). In the Piked Dog-fish, Acanthias, Kidder found a temperature of 12–16° C. above the surrounding water and even more than this in the embryo.

The brown colour may be due in part to the infusion of blood into the tissues, but in any case it is a sign of greater metabolism, and it is specially significant that the Elasmobranchs, and even the embryos, should display this greater activity. As noted in the previous chapter, they are also remarkable for the large amount of urates in the blood, which may also be taken as a sign of enhanced metabolism. On the other hand, the pink colour of the Salmon muscle is due to the presence of a special pigment, lipochrome, derived apparently from the Copepoda on which the Salmon feed (Newbigin, 1900).

By taking a series of microphotographs Hürthle has been able to follow the physical changes occurring in the muscles of living insects. The fibrillæ or striated threads of a muscle cell are embedded in a clear layer of liquid (sarcoplasm) and the same clear layer is present at the ends of a row of the striated cells. When the elastic fibrillæ contract, the liquid of the sarcoplasm is forced out into the end layers. The fibrillar portion may contract to half of its previous length or even more, but its total volume remains constant.

One of the most remarkable phenomena in nature is the way in which certain fishes have converted the ordinary, living muscle cells into powerful electric batteries. It illustrates how closely the various forms of energy are connected, and we cannot but marvel at the possession by the fishes of a power undeveloped in higher animals, even in man. Their conversion of potential into kinetic energy in the muscles with little or no waste or heat has been noted above; here it is converted directly into a form of even higher tension. The fishes possessing these organs are mostly of a sluggish nature and it would seem as if the potential energy stored up in the muscle cells, and not used in active work, had caused a reduction of the elastic, internal fibrillæ and concentrated on the outer and opposite margins, which thereby have become like positive and negative poles towards one another (Fig. 35).

The close connection between the electric organs and the muscle cells was already evident to Robin as long ago as 1847, but it is to Babukhin (1870) that we owe its definite proof. The ordinary Rays have two small organs of this kind near the end of the tail, which grade into the striated muscle fibres in front. These fishes are but feebly electric and probably expend the energy when captured, otherwise the fishermen would know of it. But in the Torpedo, the Electric Eels, and others, the transformation is complete and there is no sign of striation. In the Torpedo it has been calculated that each organ contains as many as 500,000 end-plates or terminals of the motor nerve fibres. In the Electric Eel the organs are very large and extend the whole length of the fish.

The constitution and mode of working of these electric

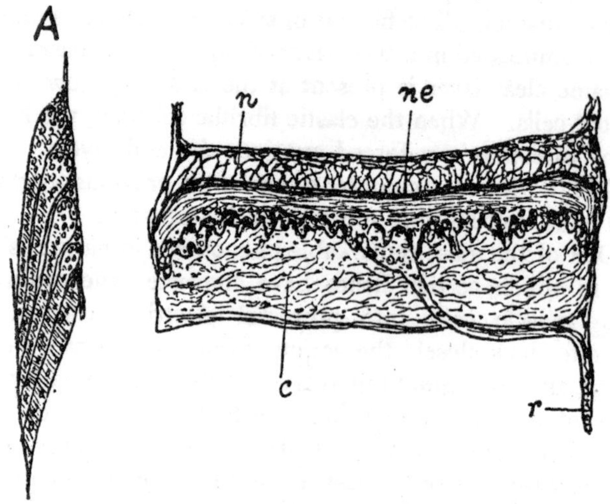

FIG. 35.—Conversion of muscle cells into electric batteries in Skate (*Raja batis*). A., cells becoming club-shaped in front, though still striated ; B., battery with stalk (*r*) still showing striation ; *c*., connective tissue ; *n*., nerve fibrils ; *ne*., nerve end-plate. (From Cossar Ewart.)

batteries are not precisely known. In Torpedo the muscle cells contain only 78 per cent. of water, the electric cells over 91 per cent., whilst the former contain more albuminoids (15 per cent.) than the latter (2 per cent.). The most remarkable thing is, that the electric cell contains a larger proportion of sodium, but a smaller proportion of potassium than the muscle cell (Baglioni, 1906). We have not yet learnt how to make these organic batteries.

Each cell possesses an electromotive force of 0·02–0·05

volt, and a "shock" from a torpedo represents about 35 volts, that of Gymnotus over 300 volts, and several hundred shocks may be given in a second. The power is thus considerable, and visitors to Anton Dohrn's famous aquarium at Naples used to be entertained by touches of the Torpedo—after some of the power had been drawn off in other ways.

The suddenness of the discharge is just as phenomenal as the battery itself; it reaches a maximum within two-thousandths of a second, but a great deal depends upon the temperature (Garten, 1910). At lower temperatures (15° C.) it takes about eight-thousandths of a second. From these and similar experiments some are inclined to believe that the nerve end-plate is the real electric element, the solution in the cells merely a conductor.

3. Mucus Glands and Radiant Energy

Mucus is a gelatinous secretion formed in cells of the epidermis and by the internal epithelium. It is present wherever there is friction or rubbing surfaces, and is not very different from the lymph fluids which occur in quantities under the skin of many fishes. The Hag-fish, Myxine, is able to surround itself with a thick glutinous coat when disturbed, and many fishes are quite slippery with mucus. With the differentiation of the body and development of a scaly covering, the secretion is restricted to single spots, and then the massing and subdivision of similar cells give rise to a definite glandular organ with an opening to the exterior.

In some cases glandular organs are developed in the very early stages, long before the scaly covering is present, in the form of so-called "larval organs." These arise during the embryonic life and disappear later, for the most part. They indicate perhaps some imperfection of the regulatory system; some of them at any rate have no apparent use and are not connected in any way with anything in the past history. Among these are the so-called suctorial discs which appear in the embryos of the Sturgeon, Amia, and

Lepidosteus (Ganoids). They are not really suctorial; according to Bashford Dean they belong to the mysterious class of pit organs or mucus organs, sense buds in some cases, from which have arisen the lateral line sense organs, the mouth in some cases perhaps, certainly the nasal organs and the ear. Lateral pores belong to the same class. We find them everywhere from the Hag-fish to the Eel, and on all or any parts of the body; at first a few cells or groups of cells simply with a secretory or excretory function, then the in-sunken epidermal cells at the base become sensory and a sense bud or taste organ develops. From these may come a light organ, an electric organ, a connected system of small sense buds or a more concentrated organ, such as the pineal gland, perhaps even the eye. As Bashford Dean remarks, there is no limit to their possibilities.

Some have thought that these organs represent a primitive condition; it would seem better to call them elemental. There is a mysterious force behind them which is difficult to understand, and which later Vertebrates have never developed, at least not retained. It may be suggested that the endocrinal organs, thymus and thyroid and probably the others, were originally of this nature.

The mucus secreted is of a peculiar nature and may even be poisonous, though we know nothing of its effect on other fishes. The most interesting thing is, that when confined within the cells or glands it leads to the production of various forms of radiant energy, like the muscle cells under similar conditions. There are two ways in which this phenomenon may be regarded: the radiant energy may come from the metabolism of the fish itself, or it may come from the presence of foreign bodies, which live in the mucus cells, perhaps even cause their production, and they alone give rise to the emanations, not the fish. The former view has been the one most generally maintained until recently, and the phenomena may be described from this point of view first of all.

According to this the radiant energy is produced by the slow oxidation of various organic substances, called "photogens," which, by the way, have no necessary connection with phosphorus. The presence of halogens or even of ammonia in the water may have the effect of stimulating these substances to radiant activity, but they are more active under certain conditions of temperature and surroundings than under others. The larvæ of fishes, though without any definite mucus organs, may become luminous simply from sudden impulses given to the water (Ryder, 1880).

The production of these substances is well-known among the lower forms of life, Bacteria, Protozoa, worms, insects, etc., whether they live in air, water, or other media. Among fishes it was formerly believed to occur only among those living in the abysses of the oceans and the luminosity was supposed to be useful, in place of the light rays, in attracting mates and repelling enemies, etc., but this view has proved erroneous. On the Valdivia expedition, Chun took every deep-sea fish as soon as it was caught into the photographic dark room, and was in most cases unable to detect any light emitted from them.

As a matter of fact, the luminosity seems to be commonest in fishes which live at no great depths. According to Burckhardt (1900) many species of Sharks, including Laemargus, which are not deep-sea forms, and even the common Haddock may display luminescence. The remarkable Harpodon, which when caught becomes brilliantly luminescent all over the body, lives in the rivers and estuaries of India. Photoblepharon, a still more remarkable form, lives in freshwater quarries and ancient craters of the Malay Archipelago. In his work on the Valdivia fishes Brauer found that the luminous organs were not peculiar to the deepest and darkest water-layers.

Brauer's conclusions have been confirmed and extended by Hjort from the results of his expedition across the Atlantic in 1910. He states that many bottom animals were brought up from the abyssal region in a luminous condition—without any special luminous structure, the luminosity being connected simply with the surface epithelium.

Only in a few fishes supposed to live on the bottom do ordinary glandular, luminous organs occur (Spinax), and only as isolated organs, not in such numbers as in the genuine luminous fishes. The oceanic Scopelids are the forms which have the most and best developed organs, and among these the surface forms of the genus Myctophum possess the largest, whilst Lampanyctus (from 600 and 800 metres) has very small organs. The deepest forms of fishes (Aceratias, Melamphaes, Cetomimus) have no luminous organs, and it appears that these are specially characteristic of the fishes which live in the upper 500 metres in warm oceanic waters.

The last point seems of great importance in the consideration of the occurrence and usefulness of these organs. A large black fish, *Paraliparis bathybii*, that lives in the depths of the Norwegian Sea, possesses well-developed eyes but no luminous organs. The temperature has thus something to do with the phenomenon. The water in oceanic depths is much colder than in the upper layers, down to 500 metres, and it may be recalled that the red rays of light penetrate only down to about 100 metres, whilst the ultra-violet and blue rays can be detected still at 1000 metres. We may say, therefore, in general, that both light and warmth are necessary for the production of luminant energy, and it may be that the luminescence of the skin in abyssal forms when brought to the surface is due to the latter fact. We cannot be certain that the fish were luminescent in the depths.

The luminescent organs may be on any part of the body, on barbels connected with the lower jaw or on tufts on the top of the head. Frequently they are arranged in rows along the body or in transverse bars between the muscle segments. And they vary greatly in structure and size even on the same fish. In the simplest cases they are just like mucus glands ; in Porichthys there are some 300 of these arranged over the body, but they do not appear to be luminous. In some the cells are arranged in the form of a number of tubules, sunk deeply in the skin, with the secretion forming a dense cover externally. Or the tubules may be still deeper with long slender cells forming a dense lens-like layer on the surface. Still more complex forms (Fig. 36) have a layer of black pigment surrounding them, and the glandular tubules are arranged irregularly inside this; further in there is a dense layer of spicules forming a kind of reflecting mirror (*s*); then follow radially arranged cells or tubes and the whole is covered by a lens-like structure. In some cases the skin at the sides of the organ projects over the organ and forms a kind of iris. In Photoblepharon this can be closed right over the organ.

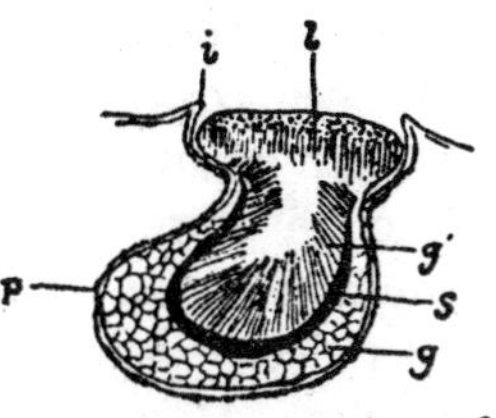

FIG. 36.—Conversion of glandular cells into a luminous organ in Pachystomias. *g.*, irregular gland tubes; *g'.*, radial gland tubes; *i.*, iris-like part of epidermis; *l.*, lens-like body; *p.*, pigment sheath; *s.*, layer of light-reflecting spicules. (From Bridge, after Lendenfeld.)

The nature and function of these enigmatical organs have been closely studied by Lendenfeld (1905), who has summarised his conclusions regarding them in the following manner. "It is most probable—although very far from scientifically certain—that: (1) The organs produce a radiation. (2) In the simple ocellar organs and the radiating discs this radiation is an ethereal wave movement directly emitted into the surrounding water. (3) In the compound ocellar organs this radiation is either also such a movement or an emission of corpuscles (electron bombardment) and originates in the inner part. It is here not emitted directly into the surrounding water, but acts on the outer part and induces this to phosphoresce and emit ethereal waves. (4) The length of the ethereal waves thus directly or indirectly produced by one and the same organ is always the same; that of the ethereal waves produced by different organs is often different. (5) In some cases the wave-length lies within the visible part of the spectrum; in other cases it may be smaller or greater, and then the radiation is invisible. (6) The ethereal waves of the invisible radiations are probably of greater length than those producing the red end of the spectrum, because such probably penetrate the water to greater distances and because the remarkably telescopic eyes of some deep-sea animals seem to be peculiarly adapted to their perception. (7) These long ethereal waves may (if not too long) be of the nature of (ultra-red) light, or (if longer) of the nature of electricity."

That the radiation from transformed glandular cells may be of the nature of electricity is clearly proved by Malopterurus, the electric Cat-fish of the Nile and other African rivers. Here the glandular system of the skin, which surrounds the whole body like a gelatinous coat, has been converted into a powerful electric organ with a strength of over 200 volts. The interesting thing is, that although formed from a seemingly different tissue the structure of the organ is precisely the same as that in the other electric fishes, where it has developed from the striated muscles. Here also the fish is of a sluggish disposition.

We may take it as certain, that the organs emit an emanation of some kind, but we can no longer believe that they are an adaptation to the abyssal and dark depths of the ocean. On the contrary, as mentioned above, their development and production of energy seem to depend on a certain amount of light and heat, which are obtained nearer the surface. Whether they serve any purpose to the fishes and other animals possessing them, must remain an open question.

This negative conclusion with regard to the function of these organs is strengthened by the discovery of Buchner, that the luminescence is really due to the presence of bacteria. In Photoblepharon, a Percoid described by Max Weber from the Malay Archipelago, Buchner found that the gland cells were filled by bacteria, and he suggests that the presence of these has stimulated the surrounding tissues to develop in the extraordinary manner described above. There is nothing impossible or even improbable in this, for we know how cells have a way of growing over foreign bodies, forming cysts and hardened layers of skin. And this view is certainly in accord with the observations of Hjort, that these organs are restricted to the upper layers of the warmer waters.

The presence of bacteria has only so far been demonstrated in the case of Photoblepharon, and it is hardly possible to go all the way with Buchner, however probable his theory sounds. For if it is true in the case of fishes it might also apply to other animals, Noctiluca, for example, and the Peridineæ generally, not to speak of the seasonal and sexual luminescence of insects. But the bacteria may only be agents stimulating the glandular cells to the production of "photogens" under particular conditions.

Apart from a general reluctance to accept a theory that rests on too few observations, it must be confessed that the phenomena of luminescence in fishes receive a clearer explanation from Buchner's standpoint than from any other. As an illustration of this the peculiar condition in the Scopelid Ipnops may be mentioned. This remarkable form has no eyes, though living among other fishes which have eyes. Instead, there is a large luminous organ inside the skull under the frontals, as demonstrated by Moseley. There seems no other reason to account for this, but that bacteria have entered the organ when in process of formation and completely transformed its character. Perhaps the early development of these forms when known will throw some light on the mystery.

There is one other remarkable feature in Ipnops that deserves to be mentioned. The luminous organ is apparently innervated by the optic nerve. This would indicate that the development had proceeded some distance before the transformation took place, and it may be that Ipnops really represents some individuals of a species going by another name.

4. Sensory Nervous System

The transformation of cells from the production of one kind of energy to that of another leads to a consideration of the nervous system. The emission of power waves of some sort was probably the first manifestation of life, and one and the same cell received and gave out messages. The differentiation of the body into many varied cells led to the receiving and emitting functions gradually becoming separated and confined to distinct organs and places. The sensory system receives the various kind of messages, the motor system, represented by the muscles and structures developed therefrom and directed from the central control, gives the response of the organism.

The separation appears to be most complete in the Higher Vertebrates and man, and our senses only register certain kinds of power waves. It is therefore somewhat difficult to realise that other forms of life, and particularly the fishes, may be tuned in quite different ways. Our range is very restricted; without special appliances we know comparatively little of the world around us; even within this range, as every one is aware, the senses are not to be trusted. On the other side, our responses go almost entirely through the muscular system, yet this is not the case among fishes, as we have just seen. And if we close our eyes and press lightly on the eye-balls, the field of vision within gradually becomes filled with light-waves—a sudden blow produces a flash. The radiant energy for the purpose of emission is there, but we do not know how to use it, or perhaps we use it without knowing we do so.

The nervous system was originally on the surface and still arises from the outermost layer of cells. The withdrawal of the central axis into the interior has led to the development of a most complicated network of communications. Through the sense organ on the surface which transforms in some way the received stimulus, a message passes inwards along a neurone to the central system, where in the higher animals it may pass through three or more neurones, closely interlinked together by dendrites or branches of the nerve cell, before the response is sent to the muscles or organs by a motor neurone (Fig. 37). The mode of transport is perhaps like that of the low-tension electric current, or there may be movements of the protoplasm, but very little energy is used in the process. Large quantities of oxygen, however, are required, as may be judged from the rich supply of blood-vessels in every sense organ.

The simplest form of sense organ is seen in the end-buds which occur anywhere on the skin of fishes. In Higher Vertebrates these become restricted to the mouth as taste-buds. Sometimes they are raised on eminences, but most often they sink deeper into the skin and the nerve-endings become embedded in mucus or glandular tissue. Whether they are all connected with reception only, we do not know, but it is possible that some papillæ-like forms may be able to send out messages (Fig. 38).

In addition to the diffuse end-buds, fishes possess an extremely characteristic system of well-developed skin-organs sunk beneath the scales, in one or several lines, and linked up together by means of a single nerve (lateralis) or two branches of the same nerve (Fig. 39). These lateral lines are so universal, that their absence from the body of the Clupeids is worthy of special note. The auditory organ,

FIG. 37.—A single neurone. D., dendrites of the nerve cell; E., endings in the skin. (From Stempell u. Koch, after Verworn.)

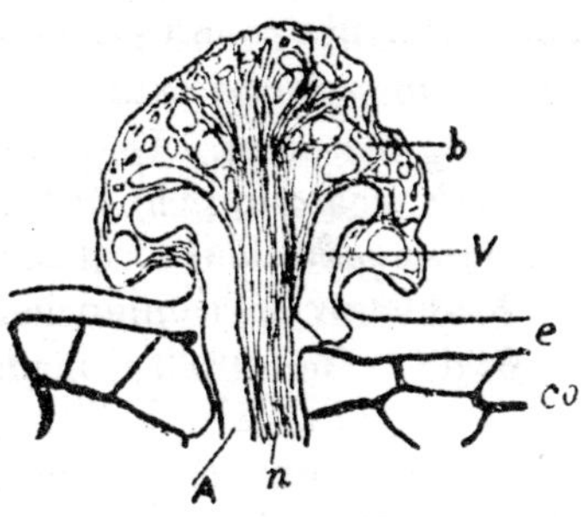

FIG. 38.—Bud-like organ on deep-sea form Malthopsis spinulosa. A., artery; *b.*, capillaries; *co.*, corium; *e.*, epidermis; *n.*, nerve; V., vein. (From E. Trojan.)

or statocyst as it should be called, is the specialised anterior portion of this system, the innervation of both coming from the VIII. cranial nerve (Mayser, Hofer, 1908). The olfactory or smelling organs have developed from similar pits on the front of the head and are supplied by the I. cranial nerve. In addition, there are several other branches of the lateral line system on the head, above and below the eyes and on the lower jaw, the principal innervation of which seems also to come from the VIII. nerve.

The part played by these organs in the life of fishes has been the subject of much discussion and research. The diffuse end-buds or pit-organs as well as the olfactory organs seem able to detect differences in the chemical nature of the surroundings. Whilst the majority of fishes seek their food by sight, Bateson found in his experiments at Plymouth that many (Eels, Flat-fishes, etc.) do not start in quest of food, when it is first put into the tanks, until the scent has become diffused. Then they begin to swim vaguely about and appear to seek the food by examining the whole area pervaded by the scent. Herrick (1903) came to the conclusion, with regard to the Cat-fish, that the organs of taste or smell must pervade the whole skin. The fish will seize unseen food with great precision provided it is brought near the skin. Here the end-buds largely replace the eyes as a means of discovering food.

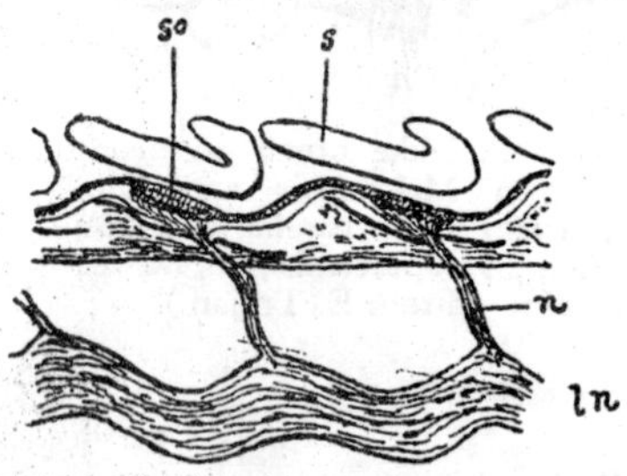

FIG. 39.—Lateral line organs. *n.*, nerve; *ln.*, lateral nerve; *s.*, scale; *so.*, sense organ. (From Bridge, after Allis.)

On the other hand, the organs of the lateral line have quite a different function. They have been closely studied by many workers and the experimental results obtained by Parker (1905), which have been corroborated on the main points by Hofer (1908), are very conclusive. According to Parker these organs are not stimulated by light, heat, salinity of water, food, oxygen, carbonic acid, water pressure, currents, or sound. They are only stimulated by, that is the fish responds to, gentle vibrations in the water of low frequency, six per second. Hence we can understand how the blind fishes in the caves of Kentucky can avoid obstacles. The reflex waves set up by their own movements inform them when they approach too near the sides. The nocturnal bats seem to have a similar sense.

Fishes living in a tank in an aquarium are guided by the same means. But if the water is violently disturbed, they lose all sense of distance and dash themselves against the sides. In disturbed waters, therefore, the lateral line organs are of no use to the fishes, but the statocyst or hearing organ may be of assistance since it responds to stronger blows than the lateral line.

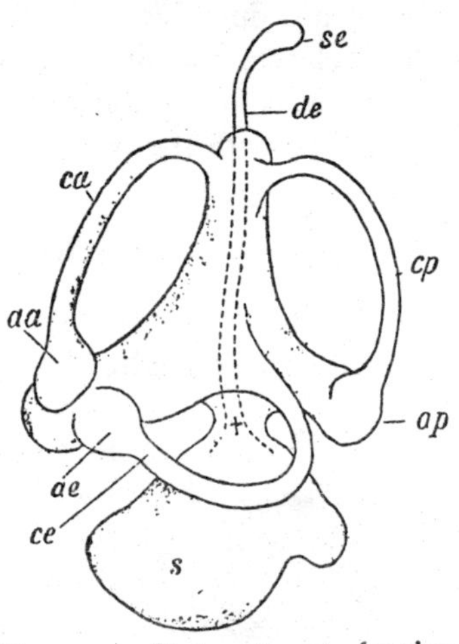

FIG. 40.—Statocyst or hearing organ. *s.*, sacculus containing otolith and sense-cells; *aa.*, *ae.*, *ap.*, ampullæ (containing sense-cells); *ca.*, anterior semi-circular canal; *ce.*, horizontal semi-circular canal; *cp.*, posterior semi-circular canal (semi-diagrammatic, from Wiedersheim).

The lateral line organs seem also to be of some use both in orientation, according to the different strengths of currents on the two sides of the body, and in balancing. Lee found that when the pectoral and ventral fins were removed, the fish (Opsanus) showed no lack of equilibrium so long as the lateral line was left intact. But the balancing of the body seems to be the main function of the sense-organs in the labyrinth of the statocyst (Fig. 40), to which the impressions are conveyed directly or by means of the air-bladder. The latter seems to be the main channel for conveying notice of differences of pressure.

The statocyst itself consists of a number of chambers

embedded in the cartilage and bone on the side of the skull. The whole is filled with a fluid, endolymph, and surrounded by a similar fluid, the perilymph. Disturbances in the external medium set up waves in the endolymph which beat upon the numerous groups of sense-cells and thus convey impressions of the strength and direction of external pressure to the central organ. In the Elasmobranchs the endolymphatic duct (*de* and *se*) is open to the exterior; but in the Teleosts the organ is completely enclosed. In some forms a continuation from the air-bladder is indirectly connected with the endolymph, as will be described later.

5. Eyes of Fishes

As with the lateral line and hearing organs, an organ for the reception of light stimuli is only present in bodies that move. We may say that it has arisen in consequence of the movement, and the latter is probably a factor in its formation. Later, it is connected with the regulation of the movements.

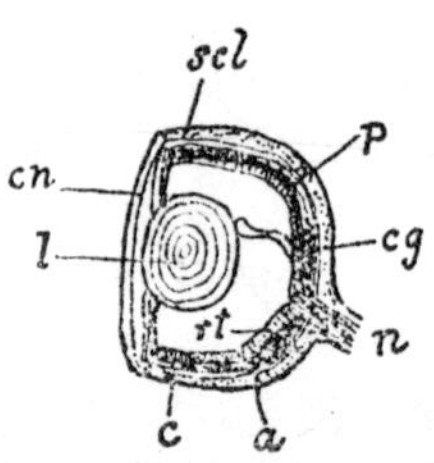

Fig. 41.—Eye of Trout (semi-diagrammatic). *a.*, argentea; *c.*, choroid coat; *cg.*, choroid gland; *cn.*, cornea; *l.*, lens; *n.*, optic nerve; *p.*, pigment layer; *rt.*, retina; *scl.*, sclerotic. (From Bridge, after Parker and Haswell.)

The eyes of fishes have also come in part from glandular organs which have sunk into the tissues and obtained a direct and important connection with the brain, through an outgrowth from the latter. Its structure is similar to that in Higher Vertebrates for the most part, but the globular lens is of a fixed size and is thus not adaptable to vision at varying distances. How this regulation is made, is still not clear. Even if the lens can be moved backwards and forwards to and from the retina, the image printed on the latter cannot be very sharp or distinct. Add to this, that in most fishes the two eyes cannot be focussed on an object at the same time, so that perspective or sense of distance is imperfect, and we see that the vision obtained by fishes must be different from ours. It is perhaps little more than a vague discrimination of light and shade, with a varying perception of colour.

In some fishes, however, the formation of the skull brings the two eyes closer together and the development of a synovial bursa or " recessus orbitalis " behind the eyeball aids in its protrusion and rotation, as in Flat-fishes. We have but to contrast the behaviour of a Cod and a Plaice in an aquarium to see what a difference this makes. The eyes of the former have a fixed stare, those of the latter can be cocked up and quickly rotated, as if the fish were alert and constantly on the watch. When the deformation of the skull gives the eyes a common field of vision in front or above, the precision of the image and perception of distance are greatly intensified. These cases are very rare, but we may well believe that such an enhanced power of vision is of assistance to Periophthalmus when it wanders about among the bushes (*cf.* figure on title-page).

The stalked or " telescopic " eyes of some deep-sea fishes may also serve for binocular vision. They are like parallel tubes with a convex cornea in front and thus seem to be adapted to the concentration of rays on a definite spot; but it is all very mysterious, since there is no light where these fishes live. And the protrusion of the eyes may be in great part artificial. As mentioned previously, when the eggs of Trachypterus are brought up to the surface, the eyes of the embryo are distinctly telescopic, but they gradually sink back to a normal condition after two or three days when the internal pressure has become regulated to the external. This may, in part, explain the extraordinary appearance of the Stylophthalmus obtained by Chun from the depths of the Indian Ocean (Fig. 42), and the young forms of deep-sea species when brought to the surface.

A great deal has been written about the extraordinary eyes of deep-sea fishes and their presumed adaptation to the light or no light prevailing in the depths, but the fact that large eyes, telescopic, normal eyes, or no eyes at all are found in these fishes, directs our attention more to the causes of the phenomena. Very small eyes like those of the Gastrostomids (Fig. 10) may react to the violet rays of the spectrum, which are still appreciable at depths of a thousand metres, but the large eyes are in the gray of darkness long before that depth is reached. It is probable that the fishes which live at these and greater depths are dependent on their other organs and not on the eyes, which in all probability have become functionless partly from lack of use and partly from the great pressures acting upon them. The retention of eyes at all by these forms is due perhaps to their comparatively recent emigration out from the littoral waters.

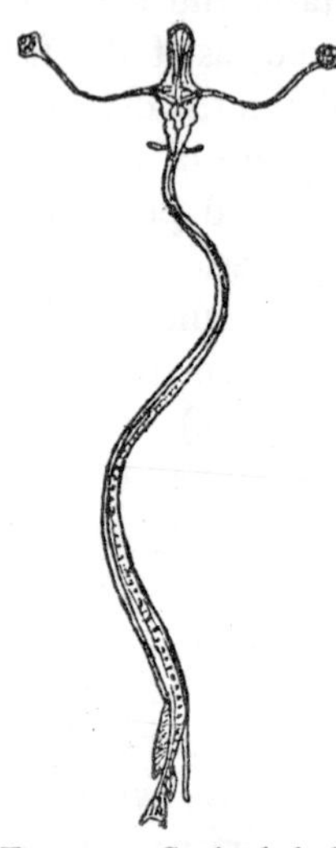
FIG. 42.—Stylophthalmus paradoxus (from Brauer).

The peculiar structure of the eyes, nature of the lens and cornea may be of service in the refraction and dispersion of the light rays. In those fishes which live near the surface or in shallower waters, the image on the retina may be a coloured one and the sense of colour in these fishes has reached a fineness and delicacy unsurpassed in any other group of animals. That their perception of colour is controlled by the eyes was conclusively proved by Pouchet (1876). On cutting the optic nerves the fish ceased to appreciate changes in the colour of the surroundings. But with their organs intact the precision and rapidity with which fishes can alter their own colours according to the different shades about them approaches the marvellous. We see a little of this faculty in the common Plaice of our shores, when they change the colour of their skin and spots on mud or bright sand; but all observers agree that the phenomenon is seen at its best in the tropical forms which live among the more variegated surroundings of the coral reefs.

From direct experiments made by Frisch (1912) and

more recently by Schiemenz (1924), we know that some fishes at any rate are able to distinguish colours. Thus, Frisch gave some specimens of Phoxinus variously coloured meats and found that red was not favoured, the yellow and grey colours being preferred.

The production of coloured pigment within the tissues of the fish is a separate, yet to some extent closely related problem. Pigment is a product of the radiant energy of the sun, on the one hand, and of the living processes on the other. The silvery iridescence comes from guanin (a waste product), or this and lime, which in places is packed tightly into a compact layer in the body, as in the retina of the eye; but the colours red, yellow, black, and blue come apparently from the oxidation of organic substances within certain connective tissue cells (chromatophores). These cells are branched just like the dendrites of the nerve cells and the protoplasm within seems to retain the amœboid power of expanding and contracting, thus spreading out or narrowing the pigment. These stream movements, Pouchet found, were affected by electric currents, the nervous system, condition of the fish, and thus by death or the loss of central control. In the latter condition the chromatophores remain in a state of moderate contraction, like the muscle cells. On the other hand, the presence of solar energy enhances the metabolism within the living cells, causing an increase of pigment formation, and a more lively play of colour ensues.

The negation of colour (black or white) is found in deep-sea fishes or cave fishes which live in darkness. But whilst the paleness of the cave fishes may be ascribed to enfeeblement of the whole system and especially to the poverty of the blood, this cannot be said of the deep-sea fishes. Here the intensely black pigment probably comes from excessive metabolism and the disruption or restricted development of the tissues, whereby the substances that make for colour are all mixed together.

Returning now to the part taken by the eyesight in regulating the play of colour, it is natural that this power should develop to a greater extent under conditions producing a greater variety of colours, as in the tropics and inshore areas. It is the law of action and reaction in the organic world which we have already seen exemplified in the skeleton. The dents or impressions from without lead to a change in the tissues, which is at first non-utilitarian, but gradually becomes responsive only to particular impressions when constantly renewed. The absence of these impressions, as in the case of the offshore fishes under more uniform conditions of light, means that the apparatus remains undeveloped.

The question whether colours are adaptations for the sake of protection, and have been evolved through and for this purpose, will be referred to later; but the opinion of Jordan and Seale with regard to the brilliantly coloured fishes of the coral reefs may just be quoted here: "It is not easy to explain the reason for these vivid hues, nor for the elaborate and striking markings which accompany them. It is clear that protective coloration is needless, for these species are exceedingly active and when disturbed move through the water like animated lightning—and the idea that recognition has a high value to the species has never been received with favour by naturalists."

The eyes of fishes have no lids, but many forms among the Teleosts have a gelatine-like covering over the eye. This is known as the "adipose" lid from containing at times a considerable amount of fat. In the Mackerel "it extends almost entirely over the eyeball leaving only a narrow slit in front of the pupil" (Ehrenbaum, 1914). It may be of service to these swiftly moving fishes in presenting a smooth surface to the water, but the origin of the structure can probably be referred back to the lymph sac which lies above the skull in the larva when the bones are being formed. In the adult Clupeids the adipose membrane extends round the eye, over the head and even above the pectoral region, thus recalling the distribution of "bone" in the early Ganoids. A somewhat similar structure or organ is found in the Sharks in the form of a "nictitating" membrane. It forms a continuation of the skin on the lower margin of the eye and can be drawn up like a shutter. Here it may be used as a protection against the rays of light when the fish comes up near the surface. And this is also the case with the well-developed nictitating membrane of the Sun-fish (Mola). According to Smitt it has a sphincter muscle with five radiating, opening or retractor muscles.

In the peculiar form Anableps tetrophthalmus of Central and South America the eyes are divided into two parts, an upper and a lower, by a cross-band of conjunctiva, and the pupil beneath is likewise constricted into an upper and lower part (Fig. 73). By this means, it is supposed, the fish, which lives on the surface mostly, has two kinds of vision, one for the air and the other (lower) for the water. How this peculiarity could have arisen, except under the law of action and reaction, it would be difficult to imagine.

6. Central Nervous System

It is perhaps difficult to realise, that the brain and spinal cord really belong to the outer skin or epidermis. This from the beginning has been the layer that received impressions from without and responded by becoming sensitive towards those impressions. In the course of time, owing to the working of some mysterious law, brain and cord have sunk in from the surface and are now formed, like the sense organs, from an inpushing or depression of the outer cell-layer in the embryo. Instead of being in separate parts, however, like the muscles and organs, the depression forms a continuous groove along the mid-dorsal line, and this groove is turned into a canal or solid column of cells by the growth upwards of the sides. Then the muscles of the sides grow in across the top and shut off the cord from the surface. Among the fishes the spinal cord and brain lie nearest the surface in the Clupeids, that is, in the early stages.

The spinal nerves connected with the different segments are arranged in the same way as in Higher Vertebrates and there is also an internal sympathetic system. The cranial nerves are for the most part the same, with some variations in their development and importance. The VIIIth nerve, for example, is specially large and is connected with an independent centre in the medulla, the tuber acusticum. This is supplied from the general cutaneous nerves, lateral line system, as well as from the auditory organ or statocyst. The surface of the brain is smooth and in some places simply covered with epithelium.

The peculiar features of the brain in fishes are the great elongation and separation of the principal parts from one another. These may be considered a sign of its more primitive condition, but it is also advisable to take them in connection with the mode of development and the forces acting upon the brain. It is certain that the shape of the brain is determined by the form and capacity of the cranial cavity and the latter is determined by the forces acting on the skull, external pressure, balance of the body, etc.

This can be seen very clearly on comparing the brain of a Dog-fish with that of a Herring (Fig. 43). The head of the adult Dog-fish appears to be narrow and elongated, but on examining the brain we find that the anterior parts are expanded well out to the sides to form the olfactory lobes, whilst the whole forebrain is distinct, with indications of two cerebral hemispheres (*fb*). This great expansion can be referred back to the movements of the embryo in the egg-capsule and stands in marked contrast to the compressed condition of the forebrain in the Herring where there is no attempt at forming the cerebral hemispheres. When we recall the conditions under which the brain of the latter develops, the free-swimming mode of life and the pressure of the water on the skull, it is not difficult to recognise the sources of these great differences.

In the Teleosts the optic lobes or mid-brain (*mb*) lie just in front of and above the articulation of the lower jaw with the skull (sphenotic) and are thus pressed upwards to a more marked degree than in the Elasmobranchs, whilst the cerebellum behind (*hb*) is still more elevated (Fig. 43), the line of the upper surface of the brain corresponding to the slope oi the skull (Fig. 24). In the Dipnoi where the lower jaw articulates directly with the skull opposite the optic lobes, the latter are pressed together and fused into one. In the Herring again, it can be noted how tightly the brain is packed

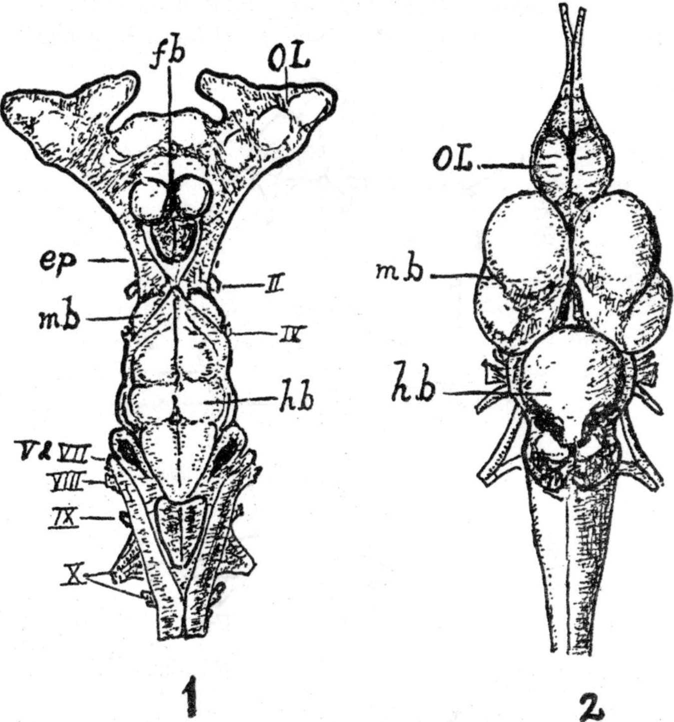

FIG. 43.—Comparison of 1, Elasmobranch (Scyllium) brain with that of 2, a Teleost (Herring). *fb.*, forebrain ; *mb.*, mid-brain ; *hb.*, hind brain ; *ol.*, olfactory lobes ; *ep.*, origin of pineal body ; II.–X., cranial nerves. (1, from Wiedersheim ; 2, from Lissner.)

into the cranium ; the optic lobes are furrowed by the frontals above and the sphenotic at the sides (anterior semicircular canal).

On the other hand, it is not easy to understand how the optic nerves below have become crossed as they proceed to the eyes, nor why the crossing should be simple in the

Teleosts, whereas in the Elasmobranchs the nerve fibres from each side get mixed up together in a chiasma. In the normal Teleosts sometimes the left optic nerve is the upper one, sometimes the right. In the Flat-fishes the upper nerve is usually the one belonging to the upper, migrating eye; but in some cases (Symphurus and Psettodes) it may be either the one or the other, and in abnormal specimens

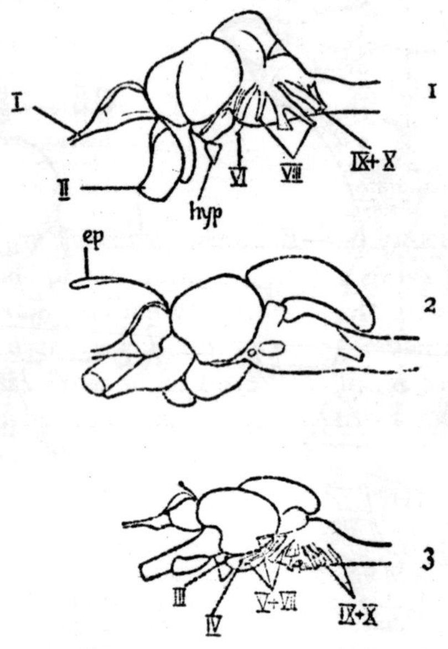

FIG. 44.—Brains of various Teleosts. 1. Herring; 2. Salmon; 3, Mackerel. I–X., cranial nerves; *ep.*, epiphysis or pineal body; *hyp.*, hypophysis or pituitary body. (From Lissner.)

(sinistral instead of dextral Flounders, for example) the nerves are twice crossed (Parker).

Among the Teleosts the greatest variation is found perhaps in the cerebellum (*hb*), which lies against the posterior wall of the cranial cavity and thus comes under the influence of the body pressure against the skull. This is greater, for example, in the forms which have the mouth inclined upwards than in those with the mouth horizontal. Thus, in the Herring the cerebellum is small and compact, where in the Mackerel it is elongated, and in the Salmon and Pike very long and curved over the medulla. This last condition has been associated with the fact that these are predaceous fishes and probably there is some connection, but both the predaceous character and the peculiar cerebellum almost certainly come from their having large mouths. These fishes also have a conspicuous pineal body (*ep*), which is hardly developed in the Herring, or Mackerel.

Whilst the shape and arrangement of the different parts of the brain are determined by the influences operating during the early development, when the structures are still plastic, it is probable that the habits and mode of life of the adults affect the growth of the cell tissue internally in the brain substance. Thus, although the eyes and optic lobes of the Dog-fishes are much smaller than those of the Herring, we cannot say that the former has less keen eyesight than the latter. The relative abilities, in fact, can hardly be judged from the external appearance of the brain, though it is generally accepted that the Elasmobranch brain with its developed cerebral hemispheres is superior to that of the Teleosts.

CHAPTER VIII

VARIATION AND DIFFERENTIATION OF FISHES

IT is a common saying, that no two peas are ever quite alike, and in no group of Vertebrates is this fact more manifest than among the fishes. The characters that later become more or less fixed, such as the head and number of vertebræ, show here a remarkable amount of variation. Fishes stand at the beginning of the Vertebrate series, and their greater variability is connected with their low constitutional level and closer contact with the environmental influences. For these reasons the study of variation in fishes is of peculiar interest.

1. NATURE OF VARIATION

Fishes differ greatly in size and shape, and according to their habits and habitats, but by variation is meant something more precise than these differences. The term is used here as Darwin used it, to express the small differences between individuals of the same species or race. These variations provided the materials, Darwin thought, for Natural Selection to work upon. If any advance has been made since Darwin's time in our way of regarding variations, it has come from endeavours to trace their causes where he had to be content with accepting them as chance or fortuitous occurrences. In this way the study of variations has become an exact science, as precise in its methods and deductions as any of the physical sciences. Its value for the solution of biological problems has perhaps been estimated too highly, but there can be no doubt that it is indispensable in their consideration.

The characters of a fish are of two kinds, those that can be counted and those that have to be measured. The former, such as the number of vertebræ, remain constant for the most part throughout life, whereas the measurements, for example of the head and body, change somewhat with the growth. The former can also be more exactly determined, with a smaller personal error, than the latter. For these reasons there has been a tendency of recent years to neglect the measurable characters altogether. This seems a mistake. It means that the form or shape of the fish is ignored, a character of the greatest biological importance, whilst the conclusions drawn from the numerical characters alone become one-sided.

Some of the characters of fishes vary only a little. The caudal fin remains very constant both in form and number of rays, not only in a species but even within wide groups. It is the most specialised part of the body, being adapted to a particular mode of movement. Even the number of vertebræ may become very constant, as will be noted later. But apart from these special cases, it may be said that all characters vary a good deal, so that one individual cannot be taken to represent the species or race. The number of vertebræ in the Plaice, for example, varies from 41 to 44; it would not be correct, therefore, to say that the Plaice has 42 vertebræ. An average has to be taken, and then we find that this average varies from one region to another and even from one year to another. Thus, by a simple arithmetical calculation we obtain a deeper insight into biological problems. The variation of one or two individuals might be called fortuitous or casual; the fluctuation of a large number at once indicates the presence of regular causes.

In the study of variations it is necessary to work with large numbers and their averages, so that the cumbersome masses of details may be made to reveal their essential secrets. Where measured characters are taken, for example, length of head or total length and so on, the measurements are expressed as percentages of a common denominator, for example, the length of the caudal region, the motor part of the fish. In this way comparable numbers are obtained for the measurable characters and they can then be dealt with in the same way as the numerical characters.

When the numbers have been obtained, they can be arranged in the following way. The "variants" are the different observed values of a character, *e.g.* number of vertebræ; the "frequencies" are the observed number of individuals at each value. As will be seen, the frequencies group themselves about a central value, becoming fewer on each side away from this value. The latter is called the "mode." The simplest way of calculating the average and higher moments is to take this mode as the central point and sum up the total differences or "deviations" from it. The mode is then 0 and the deviations from it are plus or minus according as the variants are above or below the mode. We thus obtain the following scheme.

Variants.	Frequencies. f	Deviations. x	Sum of deviations. xf	Sum of squares of deviations. x^2f
110	1	−5	− 5	25
111	17	−4	− 68	272
112	89	−3	−267	801
113	352	−2	−704	1408
114	719	−1	−719	719
115	865	0	0	—
116	500	+1	+500	500
117	192	+2	+384	768
118	36	+3	+108	324
119	4	+4	+ 16	64
	$n=2775$		$\Sigma(xf)=-755$	$\Sigma(x^2f)=4881$

Dividing the sum of the deviations by the total number (2775) we obtain the average amount of deviation −0·27, which means that the average of all the frequencies lies by this amount below the mode. This gives 114·73 as the average of all the frequencies.

The data given above represent the number of vertebræ in a very large collection of Eels (*Anguilla vulgaris*) made by Johs. Schmidt. The calculations to be mentioned in the following account have been made by Georg Duncker, who has made a special study of the variations in fishes (1899,

1908). The curve of the variations has been drawn by E. Mohr.

It will be noted, that the observed frequencies arrange themselves very evenly above and below the average, decreasing gradually to 1 at 110 vertebræ and to 4 at 119 vertebræ. This arrangement of the frequencies about an average, though not always so regular as in the case of the Eel, is a well-marked feature in the variation of fishes. To use Heincke's comparison, it is as if the variable forces were always aiming at a target, with the average in the centre,

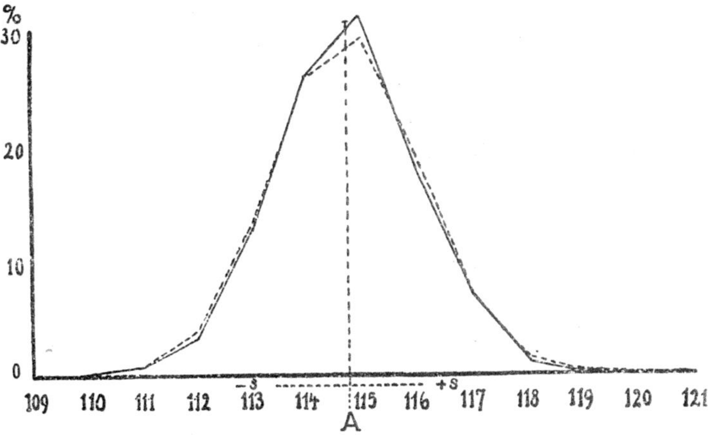

FIG. 45.—Variation-curve of the number of vertebræ in the Eel. The frequencies are expressed as percentages. A=average.

and their efforts resulted in the great majority of individuals being near the mark, over 2,000 are here between 114 and 116, whilst the remainder tail off on each side. We can see this more clearly on arranging the numbers in the form of a curve.

This grouping of the frequencies resembles very closely the occurrence of chance events. For example, if an astronomer or physicist makes a series of measurements of an object, his results group themselves about an average in the same way, and he then proceeds to calculate the probability that his average is correct, or rather, how far it is correct.

In the same way we can estimate how far the average of the observed sample of individuals is likely to represent the real average of the species in nature. For this purpose some well-known rules and formulæ are given by mathematicians.

When the frequencies have been plotted out on a curve, the latter forms a polygon whose characteristics can be calculated and expressed by means of certain factors or moments. The average is the first moment. The second moment is obtained by calculating the error of mean square or standard deviation, as it is called by biologists. Taking σ to represent this moment, its value is obtained from the formula,

$$\sigma = \frac{1}{n}\sqrt{n\Sigma(x^2f)-[\Sigma(xf)]^2}$$

where x is one of the deviations and f its frequency as before.

Hence in the case of the Eel,

$$\sigma = \frac{1}{2775}\cdot\sqrt{2775 \times 4881-(755)^2}$$

$$= \frac{1}{2775}\cdot\sqrt{13{,}544{,}775-570{,}025}$$

$$= \frac{3602{\cdot}05}{2775} = 1{\cdot}298$$

The probably true value of the standard deviation is somewhat less; that is to say, if we examined a very much greater number of specimens, they would group themselves still more about the average and thus reduce the amount of deviation. This probable value is got by multiplying the standard deviation by 0·6745, a factor obtained from the curve of probability; hence

$$P\sigma = 0{\cdot}6745\sigma$$

Then dividing by the square root of the number of individuals we obtain the probable error (r) of the observed average.

In the case of the number of vertebræ in the Eel, the standard deviation is 1·298 and the probable error of the average works out at 0·017. Hence we may say, that the probable average for the species, at the time when these observations were made, would as likely as not lie between the values $A \pm r$, that is, between 114·711 and 114·745, which means a very small range of variation, corresponding to the large number of specimens examined. If we take three times the probable error, then the probability is about 20 to 1 that the true average lies between 114·677 and 114·779.

It is by means of such calculations, if the material is sufficiently large to be representative, that races or varieties of a species can be distinguished. Thus, if similar samples of Eels from different parts of Europe gave averages lying outside these probable limits, we should be entitled to regard them, even though the frequencies might overlap, as separate races which had grown up under different conditions. Johs. Schmidt has made such a comparison between Eels from the Mediterranean and Eels from northern waters, and found that the probable average was the same in each case. He was thus entitled to come to the conclusion, that all the European Eels belong to the same stock born under the same conditions, that is, out in the Atlantic, and not as Grassi supposed, partly at any rate in the Mediterranean.

When the standard deviation has been obtained, its usefulness lies in the fact that from it we can calculate, by the factor $\frac{x}{\sigma}$, the real frequencies which should occur according to the normal curve of frequency (Gauss' curve). In this way the observed and normal values can be compared as follows:

Variants	110	111	112	113	114	115	116	117	118	119
Observed freq.	1	17	89	352	719	865	500	192	36	4
Normal freq.	1	16	101	358	716	815	527	193	40	5

It will be seen that, except at the two adjacent frequencies, 115 and 116, the agreement is extremely good. Indeed, if we allow for the personal equation, the agreement is as perfect as one can hope for. Whatever may be the causes of the whole variation, the results conform in a remarkable degree to the laws of chance occurrences. Such a close agreement has not been observed in the case of other species, mostly coastal species, and one is tempted towards the conclusion, that the conditions out in the Atlantic are more evenly balanced and more uniform than those nearer the coast.

The precise agreement of the observed curve with the normal curve can be seen from the calculation of the third moment in a similar manner as before $\left(\beta_3 = \frac{\Sigma(x^3)}{n\sigma^3}\right)$. If the agreement is perfect, this value should be equal to 0. The fourth moment $\left(\beta_4 = \frac{\Sigma(x^4)}{n\sigma^4}\right)$ shows whether the extremes meet the abscissa gradually ($\beta_4 > 3$) or at a sharp angle ($\beta_4 < 3$). In the case of the observed frequencies for the Eel, Dr. Duncker has obtained the values $\beta_3 = -0{\cdot}00244$ and $\beta_4 = 2{\cdot}03901$. These represent the closest agreement between observed variations and the normal law of chance occurrences yet discovered. If the variability came from the inheritance, as some think, then this agreement with the normal curve should be the same for all species everywhere. Since this is far from being the case, we are obliged to think that the outer conditions are responsible for the variations.

Sometimes, it is true, the observed frequencies conform more to other types of probability curves, the asymmetrical types; but for practical purposes, so far as the variation of fishes is concerned, it is better to assume that the frequencies should be normal. This directs our attention to the necessity of obtaining homogeneous material, that is, from the same place and time. We know that species vary from one locality to another and, if the material is composed of specimens from different localities, the resultant curve is almost certain to be asymmetrical. Again, the material though obtained from the same locality may be composed of individuals from different year-groups or ages, thus born under different conditions, and these have to be sorted out from one another until only a pure group is left. The asymmetry of variation-curves, derived from the study of fishes, comes far more from these material errors than from the normal variation of the fishes themselves. A pure group therefore is very difficult to obtain, but it should be aimed at.

The importance of the use of the mathematical method of treating variations lies mainly in the training it gives in drawing conclusions. One is obliged to think that the characters are not constant but vary up and down about a mean, and in this way one gets into the habit of balancing the possibilities. The mathematical method has also been extended into other problems, the calculation of the correlation between two varying characters, the death rate or

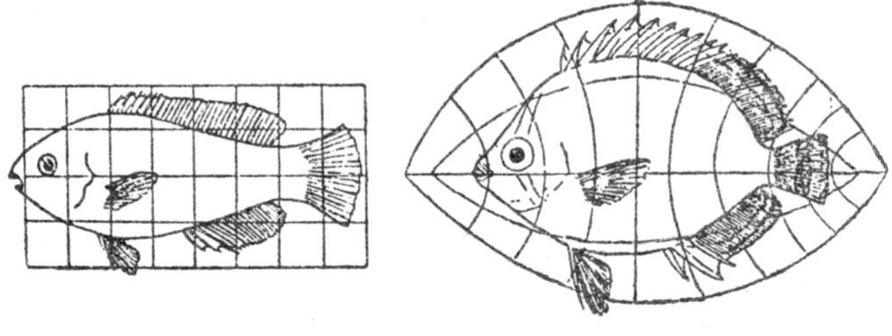

Scarus sp. *Pomacanthus*

FIG. 46.—The form of fishes expressed by co-ordinate lines. (From d'Arcy W. Thompson.)

selection of individuals and the relation between parents and offspring, and has everywhere proved its usefulness.

Of great interest and promise also is the graphic method of studying larger variations in form which has been introduced recently into the study of fishes by Prof. d'Arcy W. Thompson. In this method the form of a fish is represented by a system of co-ordinate lines and we can see how from one form many others may be derived by a simple re-arrangement of the lines, either by changing the angle of the axes or by other curves.

The importance of this method will be realised on recalling Houssay's experiments. If the anterior region of a fish varies according to the rate of travelling through the water, the direction of the axes and curves would give a measure of the relative speeds. Thus, a sharp-nosed fish darts through the water, a blunt-nosed fish is slow in its movements, and between these lie many degrees. Hence the angle of the co-ordinate axes increases with a decrease in the speed.

An example of the use of the method is given in the accompanying figures. Changing the straight lines of Scarus into appropriate, co-axial circles, we obtain the form of Pomacanthus. But if we could express these relations dynamically instead of statically, thus, by starting from the centre of movement near the tail (see Pl. X*a*.) and representing the lines of force acting through the body forwards and backwards, with a corresponding system of lines representing the pressure and resistance of the medium, we should probably obtain an even better picture of the form-variations in fishes (compare Fig. 21).

2. Heredity and Circumstances

The small variations shown by the characters of a fish thus agree more or less closely with the laws of chance occurrences. The question is now, whether we know the ways of nature sufficiently to be able to eliminate the word "chance." A stone may be rounded or pointed and nine people out of ten would give a similar explanation of the difference; in the case of biological variation it would be difficult to find two people in ten who would agree as to the causes. Not that these cannot be determined, it is rather because there are so many possibilities, and zoologists, just as superstitious as other people, like to make allowances for the presence of the "unknown" factor.

The power to vary, or variability, is inherent in the nature of organic substances. We may mix the same ingredients together in the same way, yet a slight difference in the temperature during the mixing will yield different compounds; the molecular arrangement is not the same. Even if we have the same compound, it will not behave in the same way in different media, for example, blood or water. An organism is built up of such unstable components and the marvel is, that it should remain at all constant, not that it should vary. With the formation of structures composed mainly of inorganic materials, more fixity or permanency has been obtained, but until these are formed in the early stages of each individual the organism is plastic and ready to change under the influence of the physical and chemical surroundings.

Starting from this basis Heincke in his great work on the Herring sought to establish two parallel series ; on the one side the variations in the characters, on the other the variations in the external surroundings. From the former he hoped to obtain an average condition representing a race, or it may be a species, from the latter an average of all the influences, physical, chemical, and biological, affecting the young larvæ. The racial characters and environment are considered inseparable.

On general grounds one would think that this should give a good working basis for the study of variations and their causes, but the practical difficulties in the way of forming such parallel series are immense, and no one as yet has tried to give any idea of the average of the external influences. During the past twenty years, moreover, the tendency has been for workers to leave these almost entirely out of account and to concentrate attention on the nature of the inheritance. It is conceivable, indeed, that organisms are now so well-adapted to their respective environments, right from the earliest stages, that the external influences can effect little or no change during the orderly progress of development. On this view the inherited constitution is everything and the material base of heredity has been traced to the chromosomes of the germplasm, which play such an important part in the structure and building up of the cells. It is even believed by some, that the sex and definite characters of the adult can be referred back to the presence of particular kinds of chromosomes in the germ.

Before the causes of variation in fishes can be discussed

we have therefore to consider, what is meant by heredity. That a connection exists between parents and offspring is one of the most obvious things in the world, but are we justified in thinking that this is fixed independently of the environment? Are the inheritance and environment anything but abstractions, unreal when separated from one another?

The great attraction of the heredity studies lies in the simplicity of the method of working and the ease with which it lends itself to definite experimental investigations. One or two characters readily observed are studied and compared in parents and offspring. From the results of the comparison, it is assumed, the proportionate influences of the parents in the offspring can be determined. The mode of origin and development of the characters are not considered. So far as the fishes are concerned, the only worker who has made use of this method has been Johs. Schmidt, and his investigations into the nature of heredity and the causes of variation merit careful study.

The small Cyprinodont *Lebistes reticulatus* from the West Indies and tropical America has many remarkable characters. It is viviparous, and the young when born are just like the adults. It becomes mature in three months and is able to produce no fewer than seven broods within six months, the young numbering from 50 to 70 each time. More remarkable still, one pairing with the male is sufficient for the whole lot, the spermatozoa being apparently retained in the oviduct and remaining sufficiently capable there to fertilise the successive broods, in spite of the fact that the young are meanwhile reared in the oviduct. It thus presents very favourable opportunities for experiment.

To examine the importance of the inner factors (inheritance) Johs. Schmidt kept two pairs with different characters under precisely the same outer conditions. The character studied was the number of rays in the dorsal fin. In the species this varies between 5 and 8, with an average about 7. The one pair chosen had the number 6, the other pair 8; and the experiment was repeated with two other similar

PLATE XI

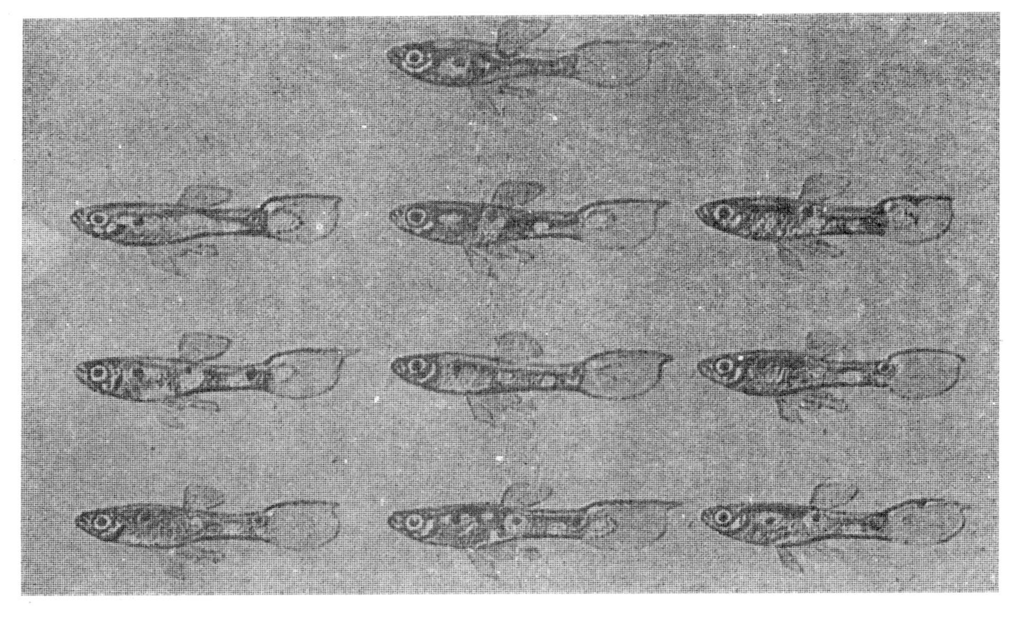

V/ RIABLE OFFSPRING FROM THE SAME MALE (UPPERMOST.)

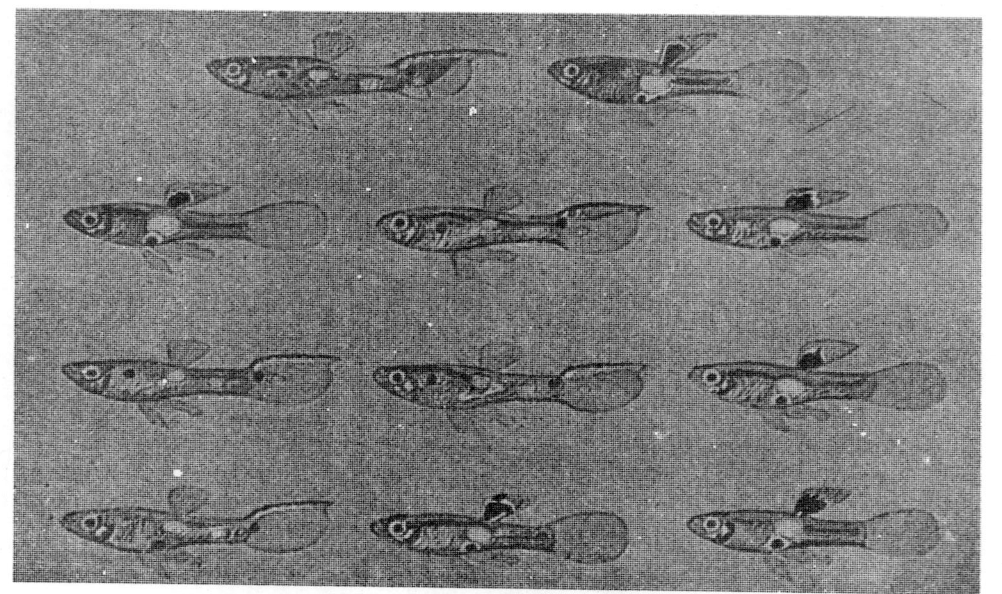

OFFSPRING FROM SAME FEMALE, BUT FROM TWO DIFFERENT MALE PARENTS (UPERMOST.)

COLOUR VARIETIES OF THE MALE LEBISTES.

(From Johs. Schmidt.)

pairs. The result was, that the offspring from the parents with 6 rays had an average of about 6·7, those from the parents with 8 rays, an average of about 7·9.

The offspring, especially the latter, thus retained the character of their parents to a very large extent, and the variation from the parents might be accounted for by the tendency of offspring to revert to the mean of the species (Galton's law of regression); but it is apparent that the results do not agree with this rule. In both cases the value obtained is much higher than we should expect. And one may doubt, therefore, whether the "outer" factors have really been excluded. For Johs. Schmidt the outer factors seem to be only the temperature and salinity; yet one would think, that all the external surroundings should be taken into account, for example, the size or capacity of the female fish. As we shall see, this is undoubtedly a factor in producing variations. Hence we can hardly assume that a female with only 6 dorsal rays presents the same opportunities of development to its offspring as one of 8 rays.

This little fish Lebistes has other characteristics which lend themselves to experiment. The female is plain and uniform in colour; the male has a brilliant coloration, and the colours differ according to the different places whence it comes. One form (A) has no black pigment but a great deal of red, the other (B) has several black spots, one handsome spot in particular on the dorsal fin, though there is a considerable amount of variation (Plate XI). By crossing these different male forms with the same kind of female, Johs. Schmidt obtained a series of interesting results.

In the first set of experiments the male B was crossed with an ordinary female from the A group; all the males without exception (998) gave the B spot. One of the male offspring was then paired with another female of the A group; here again all the male offspring had the B spot, and similarly with later generations. Then a female of the AB crossing was paired with an A male (without spot), and once more all the males (332) were of the same type—without spot.

As pointed out by Johs. Schmidt, these results are strongly at variance with Mendelian expectations. We can imagine, of course, that a special sex chromosome is present in the male and is a dominant, no matter what the female may be ; but one may reasonably hesitate to be led into the labyrinth of abstract possibilities when a simpler explanation may be available.

It is a question of the origin of the pigment. There may be a determinant or id in the germ which goes to form pigment, but it is a general rule that pigment does not appear on the body of the embryo or larva until the red blood-corpuscles are formed. It is thus apparently an oxidation product depending on the metabolism of the organism and on its movements. On comparing now the male forms A and B of Lebistes, we find that the B form has a plain homocercal tail along with the black spot on the dorsal fin. The form A has no black spot, but it has a peculiar tail. This is not quite symmetrical ; the terminal portion of the notochord (urostyle) persists and a great deal of variegated pigment, mostly red, is collected round it and over the root of the tail (Plate V).

From what has been said previously in Chapters IV and V, it seems probable that the form A represents a younger stage of development than form B, thus a slightly different balance. As this species develops at a phenomenally rapid rate (the fully developed fry escape from the mother-fish in about 3 weeks after fertilisation), the difference between the two forms seems to be simply one of time and temperature. The form B is the more advanced, and higher temperatures would probably in the course of a few generations change it into the more precocious A form.

As this possibility cannot be excluded, we can hardly say that the difference between A and B depends upon the inherited substance, chromosomes or the like. If it does so, as Johs. Schmidt himself points out, we should have the remarkable phenomenon of two separate races or forms depending upon a character inherited only by the males.

The most suggestive of Johs. Schmidt's experiments

are those in which he endeavours to find a definite expression for the inheritance. In an abstract manner we may speak of "the generative value" of the individual, that is, the value it imparts to the offspring, as distinct from "the personal value," that is, the actual appearance or character it possesses. To accomplish his aim, he has devised a most ingenious method of "diallel crossing" whereby a number of males are each crossed with the same females. When the eggs are fertilised externally, as in the Trout (Schmidt, 1919), this method can be readily carried out. The eggs of any one female are divided into lots and each lot fertilised from a separate male; similar lots of other females are treated in the same way by the same males. Hence, from any one male or female several different offsprings arise and on comparing these with regard to any one character, the generative value may be obtained. The details of the method, as worked out by Johs. Schmidt, can be seen from the following.

The character examined was the number of vertebræ, and the parents had the following values :

x	59	*y*	60	*z*	59			males
a	61	*b*	59	*c*	57	*d*	58	females

Fifty specimens of the offspring were examined in each case and the average number of vertebræ calculated :

xa	*xb*	*xc*	*xd*
61·14	59·06	58·29	59·03
60·0	59·0	58·0	58·5

ya	*yb*	*yc*	*yd*
61·35	59·22	58·59	59·28
60·5	59·5	58·5	59·0

za	*zb*	*zc*	*zd*
60·65	58·48	57·90	58·55
60·0	59·0	58·0	58·5

The first row in each case gives the average found in the offspring, the second row the average of the parents. The offspring of a row have the same father, those of a column the same mother.

Assuming now that the male and female in each case are equal in potency and that the average of the offspring gives a measure of the generative values of the parents, the latter can be obtained by a process of elimination from the following equations:

$$\frac{x+a}{2}=61{\cdot}14,\quad \frac{y+a}{2}=61{\cdot}35,\quad \frac{z+a}{2}=60{\cdot}65$$

and similar equations for the other females, *b*, *c*, and *d*. Summing up by means of these the values for *x*, etc., in the offspring averages of the first table we have:

$$2x+\tfrac{1}{2}(a+b+c+d)=237{\cdot}52$$
$$2y+\tfrac{1}{2}(a+b+c+d)=238{\cdot}44$$
$$2z+\tfrac{1}{2}(a+b+c+d)=235{\cdot}58 \text{ and so on.}$$

Hence,

$$x-z=0{\cdot}97$$
$$y-z=1{\cdot}43$$
$$a-b=4{\cdot}253$$
$$a-c=5{\cdot}573$$
$$a-d=4{\cdot}187$$

These give the differences between the generative values of the various individuals. Taking 60 to represent a standard generative value ($=y$) the personal values and the calculated generative values of the 7 experimental Trout can now be compared.

Trout.	Personal values.	Generative values.
x	59	59·54
y	60	60 (assumed)
z	59	58·57
a	61	62·72
b	59	58·47
c	57	57·15
d	58	58·54

It will be seen from this table, that the values for the parents obtained from the offspring do not agree with their actual or personal values. In the case of *x*, for example, the generative value is half a vertebra more than it actually possessed, whilst *z* shows half a vertebra less, and the female *a* gives nearly 2 more vertebræ to her offspring than she herself possessed.

The above data from Johs. Schmidt's work will give some indication of the extremely able way in which he has approached the problem of heredity. If the environmental influences could be completely excluded, the method would yield important conclusions regarding the "generative" value, as Johs. Schmidt understands it, namely, the potential value of a character contained in the germplasm of an individual. But, just as in the previous cases, one may doubt whether the environment has been or can be excluded. The generative values given by him seem in reality complex not simple. They represent not merely the inherited substance, but also the influences under which the offspring were first of all created and then developed, and we cannot assume these to be uniform.

We are thus back to the old question, whether the inheritance can be restricted to the substances contained in the germplasm. Is a given number of vertebræ inherited or not? The complexity of the problem was clearly indicated long ago by Adam Sedgwick (1899). Whilst the individual may be regarded as the unfolding of what is contained in the germ, yet "every feature which successively appears in an organism in the march from the uninucleated zygote to death is an acquired character." The offspring, we may say, receive through the germ not any particular feature, but the tendency to repeat the same experience and characters under the same conditions. One has to remember also, as pointed out by Sir John Murray, that "the chemical composition is just as likely to be handed down as morphological structure."

The inheritance consists of two, if not three elements. On the one hand, we have the organic and inorganic substances of the sexual products and the regulatory essences in the germ and egg. These tend to keep the characters constant.

On the other hand, we have the environment which tends to make the characters vary. A good impression of the first two elements can be gained from the following table.

This shows the sizes of the various Teleosts when the number of vertebræ becomes fixed in the post-larval stages. It might be thought that the number of vertebræ depended on the ultimate size of the fish, but this is far from being the case ; the largest fish are among those with the smallest number of vertebræ. It will be seen from the table that the latter depends upon the length and composition of the notochord, the plastic tube that precedes the vertebræ. The forms with a larger number of vertebræ than 60 are not included in the table, as there is a possibility that each vertebra is formed from a single notochordal cell, instead of two as in the Herring, or many as in the more concentrated forms.

Length of fish in mm. when vertebræ formed.	No. of vertebræ.	Some characteristic species.
3–4	17	Mola
5–6	25	Caranx, Mullus
6–7	30	Scomber
7–8	35	Flounder
8–11	40	Trachinus draco
12–14	45	Plaice, Perch
15–16	50	Atherina
17–18	55	Herring

This correspondence of the vertebræ with the size of the post-larva cannot be a casual one. The huge Mola, for example, might well have as many vertebræ as the small Herring, were it not for the law which regulates the number of vertebræ in the post-larval stages. And this law is simply the balance of the fish, head with body. At the upper end (Herring) the body and head are elongated and narrow, at the lower end larger and heavier, though shorter. We can see, also, that the vertebræ have become more concentrated at the lower end, for whilst the Herring has only about three vertebræ to the mm., Mola has about six, or twice as many. The inherited substance has thus become more concentrated both in the head and body.

In the earliest stages the notochord is not divided up

into vertebræ. In the words of Sedgwick, the number of vertebræ is therefore an acquired character after the little larva has been moving about for some time. If we think of the notochord as a flexible rod moving to and fro with the working of the muscles, we can readily understand how it becomes broken up into parts (vertebræ) according to the movements and the density of its substance. In agreement with this view, we find that the species with longer body, such as the Herring, display considerable variation in the number of vertebræ, whilst the concentrated forms show little or none at all.

The inheritance of fishes is thus one of balance as well as of substance. The bodies have been tuned through many generations and thousands of years to move in certain ways, and it is probably more due to this tuning than to the actual substance that the species and groups remain more or less constant. The balance of the body and head is the continuation of the regulating influence which can be detected already in the egg. It may be said to represent the life of the fish, and this, unfortunately, is just the element that is missed in the heredity studies which isolate one or more characters from the rest and take no account of their mode or time of origin.

3. Causes of Variation

If we examine a large number of eggs from the same species or the same specimen, we find that their size varies just like any other character. Heincke and Ehrenbaum (1900) found by careful and exact methods, that the larger fish have larger eggs, that the eggs spawned earlier in the season were larger than the later eggs, and that in the successive groups or batches of eggs spawned by the same fish the size became smaller. We thus have a most important source of variation right at the very beginning, for it is the general rule that the smaller eggs yield the smaller larva and the smaller larva has the smaller number of vertebræ. Whilst the number of vertebræ has some influence on the growth of all other characters.

The principal cause of the variations in size seems to be a difference in the amount of yolk actually stored up in each ovum according to its position relative to the blood-vessels in the ovary. The variation is thus to a great extent due to chance and obeys the laws of chance occurrences, as can be seen from the following example (from Heincke and Ehrenbaum) of the size of Plaice eggs, from the same fish. The " lines " refer to those on the micrometer of the microscope.

Lines	60	61	62	63	64	
Observed values . .	2	52	112	29	4	200
Theoretical values .	6	53	98	39	4	200

The agreement is not so perfect as in the case of the number of vertebræ in the Eel, yet sufficient to indicate that the diameter of the eggs, thus their size, varies according to the laws of chance occurrences. The causal connection between this variation and that of any character, number of vertebræ for example, has not yet been worked out in detail, but as indicated above we can readily see that there is a connection. With a difference in the size of the larva the balance of the fish is different when the vertebræ are formed, unless some other factor has come into play in the meantime.

Some authors are not content with this simple relation, however, and think that the sources of variation lie still more deeply, not in the yolk, nor even in the general mass of the nucleus, but in the particular parts or chromosomes of the latter. They would make these the bearers not only of the inheritance but also of the variation. For them the particular characters of the individual are not superficial, personal, or acquired ; that side of the matter is excluded ; and they maintain that the characters of the individual have come from those of two other individuals, these again from four other individuals, and so on. But this is pure assumption. If the " continuity of the germplasm " from generation to generation means anything, the individual characters have but little weight in the inherited substance. By prolonged selection and elimination we may " tune " the germplasm to yield only a particular character or " right line " without variation, but that only proves that under natural conditions the germplasm is very plastic. It does not prove that " right lines " exist in nature in such complex bodies as the fishes.

According to this view, the mixing of two different germplasms (amphimixis) accounts for the variations mentioned, but we have to draw powerfully on our imagination to picture the presence of a number of right lines or genotypes. In the case of Zoarces, the Viviparous Blenny, whose variations have been closely studied and described in a remarkable series of papers by Johs. Schmidt, the number of vertebræ varies from 101 to 126, and we should have to believe that any or all of the intermediate numbers might represent genotypes. Hence each individual has a possible chance of 26, and these should all be represented in the inherited substance. But all other characters, the number of fin-rays, size of head, etc., are almost as variable, and each germ should thus potentially have the makings of some millions of individuals. We may doubt whether the germ is really burdened in this way, and the mere effort to conceive what it means reminds one of the old riddle of the schoolmen—how many angels can stand on the point of a needle ?

The genotype is an abstraction which has arisen, like the determinants, ids, etc., from an endeavour to trace the share of the parents in the offspring without any regard to the mode of development of the vertebræ or skeleton. It means an isolation of the characters from one another and from the organism as a whole, precisely the reverse of Heincke's method. The vertebræ are taken by themselves as if they were self-existent ; the consequences appear, when Johs. Schmidt calls the " race " merely a statistical unit, whereas for Heincke it was a biological unit. The former may be defined by reference to particular numbers or averages, the latter eludes definition just as much as the species or even life itself ; yet we know it is there.

In his studies on Zoarces Johs. Schmidt has not taken the form and structure of the parents into account, though he mentions them. These vary according as the fish live in open or in secluded waters. In general, the specimens from the open waters have a more elongated shape with a larger number of vertebræ, but smaller number of pectoral rays, etc.; those living in quieter waters have a thicker body, thus a smaller number of vertebræ, and a larger number of pectoral rays. The correlations seem very apparent if we suppose the inherited balance and hence the movements of the young to have a determining influence on the number of vertebræ and other characters. Is it not conceivable, that the shape and movements of the mother-fish influence the activity of the offspring within—in fact, take the place of the genotypes?

In his painstaking endeavours to distinguish between the inherited and the acquired characters Johs. Schmidt has also made extensive experiments on Zoarces. Thus, two different sets of parents from different regions were transferred to the same external conditions (1920) and the number of vertebræ in the offspring compared. The latter retained the character of their parents, and Johs. Schmidt has taken this result as evidence of the dominating influence of the genotypical inheritance. But, for the reasons mentioned above, we may question whether the external conditions are eliminated in this way; surely the body cavity of the mother-fish may be regarded as an external influence on the movements of the young?

More convincing is the definite evidence he brings forward (1919, 1921) that the surrounding temperature has a marked effect on the number of vertebræ. In the case of Lebistes he found that, among offspring of the same parents, higher temperatures led to an increase in the number of vertebræ. In the case of Zoarces he obtained a similar result, but the most interesting experiment was that in which he placed three different lots of eggs from the same Trout under three different temperatures. The lowest average number of vertebræ came from the intermediate temperature and it rose at higher temperatures, and still more at the lower temperatures.

The results thus seem to be contradictory or paradoxical, and we find much the same thing in nature. The species of the same group living in the tropics have a lower number of vertebræ than those of the temperate zones, and in general the number of vertebræ increases towards the poles. Yet there are exceptions. The question is, therefore, whether the temperature does really operate in two different ways or whether other factors are concerned in the matter?

The matter may perhaps be understood in the following manner. An increase of temperature undoubtedly heightens the metabolic processes and leads to greater activity internally. Thus, embryos are hatched out sooner under high temperatures, but are not so well-developed as under low; the larva is smaller with more yolk to be absorbed and the internal tissues are less differentiated. Hence, where a difference of temperature operates on the eggs, the higher temperatures should lead to a smaller number of vertebræ. But the conditions are altered when the different temperatures operate on the hatched-out larvæ; the activity and movements of the latter have to be taken into account.

That the Trout hatched out under the lowest temperature should have the largest number of vertebræ is thus comprehensible, and it is possible that the highest temperature has so influenced the larvæ that they regained what they lost in the egg stage, that is by means of their own activity, so that they come to have a higher number of vertebræ than those under the intermediate temperatures, though they do not regain all that was lost by comparison with those that remained under the lowest temperatures all the time.

However this may be, it is certain that these experiments of Johs. Schmidt, on the influence of external conditions, begin to throw a flood of light on the complicated questions of the origin and differentiation of fishes—indeed, one may say, of species in general. It is not only temperature, however, which leads to variations. During its period in the egg the embryo is exposed to many and varied influences. It has been found in the Trout and Salmon hatcheries, that the greatest care must be taken of the eggs. They must have a constant flow of clean water with a plentiful supply of oxygen, and, above all, they must not be shaken or exposed to rough usage, otherwise weaklings of various kinds arise and even monstrosities.

It may be said that under natural conditions such causes of variation would not operate, but that is a mistake. They do occur and their production under experiment is proof of the importance of the inherited environment. Tornier (1908), who has investigated these matters both in fishes and Batrachians, has indeed suggested the conclusion, that the great variety of heads and mouths in different species may have arisen just from a change in the outer conditions, for example, from shallow to deep water, acting on a constitution not inured to such conditions. And we have no reason to believe that the weaklings would of necessity go to the wall under natural conditions; on the contrary, they might live on and form quite different species from their apparently stronger brethren. This point, however, will be illustrated better under the next section.

The chemical nature of the surroundings in which an embryo is reared, is also an important source of variation. Stockard (1907) has found that by mixing a certain proportion of magnesium chloride with the salt water, he was able to produce a cyclopean variety of Fundulus. This effect, it is pointed out, is due to the chemical action and not to the osmotic pressure. The effect of the chemical action seemed to be, that the optic vesicles were enlarged forward beyond their normal position and thus came to fuse together to a varying extent. An unusually large optic cup then resulted and the ectodermic lens was also larger. Although Stockard does not think that osmotic pressure had any hand in this result, it is possible that the difference in partial pressures externally had some effect.

As the waters of the sea and lakes seem everywhere uniform for the fishes living within them, such an experiment would appear to have no bearing on the natural conditions. But we are just beginning to learn that the chemical composition of the water is not the same in the different seasons of the year, and is apparently different in different depths and localities. These small differences may be of great importance in keeping the species constant in their differences from one another.

The movements of the post-larva which lead to the notochord breaking up into definite centra or vertebræ, cannot be left out of account as a cause of variation. The chase of food is a necessity of its existence. If the food is abundant, as in the littoral waters, the larva has not to work so hard for a living as its brethren out over deeper waters or in colder climates. In the critical stage, when the mouth closes during development, the young fish has to travel farther and faster to obtain what it needs (*cf. Fistularia*, Chap. V), so that the biological surroundings also have some influence on the variations.

In brief, we may say that whatever affects the balance and movements of the young fish, embryo, larva or post-larva, whether physical, chemical or biological, is a potential cause of variation. With more experiments like those of Johs. Schmidt, Tornier, and Stockard, it will in course of time be possible to disentangle the different causes. In doing so, we may regard the parents and condition of the parents as among the " external " influences.

4. Differentiation of Fishes

It is now a matter of common knowledge among those who have studied the structure and development of fishes, that " species " are not well-marked off from one another, but show a merging of the characters in every group that has been more or less well investigated. The following example, showing the number of vertebræ in various species of Arnoglossus, will illustrate what this means :

No. of vert.	33	34	35	36	37	38	39	40	41	42	43	44	45
A. Grohmanni	11	4	—	—	—	—	—	—	—	—	—	—	—
A. Thori .	—	—	—	—	9	62	39	2	—	—	—	—	—
A. laterna .	—	—	—	—	2	46	198	113	38	3	—	—	—
A. imperialis	—	—	—	—	—	—	—	—	—	16	60	10	1
A. Rüppeli .	—	—	—	—	—	—	—	—	—	—	1	6	2

The numbers opposite each species show the number of

specimens examined, and we can be sure that with a still larger number of specimens the overlapping would not diminish but rather increase. Other characters, number of fin-rays, shape and size of mouth and body, show the same grading.

Such being the character of fishes, may we not say with Darwin (" Origin of Species," 6th ed. p. 176), that " through the nature of the organism and of surrounding conditions, but not through natural selection," new groups have been formed and perhaps are still being formed ? For example, under the species *Arnoglossus laterna*, noted in the table above, three separate forms can be distinguished which some authors might call species. And there are few groups of Teleosts which do not furnish similar examples.

The question whether natural selection has played any part in the differentiation of fishes is very difficult to decide at present. We can imagine that it has, but there is very little definite evidence. In theory we may regard any character as useful ; why, then, should it vary or be altogether lost ? And in theory also, any character might be useless or harmful, like the spiral valve or the air-bladder ; why, then, should it persist ? The truth seems to be, that the fishes must make the best of what is given them, whether useful or otherwise, and are able to overcome their difficulties. The strange forms, frequently referred to in previous pages, the Pipe-fishes, Plectognaths, Pediculates, Flat-fishes, and many others, indicate that there is a power in the living organism beyond the reach of natural selection.

We cannot be certain either, that natural selection has marked off the species from one another, until we know something more about the elimination of intermediate forms. The fact that fossil fishes are not represented in the present-day fauna is not evidence in this regard. Nor can we take as evidence, that the weaklings may perish without producing offspring ; the reverse may also be said.

Though many zoologists believe in natural selection as one way of explaining the phenomena of life, very few have tried to realise what it means in practice. The only author who has endeavoured to furnish any weighty evidence in its favour, so far as the fishes are concerned, is Georg Duncker. He found that during the transport of living Flounders from one place to another, the mortality was much greater among the sinistral forms than among the dextral. Similar experiments with other fishes made in the United States have shown that the mortality is much greater among males than females. In Flounders from the Holstein coast Duncker found that 36 per cent. of the specimens under 10 cm. were sinistral, but in specimens over 20 cm. only 25 per cent. were sinistral. Similarly, in a large collection (1120) from Plymouth the percentage of left-eyed specimens decreased with the size (or age) and the selection seemed to affect the females more than the males.

In the case of *Syngnathus typhle* a similar investigation led to a curious result. The number of dorsal rings increased apparently in the adult specimens, that is, the smaller numbers or frequencies had been more eliminated. But this result only held good for the Plymouth and Baltic specimens; specimens from Naples showed the reverse condition, a larger proportion of the smaller frequencies.

The material is obviously too meagre to permit us to form any judgment with regard to the practical importance of natural selection. An elimination of weaklings is not the same thing as the formation of species. But if believers in the theory could provide similar evidence on a larger scale, we should be in a better position to judge whether it is more than a metaphorical way of describing the phenomena of life.

The difficulty of conceiving how species become marked off from one another in what appears to be a continuous region, is more a physiological than a physical one in the case of fishes. Each month of the year has its own set of variable conditions, food, temperature, and chemical composition of the water. Further, the conditions in the different depths are not the same. The conditions leading to separation are therefore always present, and if we consider a genus such as Arnoglossus or Pleuronectes we see how the species are marked off according to the depths over which the larvæ and post-larvæ live when their structures are being formed. Assuming that the genus was a single species to begin with, the separation would come from the larvæ settling down at different depths, the range of depths being from 0 to some 500 fathoms. In the case of *A. laterna*, we have a modern species which is separating up in this way. Similarly, if the spawning time of a species is spread over several months of the year, the conditions are again present for the breaking up of the species.

It is not only through small variations, however, that fishes have become differentiated from one another. There is abundant evidence that abrupt changes, mutations, have been of frequent occurrence. The transformation of a normal larva into a Flat-fish, a Pipe-fish, or a Plectognath is a mutation, and the main groups of Teleosts have probably arisen in this way and not by gradual variation. We have seen in Chapter V, how in one form of balance the anal fin has gradually crept forwards until the ventral fins could not develop (Notopterus). In Stromateoides the ventrals are well-developed to begin with but later discarded, and the same thing happens to the pectoral fins in Symphurus. Where this phenomenon has been closely investigated, it has been found to arise from a sudden disturbance of the balance of the young fish, and we can therefore assign the same reason for the non-development of the ventrals, even sometimes of the pectorals, in the Eels and eel-like fishes. The many shapes of the mouth, as the balance of the body changes, are examples of gradations within separate groups (Clupeoids, Carangoids, Sparoids), but the breakdown of the regulating mechanism which leads to the more or less abrupt appearance of a long snout is a mutation. The fusion of the pharyngeal teeth in different groups is of the same nature. Lastly, the twisting of the gut is a natural consequence of its growth, but it has given rise to several important mutations, the formation of coils and spiral valves and the loss of an open communication between the air-bladder and the alimentary tract.

Mutations do not break through any law of nature, but they make things difficult for the natural selectionist. Organs like the fins and open air-bladder duct, which are of proven utility, should not change or disappear, yet a very large number of the Teleosts have come from just these two mutations. In short, we can understand mutations and recognise the important part they have played in the differentiation of fishes far better if we leave natural selection out of consideration.

The phenomena of variation and mutation throw a great deal of light on the adaptations and habits of fishes. The loss of an open air-bladder communication, for example, brought with it an inability to regulate the internal to the external pressure. Hence the fishes were obliged, either to remain at a definite level, at the surface like many of the Plectognaths or near the bottom like the Blennies and many others, or to develop their swimming powers like the Scombroids, or to acquire a new regulating apparatus within the closed air-bladder like the Sciænoids, Gadoids, and others. In this way we can understand how differences in structure and habits arose, as the fishes adapted themselves to the various environments.

The importance of these mutations, which have played such a considerable part in the differentiation of fishes, is that they can in no way be regarded as advantageous. In the struggle for life their possessors have been handicapped, yet so far from being eliminated they comprise perhaps the majority of fishes at the present time. That these fishes have survived and led to further differentiation cannot, therefore, be ascribed to the power of natural selection. Their survival has come rather from their own efforts, their refusal to accept the seemingly predestined fate of elimination, and their ability to co-ordinate their structures and adapt themselves to new conditions.

Natural Selection, the metaphorical method of describing biological phenomena, lays stress on the other side of the struggle, the elimination of the weaklings and the unfit, but it has yet to be proved that this process affects the characters of a species. When the weaklings drop out of one struggle and sink to the bottom or float helplessly on the surface, according to the evidence now available they have not been blotted out of existence. They have developed other characters and formed new species and groups of fishes.

The doctrine of utility—that characters have survived owing to their advantageousness—lends itself too readily to misuse to be of much scientific value. It is better to remember Goethe's dictum, quoted by Charles Darwin: "The future question for naturalists will be how, for instance, cattle got their horns and not for what they are used."

CHAPTER IX

THE GENEALOGY OF FISHES

QUESTIONS of absorbing interest are always bound up with the beginning of things. What was the oldest fish like? Where was it born? Every one knows that such questions cannot be answered with any degree of certainty, yet our idea of the present world and the order in which we arrange the different fishes, depend on these questions and we must go on asking them. In the present chapter facts and theories will be discussed in furtherance of this aim.

1. THE OLDEST FISHES

Fishes are not quite as old as the earth; the process of cooling until a film appeared between the lithosphere and atmosphere took many millions of years. Even when ice and water appeared—and the temperatures of the Algonkian period were perhaps not so very different from those of the present day—organic life had to pass through many phases before the complex substances of a fish's body could be integrated to hold together. As the heavier and more solid materials settled down and the water rolled off into troughs and hollows, in great part of its own making, fishes were evolved in the shallows. Of depths in the sea or heights on the land in those days there were none.

The age of the earth has been computed at a varying number of million of years, according to the various estimates of the rate of cooling. With the discovery of radium and its radiant energy these estimates have been revised more recently by Strutt and others, and it is now generally accepted, that the earth is much older than Darwin and the geologists of his time could have believed. The following table (from A. Wegener, 1922) gives a summary of the most recent estimates.

Time elapsed since the beginning of the

Palæozoic period	500 million years
Mesozoic period	50 ,, ,,
Cenozoic—	
Tertiary period	15 ,, ,,
Eocene period	10 ,, ,,
Oligocene period	8 ,, ,,
Miocene period	6 ,, ,,
Pliocene period	2–4 ,, ,,
Diluvium period	½–1 ,, ,,
Post-diluvium period	10–50 thousand years

Beyond the Palæozoic period lay the undetermined age of the Archaic period during which the heavy metals, lead, gold, iron, etc., and their compounds were deposited.

From the above and the following table, showing the principal fossil fishes so far discovered in the earliest formations, it will be seen that the age of fishes may be placed at some 400 millions of years.

Formation.	Fossils.	Country.
Palæozoic :		
Algonkian		
Cambrian . .	Fish remains	Rocky Mts., Canada
Silurian . .	Placoderms (?)	Colorado, U.S.A.
	Cephalaspidæ	Scotland, Sweden
	Birkeniidæ	Scotland, Norway, Canada
Devonian . .	Arthrodira	Europe, N. America
	Acanthodidæ	Europe, N. America, Asia
	Dipteridæ (Dipnoi)	Scotland
	Palæoniscidæ	Europe, N. America, S. Africa, Australia
Carboniferous	Cladoselache (Elasmobranch)	N. America
	Pleuracanthus	N. America, Europe, Australia

Other forms occur in the Devonian and still many more in the Carboniferous periods, but those mentioned will be sufficient to give us a picture of the oldest fishes. Though Cladoselache, the first Elasmobranch to be recognised as such, does not appear until the Carboniferous, it has been generally believed, but with some opposition from Beard

(1890) and others, that the Elasmobranchs were the forerunners of all other groups. "If the earliest true fish could be found, it would almost certainly fall within the sub-class Elasmobranchii" (Smith Woodward, 1895). This view is not held so firmly now as formerly (Abel, 1916), and from time to time various objections have been raised against it; especially, that the Teleosts in their mode of reproduction, organs, and parts of their structure, are undoubtedly simpler. Emphasis was laid, however, on what seemed to be the more primitive characters of the Elasmobranchs, integument, gills, and the simpler condition of the tail, a condition that most Teleosts go through as a passing phase. According to the geological record also, the Teleosts seemed to have been much later in their appearance, and hence it was thought that they had degenerated from some Elasmobranch type.

There are two ways of searching for the primitive forms and characters, through the geological record and through the developmental history. Both have their drawbacks and, if taken separately, may lead one astray. But if kept together, and the lines of retreat to simpler types tend in the same direction, the results should be conclusive, so far as our present knowledge goes.

Cladoselache, so well described by Bashford Dean (1896), is a pivotal form in this connection. It is so far removed from modern sharks, that it lacks the most distinguishing features of the Elasmobranchs; it had no claspers and its mode of reproduction must have been the same as that in the majority of the Teleosts; the mouth was anterior instead of ventral; the notochord was persistent; the gill-arches opened separately to the exterior, but the foremost of the dermal gill-frills "appears to have been sufficiently large to have served as an operculum." The body was elongated with two separate dorsal fins and the caudal fin might be called a cross between the Elasmobranch and the ordinary homocercal Teleostean fin (Fig. 47).

In Cladoselache the distinctive characters of the Elasmobranchs have disappeared and it represents a connecting link between these and the bony fishes. Going further back into the Devonian period we find a number of forms which bring the relation nearer. In the Acanthodidæ there was only one dorsal fin, near the tail, the gill-arches were free from the skin and nearer the head but without an operculum, the mouth and caudal fin similar to those of Cladoselache. In these forms, however, a certain amount of calcification is present, over the skull and jaws and over the pectoral arch. We are thus approaching nearer to the Teleost types and the only distinctive Elasmobranchian character remaining is the skin with its exoskeleton of shagreen tubercles.

Further back along the Elasmobranch line we cannot go. If any form existed previous to these, and with still fewer Elasmobranch characters, we should be unable to classify

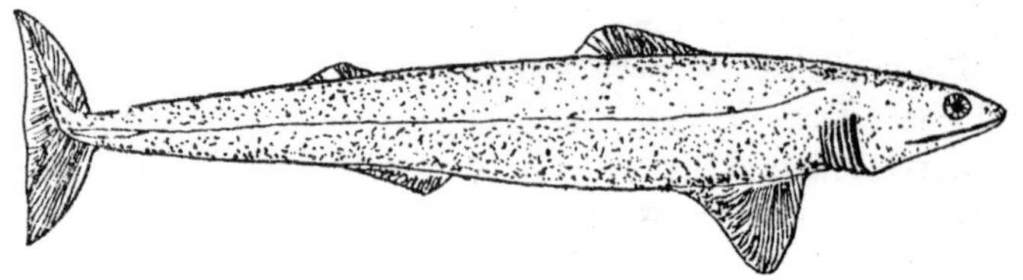

FIG. 47.—Cladoselache, a primitive Shark of the Carboniferous period. (From Bashford Dean.)

it with the latter. Some earlier forms, however, have still to be mentioned.

The ancient "armoured" fishes, formerly known as Ganoids, were more abundantly represented in the early Carboniferous and Devonian periods than the primitive sharks. They are now classed among the Teleostomi owing to the presence of distinct membrane bones, whilst their mode of reproduction was the same as that of most Teleosts and Dipnoi. An air-bladder was probably present as a rule; the gill-arches were situated under the head and covered by an operculum.

These Ganoids were exceedingly numerous over the whole of the Northern Hemisphere, even in Spitzbergen and Greenland, during the later Devonian and especially the immediately succeeding periods. The majority possessed two dorsal fins and a similar heterocercal tail to that of the primitive sharks. But the earliest of them, as well as many later, had only one dorsal, like Acanthodes.

Other forms of this early period, which can also be recognised as belonging to groups existing at the present day, were the Dipteridæ. These were Dipnoans characterised by their peculiar, limb-like paired fins; they had a well-developed opercular cover and the same kind of heterocercal tail as the forms mentioned above. Most of them possessed two dorsal fins, but in the later forms and those existing at the present day the tail was diphycercal with a long median fin continuous from about the middle of the dorsum round to the anus.

If we now review the changes in the characters of the

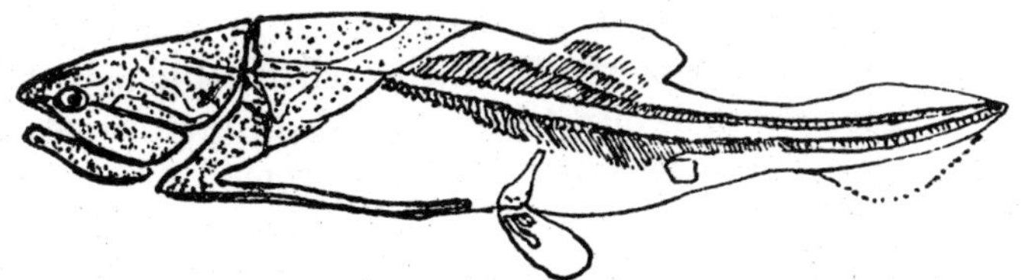

FIG. 48.—An Arthrodiran from the Old Red Sandstone of Scotland. (From Bridge, after Bashford Dean and Smith Woodward.)

above known groups as we pass backwards in time, the following will be noted. The form becomes more elongated, the tail heterocercal, the mouth terminal, the gill-arches become covered and lie nearer the head, and there is in the end only one dorsal fin near the tail, except in the Dipnoi. Lastly, an important point, the paired fins, pectorals and ventrals, arise near the ventral line or apparently are wanting.

In the Devonian period also there existed a large group of fishes, the Arthrodirans (Fig. 48), which link the Teleosts and Dipnoi closely together. They were distributed over the whole of the Northern Hemisphere and reached their highest development in North America. Only one dorsal fin is present, the tail is diphycercal to heterocercal and ventral fins were certainly present. Whether a membranous pectoral fin was developed is uncertain, yet the base for such a structure is present and not unlike that found in the simpler Teleosts like Clupea. Connected with the base we find the long pubic bar along the ventral margin which is present in many Teleosts and represented perhaps by the pterygopodia of Cladoselache. As explained in Chap. V (p. 92) this bar arises from the muscle buds along the abdominal wall and is the base from which the pectoral and ventral fins arise. Here in the Arthrodirans the connection with the ventrals is nearly but not quite maintained. The coracoid portion of the bar extends upwards on the side and one of the plates above was hinged apparently on to the side of the skull, just as the clavicle or suprascapular may be similarly attached in some of the Teleosts of the present day.

The greatest similarity of these Arthrodirans to the Teleosts lies, however, in the ossification. It is admitted, even by supporters of the Elasmobranch theory, that the bones above the skull might fairly be called paired parietals and frontals, whilst in some cases bones comparable to the elements of the third form of mouth (Chap. V), that is, maxillaries and premaxillaries, have been detected.

Many of these Arthrodirans were marine forms, where the Ganoids and Dipnoi were apparently freshwater fishes. And there seems just as close a relationship between them and the modern marine Teleosts as that above noted between Cladoselache and the modern sharks.

The Antiarchi (Pterichthys) from almost the same levels show another variant of the Dipnoi-Teleost type. The ventrals are absent, but the pectorals are present and limb-like as in the Dipnoi. The tail is heterocercal, there is only one dorsal, near the tail, and there is no anal fin.

Further back still, in the Silurian deposits of Lanarkshire and North America some forms have been discovered which, it is generally agreed, represent the earliest known fishes. Birkenia (Fig. 49) was provided with scales of a peculiar kind, a single dorsal fin, again near the heterocercal tail, and a row of ventral scutes or spines. There were no paired fins. What the head was really like, we do not know, but it was not covered with the calcified armour of the Ganoids or Antiarchians. There is also Lausanias from the same deposits, regarding which only the ventral scutes and a presumed heterocercal tail are definitely known. Lastly, some indeterminable remains, with little or no calcification, have been found in the Cambrian layers of Canada.

In these forms we are obviously approaching the limits of the fish series. The calcification is here very thin or absent. There is no armour and nothing approaching a limb, and all we can say is, that these must have been muscular bodies able to swim up through the water in a definite direction. They were not sluggish or crawling forms like the armoured Ganoids or Dipnoi.

This is perhaps a sufficient indication and description of

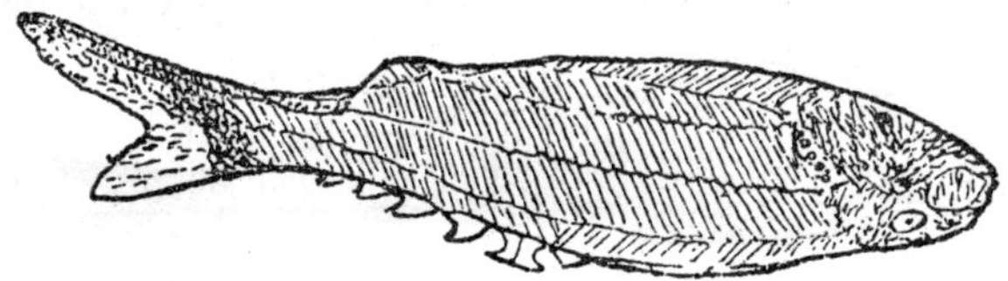

FIG. 49.—Birkenia elegans from Lanarkshire ; natural size. (From Traquair.)

the oldest fishes. The pelagic, free-swimming, muscular bodies of the Elasmobranchs have been left far behind in the Carboniferous period. The clumsy, mailed and crawling forms of the Ganoids and Dipnoi are common in the Devonian period, but some millions of years earlier than the first probable shark we have these Silurian and even Cambrian forms which are neither Ganoid nor Dipnoan nor crosses between these and Teleosts. When we say that they were pelagic forms, the nearest comparison lies with the Teleosts.

Turning now to the development of modern fishes, we can trace out a similar series from the complex to the simple. All the known Elasmobranchs must be classed among the complex, since they all show well-developed ventrals, pectorals, vascular system, etc., already in the embryonic stages. But it is worth noting, that the development of the mysterious Læmargus has so far eluded observation. If we regard Cladoselache, the fossil "shark," dispassionately without any predisposition to call it a shark, we find that it possesses Teleostean features of almost the same level as those of the Salmonidæ. The larval Dipnoan and Ganoid are just as specialised in their own way as the Elasmobranch.

The simplest larval forms are found among the Teleosts, especially those which later acquire an air-bladder with an open communication to the mouth (Physostomi). Among these also the simplest larva is that of the Clupeids, with which those of the Eels and Albula stand in close relation. Here we find the same shape of the body, single dorsal fin near the heterocercal tail, no anal and no ventral fins, just as in Cephalaspis (Fig. 50) and Birkenia. A membranous pectoral has not yet been traced in these early forms, but the

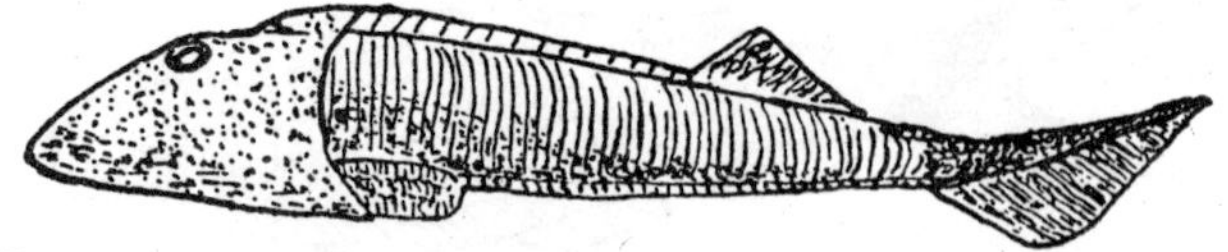

FIG. 50.—Cephalaspis murchisoni from the Upper Silurian beds of Scotland. (From Bridge, after Smith Woodward.)

basal structure of the Clupeid pectoral is acknowledged to be primitive, similar to that of the Arthrodira and perhaps Cephalaspis. A characteristic of the Clupeids also is the development of a ventral row of scutes, sometimes also a dorsal row, similar to those in the Anaspida (Birkenia).

Other primitive characters of the Clupeid larva are, the early indications of the primary mouth, the late opening of the latter, the late appearance of definite vascular and renal systems, the undeveloped condition of the sense organs, simple cranium and the continuous notochord with its fibrous, urostylic sheath. Further, the alimentary canal runs backward to the root of the tail (cf. pp. 85, 88, and 96).

No other larva, whether of Teleost or other sub-order, posesses such primitive features, and the significance of these is in no way diminished by the fact, that the Clupeid during the course of its slow development gradually changes into a highly organised fish adapted to living under more varied conditions of depth, pressure, and salinity than any other form.

The conclusion that the earliest fishes were pelagic and like the Clupeid larva leads us to picture them as elongated and muscular bodies divided into uniform segments. In consequence of their free movements the body was compressed or flattened from side to side, not rounded as in the Cyclostomes. This flattened shape of a muscular body led to the origin and development of the fins, in the order—caudal, dorsal, and then either anal or paired fins, which arose in the first place where the muscles had freest movement, near the tail. That the earliest known forms had the same asymmetry of the fins as we find in the Clupeid larva, indicates that they also had difficulties in maintaining a horizontal level, the head tending to droop downwards. Given a fairly stiff notochord, however, the movements of such a caudal region would bring the fish up to the surface, where they could take in air directly. The Ganoids and Dipnoi were certainly provided with air-sacs, probably also the Arthrodirans, to judge from the shape of their vertebral column, and perhaps all the earliest fishes. This indeed was probably the earliest form of respiration, similar to that found in the embryos of the Clupeids and other Teleosts of the present day (Chap. IV). It is not yet certain whether the earliest fishes (*e.g.* Birkeniidæ) had any gills, and these organs do not develop in the Clupeids until one to three days after the larva is hatched.

This comparison of the Clupeid larva with the earliest known fishes may be carried a step further when we take into account the enigmatical Palæospondylus, which was found in the Lower Old Red Sandstone of Scotland, thus in almost the same layers as the earliest fishes mentioned above. Traquair, who described it, thought it was the long-desired evidence of a primitive Cyclostome, and he was supported by A. Smith Woodward, both acknowledged experts on fossil fishes. But a hefty controversy raged for a long time with regard to its characters and position (see Sollas, 1903), and the fossil was variously placed among the different known orders of fishes and even outside them.

A considerable number of specimens of this form have been discovered and the following summary may be given of the principal characters. Mouth terminal but structure uncertain, the supposed cirri may have been crushed remnants of the palatines and maxillaries; the skull very broad posteriorly, narrower anteriorly, and without the calcification of the armoured fishes; no certain sign of gill-arches. Pectoral fins may have been present, but the presence of any other fins is uncertain; caudal region diphycercal or turned upwards; vertebral column consisting of distinct, ring-like centra.

Bashford Dean, who made a most thorough examination of this fossil and reviewed the available evidence from time to time, definitely rejected the idea that it represented a Cyclostome and preferred to leave the matter open; it might be a " baby Coccosteus " (Huxley) or belong to some primitive class of Vertebrate. On one important point, however, Bashford Dean was perhaps in error. He thought that the vertebræ of Palæospondylus were " apparently of a type unknown among recent fishes." As a matter of fact, they are precisely the same as the fibrous rings or tubes seen in the larval Clupeids of the same size as the fossil specimens, namely, from about 20 to 50 mm., and similar vertebræ occur in other Teleosts, but never in the Elasmobranchs, Dipnoi, or Ganoids.

The vertebræ of Palæospondylus numbered from 46 (B. Dean) to about 60 (Traquair), and of these about 40 were precaudal, to judge from the figures given by both Traquair and Bashford Dean. These are well-known proportions for the northern Clupeids (Herring and Sprat). The skull is quite characteristically Clupeid; both Traquair and Bashford Dean emphasise the great breadth of the cranium proper or " auditory capsules," and even the tympanic bulla is clearly represented in some specimens (Fig. 51, 1); the peculiar postorbital process of the frontal bone is also shown (*cf.* Fig. 51, 3). The " postoccipital plates " may have been the flattened clavicles and coracoids, since " their anterior ends (and it is interesting that they are somewhat knob-shaped) project forward below the region of the auditory capsules " (B. Dean). If so, the presence of pectoral fins, as maintained by this author, is a sound conclusion. Lastly, the notochord (Fig. 51, 2) shows the same mode of cellular division as that seen in the Clupeids (Fig. 17, p. 81) and so far not definitely known elsewhere.

Without being dogmatic on a point regarding which the most expert palæontologists have been at variance, there is sufficient justification for saying that these Clupeid affinities of Palæospondylus cannot be ignored. The great difficulty in identifying the fossil, one may believe, has arisen just from the very thin and fibrous nature of its bones, also a Clupeid characteristic.

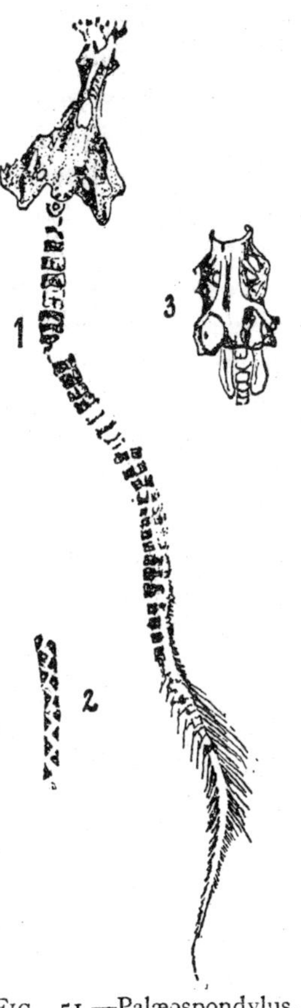

FIG. 51.—Palæospondylus, from the Lower Old Red Sandstone of Caithness. 1 and 2, from Bashford Dean; 3, from Traquair. 1 about twice the natural size.

2. ARRANGEMENT OF FISHES

The arrangement of fishes depends not only upon knowledge of the structure, past history, and the present-day development; it depends above all upon our ideas regarding the nature and mode of life of the earliest fishes. If the Elasmobranchs were really the earliest, the logical course to follow would be that adopted by Bashford Dean, to regard all the early fishes that came before and could not be regarded as Elasmobranchs—

Cephalaspis, Arthrodirans, Birkenia, etc.—as belonging, so to speak, to a different and earlier world. Bashford Dean places them even before the Protochordata, and for him the class Pisces begins with Acanthodes and Cladoselache. This is the logical course, but it cannot disguise the fact that the Arthrodirans, Palæospondylus, and many others were ossified and were fishes.

It seems better to accept the conclusion, that the Elasmobranchs came later. It may also be taken as certain, that the armoured "Ganoids" or sedentary Teleosts were not the primary forms, and the choice lies between the true Teleosts and the Dipnoi. Since the Arthrodirans show a mixture of the characteristics of both groups, we are clearly on the right lines; but the term Dipnoi refers to a special condition of the air-sacs in the modern forms, which was probably not so well developed in the earliest, and the term Teleost is comprehensive enough to embrace both as well as the Ganoids, as suggested by Bridge (1904). To avoid confusion, however, it is best to group them together under the wider term Osteichthyes.

The evidence that the earliest fishes, even Palæospondylus, possessed ossifications of some sort, cannot be gainsaid. This ossification may have been, probably was, extremely thin and fibrous, like that of the larval Clupeids of the present day, but it came as a natural consequence of a muscular body moving pelagically through the water. That these early fishes had no paired fins and no median fins except the caudal and the posterior dorsal, and that no definite gill arches have yet been found, are also features that can be readily understood from a consideration of the Clupeid larva.

Taking this larva and Palæospondylus as representing the most primitive form of fishes known, some of the changes which have taken place in the principal characters may be briefly indicated.

The form or shape of the fish has varied in two directions: on the one hand, to more elongated with the development of two dorsal fins or long single dorsal fin and paired fins in the forms which have remained pelagic and free-swimming, such as the Acanthodidæ and the primitive "sharks." On the other hand, to broader forms tending to be more sedentary with a strong development of the calcification in the anterior region and paddle-like paired fins, as in the ancient Ganoids and Dipteridæ.

The ossification of the notochordal sheath (endostyle) seen in Palæospondylus and the larval Clupeid has not been maintained in the majority of the early fishes. Instead, the cartilaginous arches from above and below have more or less completely invested the sheath and replaced it. In the modern Ganoids, Dipnoi, and Chimæroids the notochord persists with the arches pressing upon it above and below; in the other Elasmobranchs the notochord is completely invested by a complicated system of cartilages, which may often show calcified rings internally. In the Teleosts the endostylic sheath may persist, or the cartilaginous bases of the arches may grow round and replace it, becoming ossified later.

The gills and gill-arches in the earliest fishes were formed inside the skin from the lower half of the muscular segments under the cranium and had an external cover which frequently became the seat of calcareous deposits, like the rest of the head and anterior part of the trunk. There may have been several openings, as apparently in Birkenia, but usually there was only one with a well-developed opercular cover over it. In some pelagic forms, however, which did not acquire the strong head-calcification of the sedentary forms, the gill-cover became shorter as the gill-arches projected out to the open, as in Cladoselache (Fig. 47). In some Teleosts at the present day this uncovering of the gills through great development of the arches is not infrequent, and Tornier has been able to produce the condition experimentally. The retreat of the arches backwards from the head and the development of a closer connection with the skin can be followed in Acanthodes, Pleuracanthus, Cladoselache down to the modern Sharks and Rays. But it has to be noted that the Chimæroids, which also occurred in the

Devonian period, have retained the primitive opercular cover. The posterior position of the gill-arches in the other Elasmobranchs can be associated directly with the movements and mode of life of the embryos. In some Teleosts, as in the Eels, the arches also retreat behind the head during development and the pectoral arch loses its connection with the skull. The condition of the arches in the Cyclostomes cannot therefore be considered as primitive. The place where the gill-arches should form has been determined by the position of the heart and this by the position of the yolk-sac.

The superior circulatory, nervous, and renal systems of the Elasmobranchs by comparison with those of the Teleosts, can also be referred to the mode of reproduction. The greater freedom allowed to the embryo owing to the absence of a hardened chorion, and the elevation of the embryo above the yolk-sac (Fig. 14) have allowed more space for the development of the various tissues and organs.

The mode of reproduction of the Teleosts has always been recognised as the simplest, and as it also prevailed among the earliest Sharks as well as the Ganoids and Teleosts it can certainly be taken as the most primitive.

The interpretation of the structures of the mouth has always been and still is a matter of great difficulty, and it is quite uncertain what sort of mouth some of the earliest fishes had. This condition is comparable to the earliest stage of the Clupeids. There can be no doubt that most if not all of the primitive Arthrodirans were autostylic, with the lower jaw, however it may have been formed, articulating directly with the skull, and this is also the condition in the ancient Ganoids (Cœlacanths), Dipnoi (Acanthodei ?), and the Chimæroids. But Bashford Dean has noted that in the last-mentioned, the fusion of the primitive upper jaw (palato-quadrate) with the sides of the skull arises during development, and Abel (1920) has suggested, that the condition in the Dipnoi is also secondary and a special adaptation to the habits of these fishes. Nevertheless, it cannot be ignored that the Arthrodirans had this primitive autostylism along with, in some cases, maxillaries and premaxillaries, the condition found in higher Vertebrates. It seems more probable that the palato-quadrate originally formed part of the skull; at any rate, this gives a better working hypothesis to explain the differences between the Teleosts and the Elasmobranchs. In the former, the quadrate element separated from the skull, whilst the palatine remained attached to it in front; in the latter (Sharks and Rays) both descended from the skull. In each case we can represent the change as the result of the movements and mode of life of the young stages.

With regard to the skull, little can be added to what has been said in previous chapters. If we accept the skull of the Clupeid larva and Palæospondylus as the simplest and most primitive, it is not difficult to associate the cartilaginous condition of the Elasmobranchs with the mode of reproduction and development of these forms. Certain it is, that among the Teleosts, where the amount of yolk increases and the eggs are demersal, the greater is the amount of cartilage in the skull.

The changes in the alimentary canal from the simplest condition in the Clupeid larva to the complex in the Elasmobranchii have also been fully dealt with in preceding chapters, as also the air-bladder and air-sacs. It need only be added, that the primitive Ganoids like the Dipnoi also possessed the so-called spiral valve and an air-bladder. The Elasmobranchii have quite lost the latter.

The arrangement of the fishes in the accompanying genealogical tree follows in the main the lines developed by A. Smith Woodward (1891) with regard to the origin of the modern Teleosts from the Clupeoids (Isospondyli) and these from the Mesozoic Eugnathidæ and Semionotidæ (Protospondyli). It is but a short step backwards from these to the Palæoniscidæ, the early Devonian fishes. The Acanthodei were just as much Teleost as Elasmobranch and the Holocephali were of the same age. The surviving "Ganoids," Sturgeon, etc., are represented as of comparatively recent origin and quite unconnected with the ancient Ganoids, which with the ancient Dipnoi were completely blotted out in the Palæozoic and Mesozoic periods. But the modern Dipnoi, which have retained several important and primitive features of the Arthrodirans and Ctenodonts (Phaneropleuron), are represented as having maintained some connection with the earliest forms.

The principal connections between ancient and modern days have, however, been maintained by the free-swimming, pelagic forms of the Teleosts and Elasmobranchs; in both cases the lines are unbroken and unmistakable if we accept the seemingly inevitable conclusion, that the osseous forms preceded the cartilaginous.

The geological periods to the right are marked off as if of nearly equal duration, but they are only intended to give some idea of the times when the various groups arose and flourished. For the duration of the periods reference should be made to p. 202. The Palæozoic period lasted about ten times as long as the Mesozoic, the latter at least twice as long as the Cenozoic and recent together. The oldest of the living forms (Clupeids, Rays, and Sharks) go back about 20 million years, which may be taken as the longest time for the formation of the species of the present day. But with such a long period at their disposal and the numerous changes in the physical conditions during this period, we can readily understand how the fishes have become so varied and different from one another.

The principal groups mentioned are well-established, but a few words are necessary to explain the divisions given of the Teleostei. All systems of classification require a general dumping ground for the refractory forms which do not fit in well with the others. Boulenger (1904) had several of these, Haplomi, Catosteomi, Percesoces, etc., and this system had the inestimable advantage of revealing the difficulties in the way of a natural system. Tate Regan (1907 and 1913) made the Percoids the dumping ground, whilst Goodrich (1900) cut the Gordian knots by arranging the fishes according to types. There is no doubt that the last is the simplest and most practical so long as we do not wish to go very deeply into problems of phylogeny. The present scheme endeavours to combine simplicity with the phylogenetic derivation of the Teleosts. The Irregulares are taken to include a large number of miscellaneous groups,

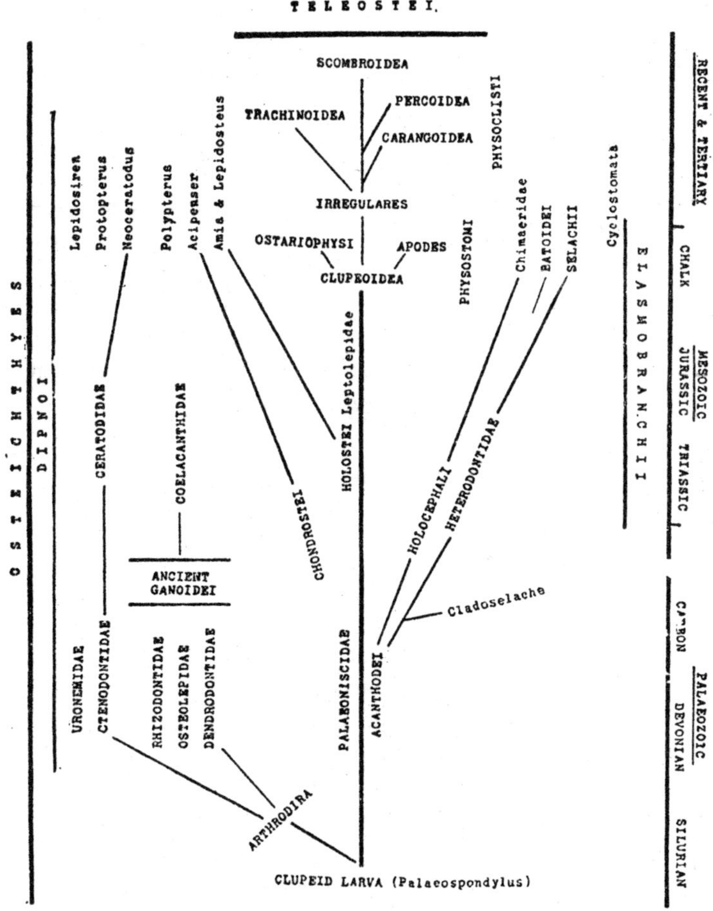

FIG. 52.—Genealogical tree of the class Pisces.

which have arisen directly from the Clupeoids or from one another, and are distinguished by the loss of the open communication of the air-bladder and the intermediate position of the ventral fins, which have not obtained a fixed, bony attachment to the pectoral arch. They include such groups as, on the one hand, the Pipe-fishes (Syngnathiformes), Plectognathi, and Scombresociformes; on the other, the Gadiformes (Anacanthini), Atheriniformes, and Anabantiformes. From these forms we can derive the four main groups of the more specialised Teleosts, the Blennies and Weevers (Trachinoidea), which live mainly on the bottom with pelagic offshoots into deeper waters, the Mackerels and Tunnies (Scombroidea), which are roving, pelagic forms, the Perches (Percoidea), which occupy the middle waters of the littoral zone and ascend into fresh water, and the more or less eccentric offshoots of these groups (Carangoidea), which have descended to the bottom (Flat-fishes) or hang about the rocks (Chætodonts and Labroids) or have wandered out to the deep-sea region. The derivation of these forms from the Clupeoids and from one another is still a matter of investigation.

3. The Drifting of the Continents

Our reading of the distribution of fishes, even of their origin, depends upon the views we hold with regard to the changes on the surface of the Earth during past ages. If any one told us, for example, that fishes came from the east to Northern Europe, and that by the shortest route, we naturally think of the land and mountains in the way and wonder whether it is easier to shift the mountains or disbelieve the story. There can be no doubt, however, that great changes have occurred in the face of the globe, not once but several times, and zoologists have had a great deal to say in the interpretation of those changes. When the same or nearly related forms are found in North America and Europe, or in Europe, East Africa, and Australia as well as India, or again in South America, Africa, and Australia, it would be contrary to our nature not to conclude, either that some one had carried them from one place to the other, or that the lands had formerly been connected.

The necessity for believing in ancient land-bridges and

waterways leads to a brief consideration of the views held by geologists regarding these matters. Formerly, indeed until quite recently, it was generally believed that the continents had always been in the same places as they are now. If land-bridges existed in ancient times, they had simply disappeared under the ocean. The fabled Atlantis lies under the waters of the North Atlantic, whilst on the other hand the moon has been torn from the bed of the Pacific. The face of the Earth assumed its present appearance from the gradual shrinking of its components as they became cooler. Here and there mountains were thrown up by violent eruptions and the land everywhere had risen and fallen at various times, all part of the general process of shrinking.

The shrinkage theory of the Earth's changes is no longer held so firmly, and a new interpretation of the ancient land-bridges has arisen. The land-masses are not so solidly rooted as was formerly supposed. From astronomical measurements made in Greenland during the past 100 years the Danish observer Koch has detected a movement of that island away from Europe, and a curious twisting of the Faeroe Isles relative to one another has been noted by the same observer. Further, it seems that the Labrador coast is moving away from Greenland in a south-westerly direction, and a swinging movement of the North American continent southwards, noticed in recent years in California, seems also to be taking place. In Europe, again, a comparison of various readings from earlier and later years points to a southward movement of that continent, though this may be due to a shifting of the North Pole towards the other side.

These observations are in keeping with the view, now an accepted belief, that the crust of the Earth (lithosphere, about 60 miles thick) lies on a plastic magma (barysphere), which, though it may have a greater resistance than steel, acts like a resilient mass or heavy fluid under the enormous pressures of the lithosphere above. It has been found also, that the magma approaches nearer to the surface at some places, under the oceans, than at others, as under the heavy mountain ranges and ice-covers. These conditions have prevailed since the beginning of the geological epochs, but the lithosphere has become probably three times as thick as it was then, whilst parts of the land surface have been rolled or folded together and other parts have separated from one another. At the earliest periods the lithosphere probably covered the entire globe and above it was a layer of water. This agrees with the general belief among biologists, that living organisms had their origins in the sea and that animals preceded plants.

This theory of the shifting of land-masses in the past has been suggested by various geologists (Wettstein, Pickering, Taylor, and others) and has been developed more especially by Wegener (1912, 1922). As an example of what the theory means, the following illustration given by the last-mentioned author may be cited. If we examine the south-west corner of Europe on the map, it looks as if the Iberian Peninsula has rotated down from the Bay of Biscay in a southerly direction, with the result that the Pyrenees have been folded together from the level land previously there. The point of the fold is in the north with the broader base in the south. Similarly, the great mountain ranges, Alps, Himalayas, Rocky Mountains, etc., have come originally from a fault or break in the magma whereby the neighbouring land, it might be for hundreds of miles around, has been drawn together with the different strata folded up against each other, not lying over one another as would be the case under the shrinkage theory. Then the magma has rolled back towards, but not quite reaching, its former level and the mountains have arisen above the plains and valleys. Again, when the great ice-sheets came down in the Glacial epochs from the Arctic, they pressed down the lithosphere so that many parts of the land were flooded and deposits were laid down. When the ice receded, the land gradually rose, but not quite to its former level, where the deposits had formed.

The sliding theory of the continents has thus an important bearing on the interpretation of fossil remains.

According to Wegener, the land-masses have not greatly altered, except for the rolling together of the mountains, since the beginning of the geological epochs. Thus we find the oldest deposits still near the surface and the marine deposits, with a few doubtful exceptions, have come from marginal floodings of the continental shelves. The principal changes, so far as the distribution of animals and plants is concerned, have arisen from the sliding away of one continent from another.

The following chart (from Wegener) shows the disposition of the land and water during the early part of the

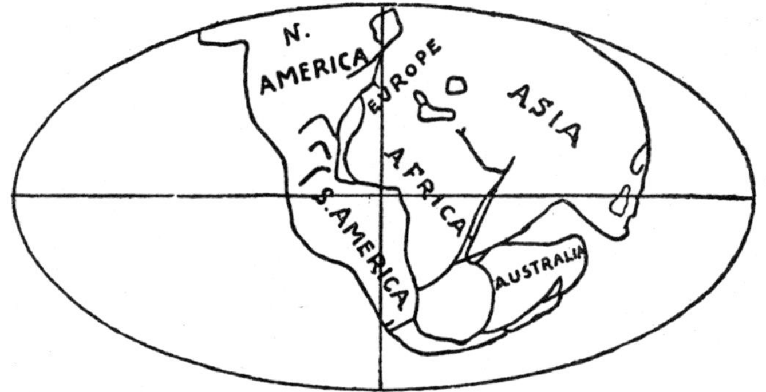

FIG. 53.—Position of the continents during the Carboniferous period. (From Wegener.)

Carboniferous period, thus at a time when fishes were already numerous in the world. The position of the water on the land is, however, merely suggested and is not meant to be representative of the actual conditions, which must indeed have been quite different, to judge from the distribution of fishes. A second difficulty arises from the view that the deepest parts in the oceans represent the earliest breaks in the lithosphere. In general, this is probably correct; the Pacific, we believe, is older than the Atlantic Ocean, and this again older than the Indian Ocean. But the deepest part of the Atlantic lies in the west, near the American continent. There probably was an early break there, as represented in the chart, with the lands still in contact to the north and south; but according to Wegener's theory this greater depth in the west should mean that the eastern masses have moved away from America, not America away from Europe and Africa, as he maintains. Some curious points in the distribution of fishes might be explained in this way.

This is a minor point, however; the main thing is, that America is supposed to have been actually in touch with Africa and Europe during the earlier epochs. Except in the middle portion and the extreme south this connection was maintained until the Tertiary period. The separation began in the south between Africa and South America previous to the Eocene and may have been completed in the Chalk period, but the separation of Ireland and Scotland from Newfoundland and South Greenland did not take place till much later, probably during the Great Glacial period. The contours of the coast lines from north to south as well as the identical lines of the coal measures, both in the northern hemisphere from Pittsburg across Newfoundland to Britain, Germany, and the Caspian, and in the southern hemisphere between Natal and the Andes, as well as many other physical data, undoubtedly speak in favour of this earlier union of the continents. The greater distance of South America from Africa indicates that the separation took place earlier there than in the north.

The similarity of the geological deposits and fossils in North America and Europe, particularly Great Britain, can thus be explained in a simpler manner than by supposing that the Atlantic covers an ancient land-bridge or continent. There are some who still maintain, however, that a considerable mass of land formerly existed in the Atlantic to the west and south of the Iberian Peninsula.

The relations between the continents on the eastern side of Africa are explained in the same way. The Lemurian land-bridge of earlier authors was in reality, according to Wegener, the direct extension of Southern India to Madagascar. This connection also persisted till the beginning of the Tertiary period, when India as it were slid away to the north with the great folding together of the Himalayas. Lastly, the connection between Australia, Africa, and South America was broken by the sliding away of Australia at a still earlier period, probably in the Lias.

These are the principal land connections of earlier times so far as the fishes are concerned, and it is now of interest to see whether they throw any light on the distribution.

CHAPTER X

DISTRIBUTION OF FISHES IN TIME AND SPACE

FROM the considerations given in previous chapters it is possible to draw a biological picture, at least in outline, of the distribution of fishes in the past and present. For this purpose it is convenient to group the geological epochs into two divisions, one reaching from the earliest times down to the great cataclysms in the beginning of the Tertiary period, the other overlapping this somewhat and extending from the Chalk period down to modern times. As central theme we may endeavour to trace the fortunes of the pelagic, free-swimming type, its changes in form and structure down through the ages and its various experiences under different conditions whilst the face of the Earth was also changing.

1. ANCIENT PERIODS

The land-masses of those early Palæozoic and Mesozoic days have to be pictured as something very different from those we know at present. The land was probably low and comparatively uniform in height. Whether the masses were all bunched together or separate as now, with land-bridges connecting the various continents, is not of much importance to begin with. In either case we must suppose that the waters enclosed by the land were in places saline, which of course is quite possible even in land-locked seas.

The significance of this point becomes apparent when we consider the actual origin of fishes. It is the general belief that all life, including fishes, began in the sea. And it has to be noted, that the nearest relatives of fishes, Amphioxus and the Tunicates, belong mainly to the coastal region within or just beyond the influence of the tides.

Again, it is also believed that the centre, from which at any rate the modern fishes started, lay somewhere along the eastern coasts of the Indian Ocean where the greatest variety of species occurs. Of the simple form Amphioxus, for example, one species is found in the Atlantic region as well as in Ceylon, another on the Pacific coast of America, and one more in South Africa, but at least three species occur in the Indian Archipelago. Of the neighbouring form Asymmetron six species occur in the Indian Archipelago, Australia, and New Zealand, but none at all in the Atlantic (Herdman, 1904).

But we must get away completely for the moment from these modern conditions. We do not know where the poles were in the ancient periods, nor the equator ; that the Earth has wobbled a great deal we can take for granted, and the great wealth of the Indo-Pacific in recent times has almost certainly come from the fact that it has lain for a long time under a tropical climate. But if we go back some 300 million years, the torrid zone may well have been several times to the north or south of the present equator and the conditions for the appearance of fishes may have been more favourable, let us say, in the north. Certain it is, that the earliest remains of fishes have been found in the northern hemisphere and far away from any sea of the present time. The fishes may have begun in fresh water. therefore, not in salt and nowhere near the Indian Ocean, Perhaps a compromise would come nearest to the truth ; like the Clupeids the earliest fishes may have been common to both salt or fresh water and probably arose on the boundary between the two and near the shore, where they obtained the richest supply of food and oxygen.

The ancient fossil fishes have also a double nature. The heavily armoured Ganoids all bear the stamp of fresh water, but the Arthrodirans and Acanthodidæ were probably marine. Since the Silurian and Devonian fossils have been found in the neighbourhood of regions where later the great zone of coal beds was laid down, thus where the land was of a marshy, freshwater nature, we have to picture the salt and fresh waters of the northern hemisphere as mixed up together in close proximity to one another. The chief centres where disturbances have been most frequent have evidently lain along this carboniferous zone and its immediate neighbourhood to the north and south, from the Rocky Mountains and Great Lakes of America across Britain and Europe to the Caspian, and further across Persia and India to the Indian Archipelago.

It is not necessary to suppose that any great fall and rise of the land occurred. The deposits were apparently laid down locally, as if extensive floodings and overflowings had taken place within the land-locked regions, perhaps due to volcanic agencies or the invasions of ice-sheets. With the presence of different kinds of water close by, these floodings would inevitably lead to a heavy mortality among the more or less sedentary fishes, whilst the active and pelagic fishes would more likely escape. We may thus believe that the extinction of the armoured Ganoids and Dipnoi was due to natural causes; other faunas were also blotted out in later periods in a similar way. Speaking metaphorically, we may say that a struggle was going on between the sedentary and the pelagic forms, and that when the physical conditions changed the pelagic forms survived. The extinction was complete and thorough though it was probably spread over many millions of years. Of all the numerous Ganoids and Dipnoi that lived in the northern hemisphere not one remains. The few modern representatives of the former, as they have been taken to be, have come from a different stock belonging to a later period.

We need not wonder that the sedentary fishes were time after time caught in their retreats and wiped out; the wonder is rather that some of the pelagic forms were also caught, and of these the Palæoniscidæ deserve to be specially mentioned.

"The Palæoniscidæ are remarkable both for their

individual and specific abundance and for their extreme range in time. Represented only by Cheirolepis in the Middle Old Red Sandstone and Devonian, the family attained its maximum development in the later Palæozoic rocks (Carboniferous and Lower Permian), became rare in the Mesozoic, finally dwindling away at the close of the Jurassic period. Their geographical distribution in the past is hardly less remarkable. In various geological formations they have been found in Great Britain and Ireland, in widely remote parts of continental Europe, and in North America, South Africa, and Australia" (Bridge). The Southern Atlantic and the Eastern Pacific are not represented; the former because Africa and South America were probably continuous, the latter because the Palæo-

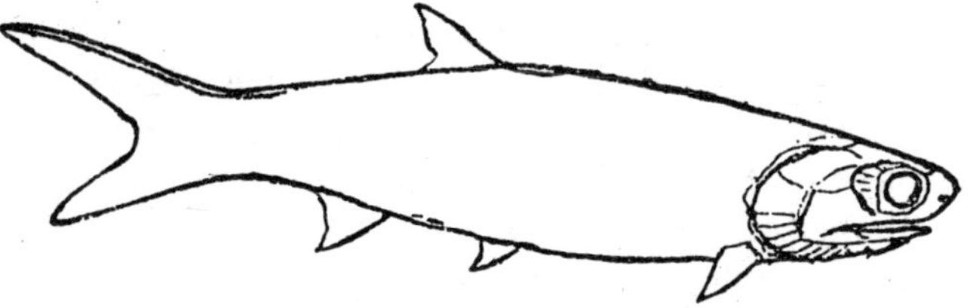

FIG. 54.—Palæoniscus macropomus, scales omitted. (From Bridge, after Traquair.)

niscidæ, like their descendants of the present day, probably kept to the coastal shelfs, though it is clear from the distribution that they were marine forms.

The characters of the family are no less remarkable. From the Clupeid shape, head, mouth and position of the fins (Fig. 54), it can be seen that they were active pelagic fishes. They differ from the modern Clupeids only in their squamation, absence of ribs possibly, and the structure of the lower jaw. In describing the changes through which the Teleostean mouth has passed (Chap. V) it was noted that the last of the bones to develop in connection with the opening apparatus of the mouth, is the interoperculum. This bone is wanting in the Palæoniscidæ, the opening of the mouth being accomplished apparently by the branchiostegal rays and covering bones which extended forward to the symphysis of the lower jaw. These forms represent, therefore, an early stage in the development of the Clupeid type.

With regard to the squamation, some of these ancient Clupeids had small rhombic or somewhat square-shaped scales which fitted into one another, usually with a peg-and-socket joint, so that the whole body was encased in a sheath. If we regard the deposit of calcareous substances in the skin as an excretion, it is probable that these fishes had an even more imperfect excretory system than the modern Clupeids, and with this we can associate the lack of flexibility in the rhombic squamation. In other words, the earlier pelagic forms were less mobile and less active and their internal organs less developed.

In somewhat later forms of the same type there is a gradual transition from the rhombic to the cycloid scales even on the same fish, the trunk having the former, the more flexible tail the latter. As Ryder and Smith Woodward have indicated, these changes—as, indeed, the whole formation of the scales—can be associated with the working of the muscles on the skin and dermis. The faster the fish moves, the more the skin is wrinkled and tends to form cycloid, overlapping scales. The rhombic squamation practically disappeared during the Mesozoic period and only persists at the present time in the isolated Lepidosteus and on the tail of the Sturgeons. Yet a close approximation to the rhombic arrangement can still be seen in some of the pelagic, slow-moving Teleosts. It should be mentioned that some of the earliest Dipnoi and most of the earliest Ganoids had cycloid scales, but their paddle-shaped paired fins indicate that they were mainly grovelling forms of the fresh waters, not marine.

We can follow this pelagic type, its distribution and even the changes in its mouth and squamation, down to the present day. During the Mesozoic period it was represented not only by the Palæoniscidæ but also by the large group of Protospondyli (Smith Woodward). Some of these are of great interest. Lepidotus, with a peculiar distorted mouth, has been found in the earlier deposits of the Mesozoic period (Trias and Lias) of Europe; four species occur on the boundaries between the Jurassic and the Chalk. Further, four species have been found in the Deccan of India (Lias), and one of the European species has been found in the Wealden deposits of East Africa (Hennig, 1914). Lastly, the same family has been found in the Chalk of Brazil. Other genera have been found in North America, the Chalk deposits of the Lebanon, and in the Jurassic of New South Wales.

The cosmopolitan range of the Protospondyli sufficiently indicates their pelagic character. We see also, that the Lemurian bridge between India and Madagascar has disappeared, at least partly, during the Jurassic period and that South America has been separated from Africa in the Chalk period, as far north at least as the Amazon.

These pelagic forms, then, were abundant during the Mesozoic period and their range extended from Australia (New South Wales) across the Indian Ocean to East Africa and India, thence across Syria to Middle Europe and North America. They are connected directly with the Clupeids of the present day through a number of forms, especially the Leptolepidæ. These had a similar range to the above, from Australia to Europe, but not apparently in North America. The changes in structure consist in the gradual replacement of the Ganoid by the cycloid scales and the acquirement of the interoperculum. The true Clupeids occur abundantly in the Cretaceous period in all parts of the world, including South America and West Africa (Guinea and Cameroon).

Going back a little, we have to note how this pelagic form has from time to time thrown out various offshoots, which branched off in several directions. Two of these may be specially mentioned. Chondrosteus was a form with Palæoniscid characters which occurred in England during the earlier Mesozoic period (Lias), but it differed from the pelagic type in that the mouth was drawn backwards under the snout and seems to have been protrusible. From this form or a form like it came, it is believed, the modern Sturgeons, which have adapted themselves to a more or less sedentary life, feeding on the bottom, with the osseous skeleton largely replaced by cartilage. Including Polyodon their range extends across the whole of the northern hemisphere, from China to the western States of America. The chief centres, however, are Eastern Europe, the Caspian and Black Sea, and the rivers flowing into these seas, and in the rivers of the Great Lakes of North America. They have been shut off there by changes in the land elevations, but they retain the habit of descending periodically to the lakes or seas. One species, the common sturgeon (Acipenser

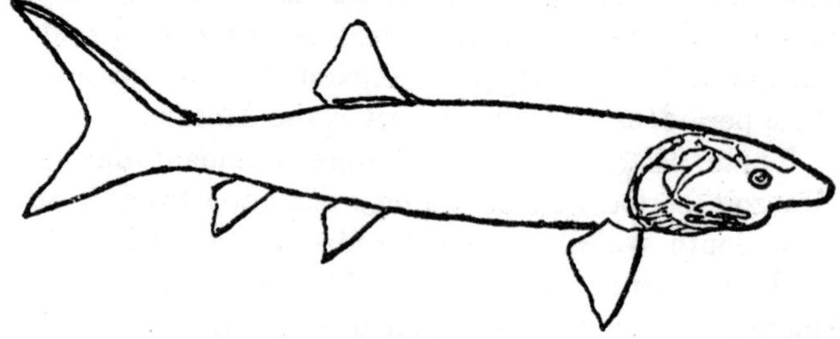

FIG. 55.—Chondrosteus acipenseroides. Perhaps an ancestor of the Sturgeon. (From Bridge, after Smith Woodward.)

sturio), belongs to the central or Atlantic area, where it has kept in touch with both coasts. Though almost fished out of recent years, it was formerly abundant in the Rhine and Mediterranean as well as in the eastern rivers of the United States.

The Polyodon family is represented at the present day only in the rivers of China on the one hand and in the Mississippi and its tributaries in the States. There also another old relict persists in Amia, the Bow-Fin, which lives in the rivers and lakes of Central and Southern North America. Representatives of Amia occur in the later Mesozoic (Jurassic and Chalk) deposits of Europe and persisted there till the Lowei Miocene period. It was about this time, according to Wegener, that the northern parts of America and Europe began to separate from each other, and it is certain that the physical changes accompanying the separation contributed to the extinction of many forms and the isolation of those remaining. Lepidosteus is another form that seems to have been abundant in Europe up to the Miocene period, but it is now only represented by some four species in North America, ranging from the Great Lakes southwards to Mexico and Cuba, and by one species in China (Jordan and Evermann).

Whilst the pelagic type was thus sending off a number of forms that settled down to a more or less sedentary mode of life in the land-locked waters of the northern hemisphere from China to America, we cannot suppose that it was less active in spreading out into deeper waters. Unless they occasionally approached the coasts, however, or were caught in a more than usually great elevation of the land, they would not occur in the geological deposits. The Chalk period seems to have been of this kind, and its deposits both in America and Europe contain several forms (Protosphyræna, for example) which appear to have been the forerunners of several deep-sea families of the present day.

In addition to the pelagic, cosmopolitan Clupeid type, the Palæozoic waters contained another which with some reason has been taken as the forerunner of the true Sharks. The long fusiform body of Acanthodes with heterocercal tail shows that it was a rapid swimmer, and this is also indicated by the cutwater spines in front of the fins. In shape of body and position of the latter it was similar to Palæoniscus, but the gills were free and the body covered with the placoid denticles of the modern Sharks. Acanthodes still retained membrane calcifications, however, in relation with the jaws and roof of the skull, even the pectoral girdle was partly ossified.

This type, so far as the fossil records go, was apparently restricted to the northern, land-locked hemisphere, and the North American Cladoselache might well be considered a later offshoot, which had lost the calcifications of the fins and head, but retained calcified arches in the caudal region. Many other " Sharks " of a similar type must have ranged over the whole northern hemisphere from the Carboniferous right on to the Chalk period, but it is curious that they have not been noted from Australia. From the latter as well as the northern hemisphere, Europe and America, we have the peculiar form Pleuracanthus, which can hardly be considered a pelagic fish. Its paddle-like fins and apparently autostylic mouth, as well as the long caudal region, show that it was a sedentary form and place it near to the Dipnoi; all the endoskeletal structures were partially calcified. As it occurred first in the Carboniferous period, it may have been a wanderer from the north before Australia separated from South Africa.

It is possible, of course, that the Pacific [1] of those early days was the home of " Sharks " just as it is so at the present time, but it is evident that they had a free range right across the northern hemisphere. Quite a large number of forms (Scyllidæ, Lamnidæ, Rhinidæ, Rhinobatidæ, and even the Rajidæ) have been found in the Cretaceous deposits of Mount Lebanon.

The impression given by both of these pelagic types, Palæoniscus and Acanthodes, as well as by the Ganoids and other sedentary fishes, is that the waters in which they moved may have been of greater density or salinity than we find in the waters of the present day.

From the distribution of the fossil fishes it is possible to obtain some general idea of the distribution of land and water on the globe during the Palæozoic and Mesozoic periods up to the formation of the great Chalk deposits (Fig. 56). It seems fairly certain that the Southern Atlantic was closed during the whole of that time, whilst Australia was near South Africa; at any rate not separated by waters

[1] On the other hand, there may have been no Pacific if the Moon had its origin in that ocean, as Sir George Darwin suggested. The departure of such a weighty mass, let us say, towards the end of the Mesozoic period would provide sufficient cause for the upheavals then and in the Tertiary. According to Smith Woodward also the deep-sea fauna dates no further back than the Tertiary or Chalk period.

of great depth. East Africa, Natal and northwards, came into touch with the Indian Ocean about the middle of the Mesozoic period (Jurassic and Wealden), since the Chalk deposits there contain the same families as the late Jura of India. On both points, therefore, the distribution is in agreement with Wegener's theory of the continental connections.

To account for the wide distribution of the early Teleosts and Dipnoi in the northern hemisphere, consequently, it is necessary to believe that a broad communication existed during the Palæozoic and early Mesozoic periods between

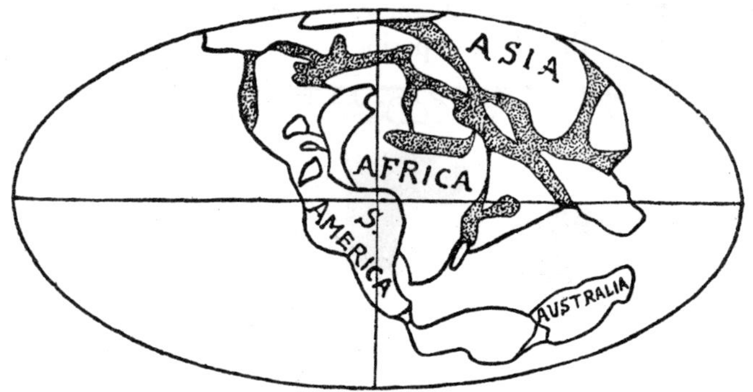

FIG. 56.—Presumed distribution of the water (shaded portions) in the northern hemisphere during the early Mesozoic period, based on the occurrence of fossils. Disposition of land-masses after Wegener.

the Indian Ocean or Pacific and the waters of Central Europe and North America. This communication, varying no doubt at different times, must have extended from at least the Great Lakes across Newfoundland, Britain, Central Europe, Black Sea, Syria, the Caspian Sea, and farther east perhaps across the region where the Himalayas now stand. It is not necessary to believe that this was one continuous sheet of water, or that the communications were of great breadth. There was probably more land than water and the different regions may only have been connected by rivers.

Some of these waterways must have run from south to

north, as we understand those terms nowadays. It is believed, for example, that a wide communication must have existed in Eastern Europe from the White Sea and Spitzbergen in the north to the Caspian and Black Sea in the south. There may have been another and separate communication farther west, from the Adriatic northwards across where the Alps now stand to Britain and Scandinavia. These and various other supposed connections can only be conjectures, however, which are helpful in explaining the distribution in the north of the ancient Ganoids and Dipnoi. The former, and especially the Cœlacanthidæ, were extremely abundant, for example, from the Devonian right on to the Chalk period, in Spitzbergen, Russia,

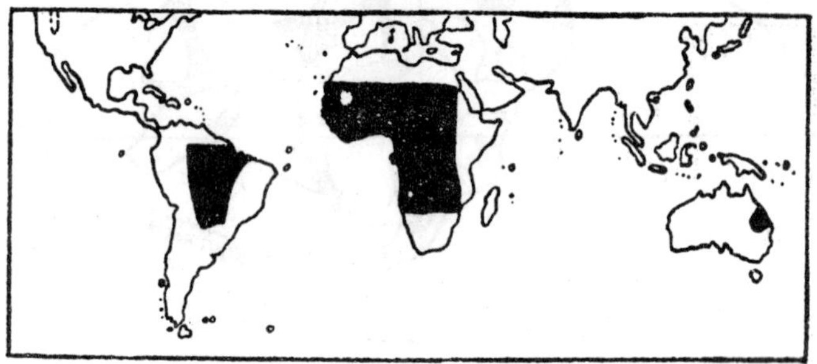

FIG. 57.—Distribution of the modern Dipnoi in South America, Africa, and Queensland. (From Bridge.)

Germany, Britain, Greenland, and North America. The Dipnoi did not extend so far north as Spitzbergen, but their remains have been found in many deposits ranging from India, Persia, across Germany and Britain to North America.

By way of contrast to these early conditions the chart (Fig. 57) is worthy of study. There are no Dipnoans in the north at the present day. The three remaining forms occur only in the southern hemisphere and in widely separated regions. One of the most interesting things connected with them is that the Queensland form (Neoceratodus) shows a far greater resemblance in habits and structure to the Mesozoic forms which were numerous over the whole of the northern hemisphere; one of the latter occurred in South Africa. It is generally agreed that the remaining forms, Protopterus in the rivers of Africa and Lepidosiren in those of South America, represent a higher degree of specialization. They are in all probability, therefore, of recent origin and their distribution affords strong evidence in favour of the view that Africa and South America were united until the Chalk period.

Reviewing now the occurrences in the northern hemisphere during the ancient periods, we see that the first groups of the Dipnoi and Ganoids were completely wiped out before the end of the Palæozoic period. One group of each, Ceratodontidæ of the former, Cœlacanthidæ of the latter, lived in North America and Europe during the Mesozoic period, the latter right to the Upper Cretaceous. From their shape and structure we may say that these were all sedentary forms living in fresh water. Their final extinction occurred during the Chalk period, when the tremendous changes in the physical appearance of the world began which culminated in the upheaval of the mountain ranges during the early Tertiary period. The Chalk deposits are evidence that the sea must have swept over large tracts of land at that time, and it is probable that the whole of the freshwater population was destroyed.

On the other hand, both of the pelagic, marine types escaped. Some of the Clupeids were trapped locally in the Chalk as also some of the Elasmobranchs. The most interesting thing is, that just about this period both types began to break up into offshoots, which have become the characteristic forms of the present day. But whereas the Clupeid type persisted as well as its descendants in the deep sea or fresh water, many of them losing the air-bladder or its open duct, the Sharks with only one apparent exception quite changed their mode of reproduction and developed claspers, whilst the Rays were only formed, probably from the Mesozoic Rhinobatidæ, during the Chalk period. The Chimæridæ of the deep seas, on the other hand, appear to have remained unchanged from the Mesozoic period.

It would appear, therefore, that the physical changes during the Chalk period had a profound influence on the mode of life and characters of fishes. They led to the appearance of new forms of both a sedentary and pelagic type, coastal and deep-sea forms, which did not exist previously. And this led, again, to the appearance of new freshwater forms.

2. Modern Periods

What happened to the Earth during the Chalk and Tertiary periods we can but faintly imagine. The land-masses appear to have been in a more or less fluid condition, rolling together in places and rising subsequently there into huge mountains, thus causing many parts to be submerged under the sea for long periods. The changes cannot be referred to ice-sheets, like those of the later Glacial epochs, but seem rather to have come from deep breaks in the lower-lying magma. Whatever may have been the causes, we have to picture the continents slowly becoming separated from one another, so that Africa is no longer connected with South America and the North Atlantic basin is open to the south. The northern part, on the other hand, is closed between Great Britain, Newfoundland, and the North-Eastern States until late in the Cenozoic period, and Scandinavia, Greenland, and Canada were probably united until the first Glacial epoch.

With the breakdown of the bridge between Africa and South America it might be supposed that a considerable immigration into the North Atlantic took place from the south, especially of the pelagic forms. This was probably the case, but it has to be remembered that the waters of the northern hemisphere were already well populated with the pelagic Clupeids and Sharks, and that there had been an eastern entrance over Syria and Russia long before the South Atlantic opened. It is not very easy, therefore, to distinguish between the two streams. For example, the Protosphyrænidæ which seem to have given rise to quite a number of modern Atlantic, as well as Pacific, forms, occurred both in Europe and North America. Their teeth have been identified from Patagonia, and the distribution of the modern forms would suggest that they came in from the South Atlantic. Yet the teeth have also been found in Egypt. Again, Belonostomus of the same period also had a very wide distribution, from Queensland to Europe and Mexico. It occurred also in the Upper Chalk of Brazil and thus seems to have come from the south. Yet it has likewise been found in the Lower Chalk of the Carpathians, though like the Protosphyrænidæ it does not occur in the Lebanon Chalk. Similar difficulties will be noted when we come to the freshwater fishes.

In some cases, however, there is a greater probability in favour of the South Atlantic entrance. Chirocentrus, a near relative of Clupea, has its home in the Eastern Archipelago of the Indian Ocean, yet the fossils have been found in the Upper Chalk and Eocene of Brazil. Ichthyodectes, another related form, has been found in the Upper Chalk of Cameroon and various Elopidæ occur in the Chalk of Brazil. At the present time several species of Elopidæ occur off the south-eastern coasts of North America, the West Indies, West Africa, and one species in India. No trace of these has been discovered in the Mediterranean region. These afford clear evidence of a southern immigration since the Chalk and there can hardly be any doubt that the whole of the Scombroids, Tunnies, Mackerels, Sword-fish, and the Sail-fishes, etc., arrived by the same route. These are the pelagic fishes par excellence of the present day and their range extends over the whole world.

Another large and valuable group that we probably owe to this southern route is that of the Anacanthini or Gadoids. Remains of these have been found in the Eocene of Europe and North America, even earlier in Denmark. As Greenland was not yet separated from Scandinavia and they have not been found in the Mediterranean deposits, there seems little doubt that they came from the south. Boulenger has suggested that the Gadoids may have been derived from deep-sea forms like the Macrurids, but it is possible that the cosmopolitan and less specialised genus Merluccius, which occurs on the west coast of South America and in New Zealand, as well as in the North Atlantic, might have been the original form.

On the other hand, the Rays and Dog-fishes, most of the coastal forms, Blennies, Cottoids, and Gobies, perhaps also the Labroids, have come to Europe apparently by the eastern entrance. Remains of the first five groups have been found in the Upper Chalk of Persia (Priem) or the Lebanon. It is probable that the Carangoids, Stromateids, and some of the Flat-fishes, Soles and Turbots, have come by the same route.

In all these cases, with exception of the Gadoids, the European or North Atlantic fauna is poor by comparison with that of the eastern part of the Indian Ocean, and there seems no doubt that after the European waters were depleted of their stocks during the changes noted above, they became occupied by new-comers from the south and survivors from the east. One of these survivors seems to have been Anguilla. This is said to be represented in the Upper Chalk of the Lebanon and it also occurs in the Eocene of Mt. Bolca. It could not have entered the Atlantic from the north and its entrance from the south is rendered improbable by the fact that the genus does not occur in the South Atlantic. Related forms occur in the rivers of West Africa (Ehrenbaum), but these may have come from the north. On the other hand, the Muranoids have their centre in the Indian Ocean and one species, M. helena, is common to the Mediterranean and the last-mentioned region. Anguilla also has its true home in the Indian Ocean, as Johs. Schmidt has shown.

It seems certain, therefore, that the Eels entered the North Atlantic waters from the east before the southern entrance was open and became acclimatised there before North America separated from Europe. This explains the close affinity of the American and European Eels and raises an interesting question connected with Wegener's theory. According to the latter, America travelled away to the west from Europe and the American Eel should therefore have the longer distance to migrate. As it is the European Eel which makes the longer journey, it looks as if Europe had travelled away from America instead of the reverse.

The view that modern fishes have come mainly from the Indian Ocean and Eastern Archipelago is supported by the investigations in the Pacific made by the American authors, Jordan, Evermann, Gilbert and their collaborators. These authors are agreed that the fishes of Hawaii and other Pacific Islands of the west coast of America have come from that Archipelago. Gilbert writes, " An Analysis of the list of species recorded shows conclusively that the bathybial fishes of Hawaii, like those of its reefs and shores, have been derived as a whole from the west and south, and not from the east or north. In its entire facies, the fauna is strikingly unlike that of the Pacific coast of Mexico and Central America, and resembles strongly the assemblage of forms discovered by the *Albatross* and the *Challenger* off the coasts of Japan and the East Indies. Some of its members find their nearest affines in the Bay of Bengal " (1905).

The distribution of the freshwater and brackish-water fishes of the present day affords many interesting examples of probable land and water connections in the distant past. They have all come from pelagic forms at different times, we may believe, and frequently still have near relatives on the coast. But deep or salt water is a barrier to their spreading from one land to another, and we have a classical example of this in the East Indian Archipelago. The island of Borneo, which belongs to the Indian region, has hundreds of species of freshwater fishes, yet Celebes of the Australian region, but a few miles away to the east, has not a single indigenous freshwater fish (Regan, 1911).

Among the most interesting of these groups are the Carplings, or Cyprinodonts, small fishes which live in tropical and subtropical countries. As the map shows

(Fig. 58), they extend over the southern parts of Asia, the whole of Africa and parts of America, chiefly in South America. In the latter, indeed, they are represented by nearly ten times as many species as elsewhere; but they are not found in the Australian region. From their occurrence in Madagascar and South Europe it would appear that they have arisen earlier than the Tertiary period, but after the southern part of South America had separated from Africa.

A remarkably similar distribution is shown by the Cichlidæ, also fresh or brackish-water fishes, but widely

FIG. 58.—Distribution of the Cyprinodonts. (From Boulenger.)

separated in structure from the Cyprinodonts and probably, for that reason, of a distinct and later origin. The majority of the species belong to Africa, including Syria and Madagascar, but a very large number live in the same regions of America. The curious thing is, that two species, Etroplus, occur in Ceylon and South India, and they are hardly distinguishable generically from Paretroplus, which occurs in Madagascar, though both genera are different from all the other Cichlids (Duncker).

On the other hand, the Galaxiidæ, a family not far removed from the Cyprinodonts, show quite the reverse distribution, being confined to the south parts of Australia, Tasmania, New Zealand, the south point of Africa and the south end of South America. One and the same species is believed to be common to New Zealand, Australia and South America (Boulenger). We can hardly suppose that this points to the earlier connection between these parts; the distribution would have been more extensive in that case. It is more probable that they have come later from some common marine species like the Clupeids, as, indeed, their form suggests.

Turning now to more typically freshwater forms, the Ostariophysi, which have practically lost all connection with marine forms, we obtain a picture of a wide dispersal of some 4,000 species of fishes (Regan, 1922) since the

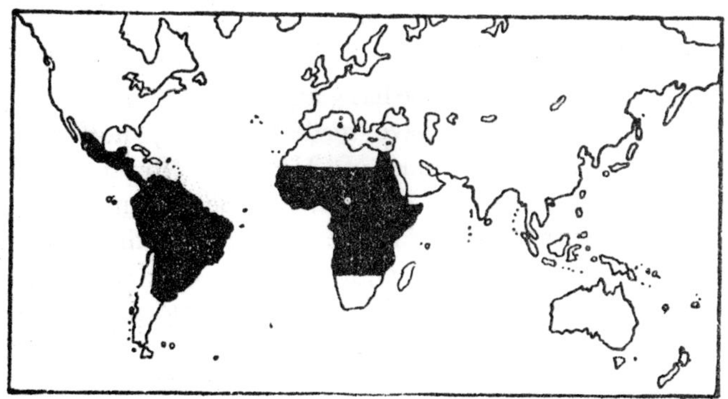

FIG. 59.—Distribution of the Characinidæ. (From Boulenger.)

beginning of the Tertiary period. Some of these may have existed in the Cretaceous, for the Characinidæ, the oldest group, are not far removed from the Clupeids of that period, but no fossils older than the Mid-Eocene have yet been found.

The outstanding character of the whole group of Ostariophysi is that none occur in Madagascar or the east (with possibly an exception in the Australian region). They originated, therefore, after these regions had separated from Africa, but whilst the connection between Africa and South America was still maintained. The accompanying map of the distribution of the Characinidæ shows a remarkable correspondence with that given above for the Cyprinodonts, and it is further remarkable that, like the latter group, the Characinidæ are ten times more numerous in South America than in Africa (Regan), yet the Characinidæ do not occur in India or other parts of Asia.

The obvious conclusion seems to be, that the Characinids and Cyprinodonts came from some wandering Clupeids (Leptolepidæ), which had reached round the southern end of Africa into the South Atlantic, when the gap was forming between Africa and South America. "The subsequent history of South America is simple; it was isolated and developed its rich and remarkable fauna of endemic families of Characins, Gymnotiformes, and Siluroids; at the beginning of the Miocene it became connected with Central America, which has been colonised by immigrants from the south, but none of these has gained access to the Mexican Plateau and very few have made their way northwards into the Atlantic streams of Mexico" (Regan). This author concludes, it will be seen, that South America was not united with Central America until late in the Tertiary period, but the Cyprinodonts, which came over earlier than the Characinids (see map above), have spread more to the north than the latter, so that the time factor has to be taken into account in the distribution of the Characins. Regan himself, as will be noted presently, has shown that the dispersal of fishes has been a slow process.

From the Characinids came the Gymnotids, the freshwater "Eels" of Central and South America but of nowhere else. It is very remarkable that the true Eels (Anguilla) are not at all found in South America. The great similarity of the larva of the latter (Leptocephalus) to that of Albula, which is exceedingly common in the West Indies, suggests that the early ancestors of the Characinids and Eels were the same (Leptolepidæ), and that whilst some were changed into the Characinids and Gymnotids on their way northwards along the South American continent, others passed on into the North Atlantic and became transformed into the Eels. This is certainly an attractive idea, but the occurrence of several species of Anguilla in the Indian Ocean and Eastern Asia makes it improbable. The more probable line of immigration for the Eels has evidently been across the plains of Syria, as already noted above.

Of the remaining Ostariophysi the Siluroids are common to South America, Africa, and the Indian region, but very few now remain in the northern hemisphere, though fossils have been obtained from the later Tertiary deposits of North America as well as Europe. Some early families also exist in Madagascar and Australia. It is probable, therefore, that this group originated in the Indian region in pre-Tertiary times.

The last group of the Ostariophysi to be mentioned, the Cyprinidæ, is only found in the northern hemisphere, from India and Siberia right across Europe and North America. The fossils have been found in the lower Tertiary of North America. It would appear, therefore, that this group has arisen after the separation of India from Madagascar and Africa, and the Cyprinoids may have found their way across Europe into North America by the earlier Mesozoic route (Fig. 56) or, as Regan suggests, partly by a more direct course over the Bering Straits, when Asia was united with Alaska. As the temperatures along the latter route were probably too low for this Indian group, it is more likely that they followed the same channels as the old Ganoids and Dipnoi. However this may be, the most interesting thing is, that the peculiar air-bladder attachment to the " ear," which is characteristic of the whole group of Ostariophysi, must have arisen separately in different groups in different parts of the world ; thus, in the Cyprinidæ of the north and India and in the Characinids of South America. This would be extraordinary, if we could not explain it by the fact that both groups have come from the same pelagic, marine and freshwater stock, the Clupeids, which already have indications or rudiments of the air-bladder connection.

Whilst the groups so far mentioned have a more or less restricted range, there are some puzzling cases of cosmopolitan distribution with apparently no means of communication between the different members of the group. As an example of this kind we have the large family of Pipe-fishes (Syngnathidæ and related forms), for the most part coastal forms in all parts of the world with not a few living in fresh water. Regarding these Duncker gives the following information. Nerophina belongs exclusively to the North-East Atlantic; the freshwater Doryichthys, Cœlonotus, and Belonichthys are Indo-Pacific forms, yet Belonichthys fluviatilis occurs in the Zambesi, in Madagascar, Celebes, and the Philippines; Microphis is mainly Indo-Pacific also, but M. lineatus and M. aculeatus, nearly related forms, occur in the mouths of rivers in the Atlantic, the former in Central and South America, the latter on the African side from Senegambia to Portuguese Congo; the genus Leptonotus again occurs on the west coast of America, in New Zealand, Tasmania, Australia, and the Fiji Islands, whilst the Pegasus family occurs in China, Japan, the Malay Archipelago, and Australia.

These are all forms of Tertiary or more recent origin, and the wide distribution may have been effected in part through drifting seaweed to which they can attach themselves, as in the Sargasso Sea, but we can hardly accept this explanation for all cases. It is more probable that they have come from some pelagic and related forms of wide distribution, such as the Flute-mouths, Fistularia, which is common to the tropical and subtropical waters of the Atlantic and Indo-Pacific and which is also represented in the later Cenozoic deposits of Europe. As shown in Chapter V, Fistularia begins life with the characters of a Clupeid larva, and we thus obtain another good example of the transformation of the pelagic, free-swimming type into a diversity of sedentary, coastal, or freshwater forms.

With regard to the distribution of more recent fishes since the Tertiary period, we find in C. Tate Regan's book on the British Freshwater Fishes an interesting account of the swaying to and fro of the fish population of these islands with the advance and retreat of the Ice-covers from the north. The glaciers that extended downwards from Scandinavia and North Russia over Mid-Europe and England as far as Buckinghamshire made a clean sweep of the earlier inhabitants of the lakes and streams. The same thing happened in North America, where the Ice-cover extended some 5° still farther south. A new population then arose, and Regan shows how the most characteristic freshwater forms of Great Britain, the Chars (Salvelinus) and Whitefish (Coregonus), belong to a more northerly fauna. They were left behind in isolated lakes as the Ice-cover retreated and have adapted themselves so well to the varied conditions that almost every lake has its own peculiar race or species. They are now freshwater fishes, but their northern relatives spend some part of their time in the sea.

These are the only species peculiar to the British freshwater fauna. The Char also occurs in isolated lakes of Scandinavia and the Alps, but not in Germany. Coregonus is represented by other distinct forms in the countries round the Baltic, but the main types of the family live as anadromous fishes in the Arctic Ocean and rivers of Siberia.

In addition to these relicts of the Glacial periods, the British waters also contain a considerable number of true freshwater species, such as the Perch, Pike, Carps, and Loaches (Cobitidæ). All these occur on the Continent of Europe, most of them even in Asia. A striking peculiarity is, that of the twenty-two species mentioned by Regan only one occurs in the Iberian Peninsula. These freshwater forms have thus come from the east and have entered the British waters from the Rhine region, when South Britain was still in touch with the Continent. Once more, therefore, the northern waters have been stocked from the east. These fishes have apparently entered by way of the Humber and Trent, and Regan is able to show that their distribution to the west, south, and north has been very gradual. Ireland, for example, has only ten of the twenty-two species.

Even more interesting are the relations with North America. The Pike (Fig. 60) and the Burbot also occur there in the Great Lakes and northwards, as well as several species of the Perch and Sander (Lucioperca). The Perch occurs in Britain and the Rhine district, but Lucioperca is found in neither; its home lies in the rivers of Eastern Europe draining into the Caspian and Black Sea. Yet it occurs in Scandinavia, Canada, and the States. Similarly, the Mud-fish (Umbra) occurs nowadays only in stagnant waters of Austro-Hungary (U. crameri) and in swamps of Canada and the north-eastern States (U. limi). The occurrence of these forms in places so far apart would indicate that Scandinavia has had an independent con-

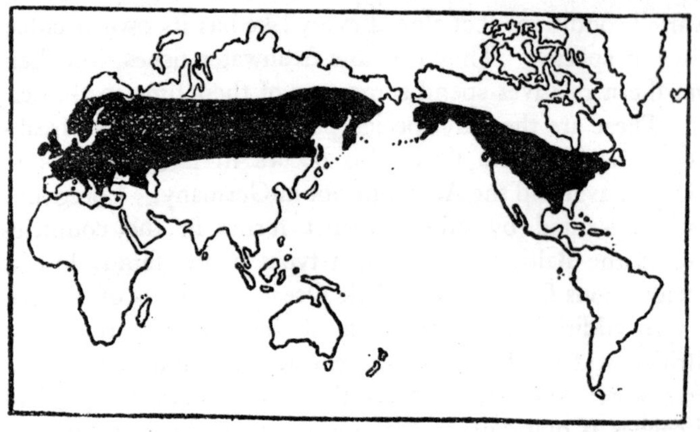

FIG. 60.—Distribution of the Pike family. (From Boulenger.)

nection with Greenland and this with North America apart from that between Newfoundland and Britain.

The distribution of the three-spined Stickleback (Gastrosteus aculeatus) also indicates this. In Eurasia it extends from Siberia across the Black Sea region to Scandinavia and Iceland. It does not occur in the Mediterranean, but is found in Algiers—an occurrence Regan traces to the influence of the Glacial Age. It is also found in Greenland and in North America as far south as California and New York. It lives both in salt and fresh waters. On the other hand, the strictly marine Fifteen-spined Stickleback (G. spinachia) is not known from Iceland or North America, but it occurs in the Faeroes.

It is generally believed that the Salmon family came to Europe from the west and not from the east (Smitt). The Salmon is an arctic and subarctic genus that has its centre in the Northern Pacific, where four to five species occur in the rivers of Asia. These are all represented on the Pacific coasts of Canada and the States and one or two even run up long distances into the land. In this way the Salmo salar of Europe probably crossed the American continent before the great rolling together of the Rockies took place in the Tertiary period, and before Great Britain was separated from America. The Trout is probably a derivative which has adapted itself more particularly to the rivers and lakes. It is of interest to note, that the small Trout transplanted to New Zealand rivers have become almost as large as the Salmon.

If we now glance back over this brief summary of the distribution of fishes in the past and present, we cannot fail to be struck by the many and great changes which have occurred. The endeavour has been made to associate these changes with known physical causes, the separation of the continents, the drying up of regions, the incursion of various ice-sheets, and the flooding of areas by salt or fresh water. It may be that the Indian Ocean and the lands bordering on it owe their great wealth of forms just to the comparative absence of these physical changes, which have blotted out from time to time the faunas in other parts of the globe. But the main thing is, that after one fauna was blotted out or depleted, its place was taken by another coming from the pelagic stock of the sea. Here the Clupea tribe, under various names, has reigned from the beginning with but slight alteration in its structure and practically none in its form. It has been preyed upon by its descendants and other animals in the most ruthless manner, beyond comparison with anything known elsewhere, yet it persists in all parts of the world, in salt and fresh water, in greater variety and greater profusion than any other group. Its constancy and persistence have come probably from its refusal to specialise. It was born in shallow waters with free access to the air, and these are the conditions which have made for permanency. The food in the neighbourhood of the plant-covered coasts is abundant, the supply of oxygen is plentiful, and it has adapted the simplest of hydrostatic organs in a rough-and-ready manner so as to be able to rise and descend in the water without difficulty. From this group, we may believe, have come all other forms, fresh-water fishes, the fishes of the deep sea, and even the Sharks. It is at once the most primitive and most modern group of fishes.

CHAPTER XI

ADAPTATIONS TO SUIT PARTICULAR CONDITIONS

It has been indicated that many characters of fishes, vertebræ, form, fins, etc., have arisen as a consequence of their movements. Each individual recapitulates the ancestral movements and develops its structure according to the ancestral balance. As they develop, some structures fix or consolidate the form of movement in a particular direction and may reach the dignity of adaptations.

The term "adaptation" has both a general and a special sense. We say that fishes are "adapted" to live in the water, their gills being "adapted" for aquatic respiration, their tail for aquatic locomotion, and so on. These qualities are involved in our idea of a fish. In a similar way we might say, a stone is "adapted" to sink in the water, a cork to float. These general qualities are for the most part *regulations*, i.e. characters that have arisen from the physical laws of action and reaction without any reference to their utility (*vide* Chapter V). To come nearer to the special meaning of the word adaptation, we may consider the opposite conditions: a fish that has practically no gills (Monopterus) and a fish that has practically no tail (Sun-fish). These manage to live quite well by adapting their remaining structures to a particular mode of life. The former lives in the mud and breathes in air somehow, the latter uses its dorsal and anal fins for propulsion in place of the lost caudal. In fact, if we picture to ourselves what a fish should be like in form, structure, and habits, we find that adaptations are in reality departures from that picture in one direction or another. As Professor J. Arthur Thomson defines it, an adaptation is a "special adjustment of structure or function to meet particular conditions of life."

This definition fits in with the conclusion stated in previous chapters, that a pelagic form of fish, a generalised Clupeid, let us say, was the original parent of all fishes, and that from this type at various times all others have been derived, just by adapting themselves to other modes of life. Starting from this point of view, we may trace the gradual attainment of some adaptations.

1. Growth of Adaptations

Like most of the Clupeids the Herring is a pelagic fish that can rise to the surface or descend to considerable depths, at least 100 fathoms. A few other forms can also do this, and we may examine how it is done. The Herring has no special adaptation of bone or gills for the purpose. Its ability to rise and sink comes from a peculiarity in the air-bladder. In addition to the communication with the blind sac or stomach, this organ has also a posterior opening direct to the exterior—the "safety opening," as it has been called, which is formed in the late postlarval stages. Being able to take in water by this means the fish can adjust its balance to different depths without any effort. But the pelagic zone is invaded at times by other fishes with a different structure. The Mackerels and the Tunnies come up to the surface to spawn, and when they have done so, in their hunger, they pursue the pelagic forms of life, both of them the Herring, and the Tunny also the Mackerel, and when they have fed they retire again into deeper waters. It seems certain that these are not really adapted to living at the surface. What enables them to do so is simply their great speed, and in developing the speed they have attained to a remarkable form and structure which are adaptations to rapid movement and secondarily to the pelagic mode of life.

In whatever direction we look, in the coastal region or in fresh water, we find many diverse forms living under precisely the same conditions, yet with quite different shapes and structures. How this has come about can be understood from their past history. Each region contains immigrants from various sources, some old and some new, with different structures to begin with, and these structures have then to be adapted to the same conditions, even the same habits. For example, the Sturgeon is an earlier immigrant into fresh water than the Loaches or the Carp. It has a ventral mouth like the Sharks but without teeth, except in the larva; it manages to feed by protruding the inside of its mouth into a spout-like structure. The Carp and Loach have no teeth either, but the former has a pointed snout of the ordinary shape, whilst the Loach has a sucker-like mouth with tactile barbels. All three grub in the mud. And close by is the Eel living in a similar way, but it is well provided with teeth. In all cases the reference is to the teeth of the jaws or mouth, not of the pharynx.

In the sea we find the same phenomenon on a larger scale. The Rays and Flat-fishes are often found together and partaking of the same food, but the structures are very different. The same may be said of the Dog-fishes and Gadoids, or the Weevers and Sand-eels, or the Grey Mullets and Wrasses—all with a choice of the same food but again with different structures. In all cases it is evident that the environment could not blot out the old structures and make new ones alike for all fishes. It could only make small modifications here and there. For example, as was mentioned in Chapter X, the Chars in the different lakes of England have gradually become so modified that each lake has its own peculiar race or even species.

Perhaps the best idea of what is and what is not an adaptation is furnished by the coloration of fishes. According to the older view this was considered a protection against enemies; the colours so closely resembled those of the surroundings that other predatory animals could not see them. But with a closer study of the phenomenon, this view has fallen into disuse, as already mentioned. The colour is a product of the metabolism of the fish on the one hand and the rays of light on the other. Where both are intense, as in the tropics, the colours are vivid; where both are feeble, as in the dark caves of Cuba and the States, colours may be quite wanting and even the pigment of the retina does not develop. Cunningham has proved also that the under-surface of Flat-fishes, normally white, becomes coloured when the rays of light are brought to bear upon it. Inshore fishes living among variegated surroundings are more coloured than offshore fishes, and the deep-sea forms are either silvery or black. If the colour is any protection it comes from the fact that the pigment absorbs the dangerous rays of light (Buchmann).

How difficult it is to regard colour as an adaptation, other than in the above sense, can be seen from a comparison of the different colours of fishes living under the same conditions. It has been supposed that the slaty blue of the back and white of the belly in many fishes, like the Cod, render them more or less invisible to enemies from above or below. Yet the Mackerel and Herring are conspicuous in the same waters and other fishes may be almost black (Cantharus) or even red (Gurnards and Red Mullet).

On the other hand, if we meet with cases where the fish appears to make a definite use of its resemblance to particular surroundings, we can say that it adapts itself thereto. In the summer months the Plaice chooses the bright, clean sand for its residence, whilst the darker Dab and Flounder prefer mud. But in the spring-time, when it has spawned and is off colour or out of condition, the Plaice haunts the muddy ground. The different species of Pipe-fishes, when they return to the inshore waters, choose the seaweed and surroundings that suit each best.

In all cases it is advisable to distinguish between the origin of a character, if possible, and the use to which it is put. Each character may have several different uses. The tail or caudal fin develops as a balancing organ, providing an equipoise to the head. Where no air-bladder is present and where the centre of gravity lies near the head, the tail is bent upwards (heterocercy) and thus by its movements serves to raise the head. That is the general purpose.

But when the upper lobe reaches to the huge dimensions seen in the Thresher-shark (Alopecias, Fig. 61), where it is as long as the body, its use as a flail in the water may well be called an adaptation.

When an air-bladder is present, however, the strain of supporting the head is removed to a greater or less extent and the tail tends to become homocercal. Yet many intermediate stages occur. The middle rays may grow out a considerable distance (Fistularia), or the upper or lower rays (Trachypterus, Regalecus)—or no rays may develop (some Pipe-fishes, Trichiurus), all depending on the balance and movements in the early stages, as explained in Chapter V. And these different tails may be put to various uses, either to hang on seaweed (Pipe-fishes), or to climb

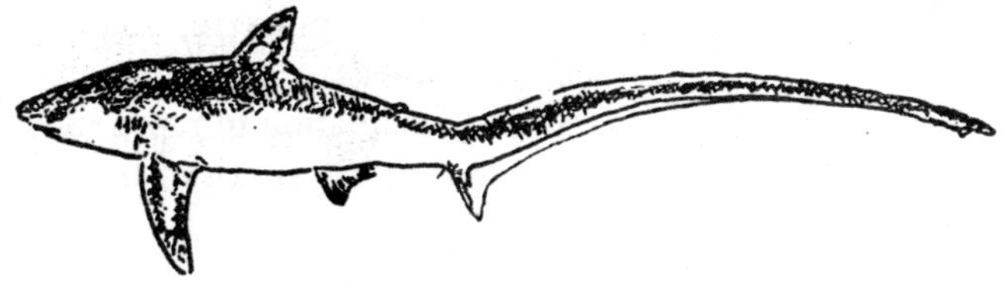

FIG. 61.—The Thresher Shark (*Alopecias vulpes*) grows to a length of 15 feet, more than half being tail. (From New York Zoological Society Bulletin.)

out of a tank (Eel), to hold on to a rock (Rivulus), or even for fighting purposes (Pyrrhulina).

But the ordinary homocercal tail in conjunction with an air-bladder does not make for very rapid swimming, the air-bladder is in the way of the movements, and a different combination of characters has developed in the Mackerel family. These are down by the head, like the Sharks, and tend to sink when at rest, but they have managed to convert the tail muscles and vertebral column into a propelling apparatus that would call forth the admiration of an expert engineer. The tail is deeply forked (Fig. 62) with the rays overlapping one another on the end of the vertebral column. With every twist of the body the tail is thus able to give a double stroke on the water. And each vertebra has become like a living cog-wheel fitting into a series of similar wheels, but not for the purpose of rotating. Extended on the one side, as in bending, each vertebra has a certain amount of free-play or side-slip, but on the compressed side the locking is rigid. Flexibility and rigidity

FIG. 62.—Sword-fish (*Xiphias gladius*) with long snout and powerful tail. (From New York Zoological Society Bulletin, after American Museum Natural History.)

at the same time combine to make a powerful shaft and screw adapted to reciprocal movements.

We may call this apparatus an adaptation to rapid movement, but we cannot fail to see that it has arisen from the

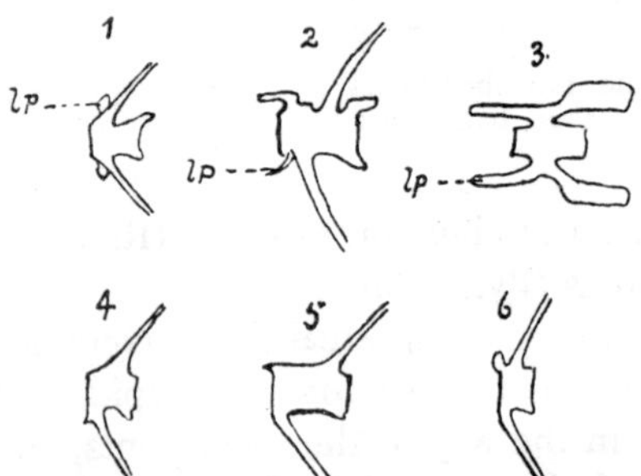

FIG. 63.—Vertebræ of fishes. 1–3, fast swimmers: 1, Cod; 2, Pilot-fish (*Naucrates*); 3, Sail-fish (*Histiophorus*). 4–6, slow swimmers: 4, Blenny; 5, Weever (*Trachinus*); 6, Perch; *lp*.=locking process.

effort to swim more and more swiftly. If we compare a series of vertebræ of different forms (Fig. 63), we can readily see that the ability to move fast is connected with the locking of the vertebræ together to form a solid axis. This locking in the Higher Vertebrates is produced by definite articular processes, but in fishes (Teleosts) simple spinous processes are the usual means. Where the body is elongated and the movements are of a wriggling kind, as in the Blennies and Ribbon-fishes, the locking spines are hardly developed; but in the Mackerel, Tunnies, and Sail-fishes, as mentioned, the spines have become huge locking processes. On the other hand, in the specialised Carangoids and Percoids, which have developed a different mode of balancing, the spines are but little developed.

Whilst the structure of the caudal region may thus be referred to the efforts of the fish in the way of movement, the head and anterior part of the body show the impression of the resistance of the water. In the most powerful swimmers (Tunnies) the head is pointed, compressed more above than below, and even the sides of the body are compressed, so that the body is no longer torpedo-shaped in section but rather a broad oval. Further, the jaws are pressed together and covered over by the skin and broad suborbitals in such a way that they present a minimum of resistance to the water. For a similar reason the fins are folded back out of the way when swimming.

It is evident that a mouth with such a structure has not been arrived at without difficulties; the opening and closing muscles have to be co-ordinated to a nicety. In other species the feeble development or absence of teeth indicates a disturbance of the growing tissues, and this has resulted in the formation of a long snout, as in the Sail-fish and Sword-fish. In other cases, apparently, the disturbance has led to a temporary closing of the mouth (as in the Pipe-fishes and Flat-fishes), and this to a great elongation of the body with increase in number of the vertebræ, loss of the ventral fins and even of the caudal. We can follow such a change in the series Thyrsites, Lepidopus, and Trichiurus. We can hardly call these adaptations.

In default of teeth, however, the Sword-fish makes a special use of its long snout (Fig. 62). It pursues the Herring, Mackerel, and Cuttle-fishes and with sharp swings to the right and left cuts them into pieces which it can conveniently swallow. It has often driven its sword into small boats and left parts of it in large ships. For this reason it has been accused of ferocity, even of attacking a whale, but Townsend, who has carefully collated and sifted the available evidence on the matter, has come to the conclusion that it is really of a peaceful disposition. When attacked by the Sword-fish, the Herring and Mackerel gather together near the boats, seeming to know that safety lies there, and in following them the Sword-fish may accidentally run its sword through a boat. And if it apparently loses its temper and attacks more than once, this is probably from blind fear rather than rage.

On one occasion the Sword-fish figured in the London Law Courts, with Huxley and Buckland as expert witnesses. An Indian clipper, the *Dreadnought*, had been rammed by a Sword-fish, sprung a leak, and the cargo, if not the vessel itself, suffered serious damage. The Insurance Company, when called upon to pay damages, pleaded that the loss should be considered " an act of God " and not a fair risk of the sea ; nevertheless, they had to pay. As showing what a deadly weapon this sword may be, it may be mentioned that its length reaches three feet and one of six feet has been recorded by Brown Goode.

When one structure fails, the fish is not beaten but adapts another to serve the purpose. The Sword-fish is an example ; but many other fishes have lost their teeth in a similar way without being provided with a long snout to transfix and cut up their food. The Pipe-fishes, for example, have no teeth, and it is impossible to regard their snout as an adaptation to anything. They do the best they can with it, however, and manage to suck in the small Crustacea of the water and even a large shrimp occasionally. Other forms, again, have taken to grubbing in the mud, and this has led to various modifications internally, as will be mentioned later.

The spines possessed by many fishes have arisen in various ways connected with the movements and balancing of the body and in so far can hardly be called adaptations, yet in some cases they are put to a particular use, which, again, has probably led to their greater development. The

Sting-rays and Eagle-rays have a strong spine in their tail which they use to cut up their victims, and if one spine is lost another is developed to replace it. The Climbing Perch makes its way overland by means of the spines on the gill-cover and fins, and other species of the same family hold on to stones in running streams. But, of course, the same end can be attained in other ways without spines.

In some cases the spines are used as weapons of offence and defence—for example, by the small Gasterosteus (Stickleback) and Osphromenidæ when engaged in fighting with one another, but we can hardly believe that they were developed for that purpose. In fact, many fishes possess most deadly spines, but are not known to use them. The Common Weevers of the shore are peaceful fishes and have no more need of spines than the Plaice they live with. If these spines are accidentally stepped on, however, they cause a serious wound made more dangerous by the poisonous mucous secretion which gets into the wound. The peculiarity of the spines which have poisonous glands or sacs connected with them is, that they are loose in the tissues, where usually they are fixed. It is probable, therefore, that the rubbing of the loose spine on the surrounding connective tissue has led to the development of glandular cells above and below—just as the eyes of some fishes have an open gland-like cushion or bursa behind.

The spines of the first dorsal fin in the Weever are looser than usual, but the glands are small and, in the rays behind the first, mainly detached cells of the surrounding epithelium. As a sign of some disruptive agent at work, the skin covering the spines is pigmented a deep black.

The Toad-fishes (Batrachus) show another variety of the poisoned spines. The agency at work is presumably the same, but the opercular spine is perforated, that is, the loose spine has worked more on the cells at the base and the acid secretion has made its way outwards through the spine. Owing to this difference the spines of Trachinus and Batrachus have been compared to the poison fangs of the colubrine snakes and the true vipers respectively.

Unless we know that these spines and poisons are deliberately used by the fishes as weapons of offence, just as the snakes use them, we can hardly call them adaptations. Spines are the common inheritance of many fishes and it is well-known that the blood and serous fluid of one animal is a deadly poison to other animals. That the latter may be so used, however, would appear from the following account of the Porcupine-fish given by Darwin in his *Voyage of a Naturalist*.

" One day I was amused by watching the habits of the Diodon. This Diodon possessed several means of defence. It could give a severe bite, and could eject water from its mouth to some distance, at the same time making a curious noise by the movement of its jaws. By the inflation of its body, the papillæ, with which the skin is covered, become erect and pointed. But the most curious circumstance is, that it secretes from the skin of its belly, when handled, a most beautiful carmine-red fibrous matter, which stains ivory and paper in so permanent a manner that the tint is retained with all its brightness to the present day. I am quite ignorant of the nature and use of this secretion. I have heard from Dr. Allan, of Forres, that he has frequently found a Diodon, floating alive and distended, in the stomach of the Shark ; and that on several occasions he has known it to eat its way, not only through the coats of the stomach, but through the sides of the monster, which has thus been killed. Who would ever have imagined that a little soft fish could have destroyed the great and savage Shark ? "

Such a Diodon clearly deserves to live, but how many escape from such an ordeal ? How the Diodon avoids being rendered innocuous by one bite of the Shark's teeth is also a mystery. But it appears that when it has run the gauntlet of the teeth and arrived in the stomach, it has the sense to give off the secretion which counteracts the acid juices for a sufficiently long time to enable it to bore its way out of the Shark. Other fishes are not so fortunate under similar circumstances. Williamson has described how the Sand-eel may bore its way through the stomach of a Cod by means of its sharp snout, only to perish in the surrounding tissues. It is evident that Diodon has some advantage over other species and makes use of it.

Diodon belongs to a curious group of fishes, the Plectognaths, which have acquired various compensations for deformed bodies, not least, perhaps, a high degree of intelligence apparently above that of their neighbours and enemies. Several possess a large and powerful spine in the dorsal fin, whose mode of working is worthy of mention. In Balistes, for example, the first spine is very large and rests on the second, which, again, is tied to the third. The peculiarity of this arrangement, which is repeated in various ways in other fishes (Sticklebacks), is that when the first spine is erected, it becomes locked at the base and cannot be depressed by any external agency, short of breaking the apparatus, until one pulls on the small ray behind. On account of this, the sailors have called it the "Trigger-fish."

A clear interpretation of this phenomenon has still to come. We know that the erect spines of the Stickleback sometimes stick in the throat of its captors and cause their death, but at times the Stickleback also loses its life, and such fishes as the Perch do not pay much attention to the spines. These are certainly nasty things to deal with. The natives of India, when they catch the Climbing Perch (Anabas), are accustomed to kill it by putting it into their mouth and biting through the spinal column, just as dogs kill a rat. But sometimes the fish in its struggles penetrates into the throat and sticks there. The only way of dealing with such a case, according to Day, is to cut off the projecting tail-end of the fish and let the remainder rot until it disintegrates and can be taken out piece by piece.

Spines are put to various other peculiar uses and in so far might be called adaptations. Yet we cannot go to the length of thinking that they are necessary to the existence of any particular species. They may or may not have saved many individuals in an extremity, but that the species with spines have persisted through such accidents—the natural selection point of view—seems too slender an hypothesis to suit the facts.

Gudger (1905) has given an excellent example of nature's disregard of theories. As mentioned above, the Sting-ray is provided with a powerful spine in its tail, which may well be called an adaptation, useful to and used by the fish. Yet even this terrible weapon cannot save it from an enemy that ignores spines. In one specimen of a Hammer-headed Shark (Zygæna) no fewer than fifty broken spines—representing about so many individuals—were found embedded in the muscles and sticking into the throat (Ehrenbaum, 1915). This Shark, at any rate, was not killed by something far more formidable than a Diodon.

It is of interest to note how the development of one adaptation leads to others. The vertebral column of the Mackerels is an adaptation to rapid swimming, and in trying to attain this end some of the family have their mouths and tails transformed in various ways. The Sword-fish and the Trichiurids are examples which have already been mentioned. A still more striking example is provided by the Sail-fishes (Histiophorus) with their high dorsal fin, higher than the depth of the body. It is clear that such a fin is out of place in a rapid swimmer, yet it has probably arisen just from the firm and solid locking of the vertebræ. Such a fin would readily be torn in pieces, if the vertebræ were flexible. Having developed, it is probably of use in preventing the fish from rolling when swimming at a moderate pace, but we may doubt whether it is of any use as a "sail."

The peculiar use of the pectoral fins in the Flying-fishes (Exocœtus and Dactylopterus) is an example of a similar kind. The fins were not made to fly; they were an old-established organ put to a new purpose. The mode of acquiring the new organ is clearly illustrated by the Flying Gurnards. The ordinary Gurnards have broad pectorals, of moderate but not excessive length. The *young* Flying Gurnards have similar fins, but as they grow older the upper part, which is used as a wing, grows longer until it becomes as long as the body (Fig. 64). In the course of time, we may believe, the aptitude will arise at earlier periods until even the young have long pectorals, like Exocœtus. How long it has taken for such a character to develop can hardly be told; perhaps a thousand years or more, for only the aptitude, and not the acquirement, of one generation is handed on to the next. Imitation and example are probably of more importance than the structural inheritance.

Again, when the Rays took to expanding laterally it was not part of this adaptation to a bottom-life that the arms or wings should become attached to the sides of the head. In the Monk-fish (Rhina) they are quite separate,

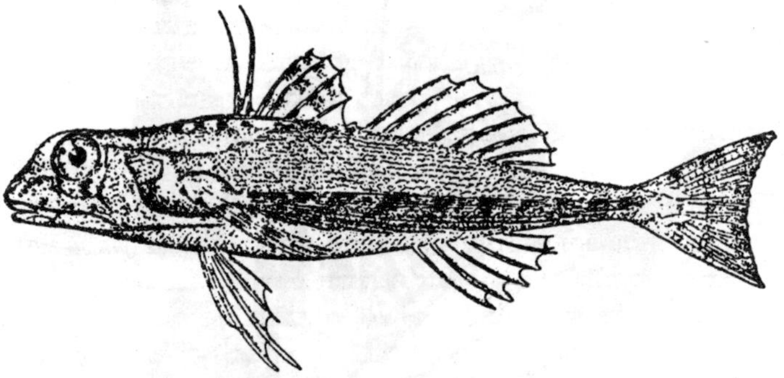

FIG. 64.—The Flying Gurnard (*Dactylopterus volitans*) does not have such long pectoral fins in its youth. It has developed them by vigorous efforts at flying. (From Gill.)

and this fish has a much narrower body than a Ray. When this regulative change was made, however, it led to a wide expansion of the wings, which in the Vampyre Ray may reach a width of twenty feet or more; with these wings they envelop their prey before devouring it.

In a thrilling story of adventures off the coast of Florida Russell J. Coles describes the immense strength and size of these Devil-fishes, or Vampyre Rays as they are variously called. One has been known to tow a hundred-ton vessel far out to sea. As Fig. 65 shows, they are very dangerous at close quarters; a blow from their powerful wings can easily smash a boat. In the fight depicted the fish happened to come up under the boat, and the issue was very doubtful for more than twenty minutes.

On the other hand, the loss of fins can in no case be regarded as an adaptation to anything and just as little as the result of disuse. If it had ventral fins, the Eel could make as good use of them as other fishes which wander out on land, and we cannot believe that it has lost them through crawling through holes and narrow pipes. The explanation

FIG. 65.—The terrible Devil-fish (*Manta birostris*) in a combat described on p. 261. The fish, boat, and men are drawn to scale. (From the *American Museum Journal*.)

of the loss has to be sought for in a different direction. Stromateoides, for example, has long ventrals in the beginning, but they are thrown off as the fish grows. In other fishes the pectoral fins are discarded, and this comes from some convulsion of the whole body and not from any shortcoming on the part of the fins. It may be concluded, therefore, that the movements and correlation of the different parts during growth are responsible for the loss of structures, not the mode of life. It would be difficult in any case to connect the life of the adult Eels in fresh water with the non-development of ventrals in the Leptocephalus out in the Atlantic.

The structures of the head arise under the influence of three main factors : the movements of the body, the working of the mouth, and the pressure of the surrounding medium. In some cases we can see how the nature of the movements

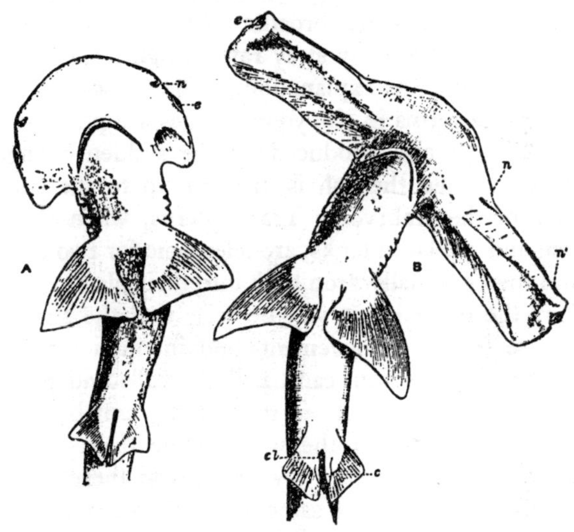

FIG. 66.—Ventral view of the head and trunk (A) of a young Bonnet Shark (*Sphyrna tiburo*), and (B) of a young male Hammer-head (*S. zygaena*). *c.*, Clasper ; *cl.*, cloacal aperture ; *e.*, eye ; *n.*, nostril ; *n'.*, nasal groove. (From Bridge.)

affects the shape. For example, the head is broadened out from side to side and likewise the mouth in the Gadoids and many other Teleosts. But the most interesting examples are furnished by the Sharks and Rays. The Sphyrnidæ have the front part of the head, including the eyes and nostrils, pressed out on each side into varying shapes, of which the best known perhaps is the Hammer-head (Fig. 66). As the young do not have these peculiarities to the same extent, it would appear that the fish, by a sort of centrifugal action, go on shaking out the comparatively soft materials of the head more and more to the sides. And most figures published of these forms seem to have caught the fish in the act.

In the deep-sea Teleosts, on the other hand, the head is usually compressed from side to side and the eyes are also compressed and large. The latter phenomenon has often been called an adaptation to the darkness of the deep sea, it being assumed that the broader surface would somehow catch and concentrate any stray rays of light, but the great pressure is a sufficient explanation. The telescopic eyes of many deep-sea forms in the young stages are possibly to a great extent artificial, produced by the sudden change to less pressure when the fish is brought up to the surface. The eyes of the embryonic Trachypterus, when the eggs are taken and placed in tanks, are telescopic for two or three days and then gradually drop back into their places.

From the foregoing it will be seen that the structures which arise from the movements and interaction with the external physical medium can hardly be called adaptations, unless they are further turned to good account by the fish itself. A stone sinks in the water while the cork floats, yet we do not say that they are adapted to these purposes until we use them. In the same way it does not seem right to say that pelagic eggs are adapted to float and demersal eggs to remain at the bottom, when the significant meaning is, that the density of the substances concerned is different. Further, to say that the Herring is adapted to a pelagic life, the Plaice to a demersal, and the Carp to fresh water, adds nothing to the understanding of the facts. When we have learnt the causes of the differences, we do not require to use the word "adaptation."

The chief difficulty about the word is, however, that in the minds of many it implies some power or combination of circumstances, external to the organism, which has made or "adapted" it to suit the conditions. If it be recognized that this view has no prerogative to the use of the word and that adaptations may arise in a different way, from the power of the organism to use and adapt its structures according to the conditions, a great deal of the difficulty disappears. Some of the following examples will perhaps make the distinction more evident.

2. Adaptations connected with the Mode of Life

The mouth of fishes is a grasping organ, the only organ of the kind they possess, and it is used for several different purposes and in several different ways for the same purpose. It has to obtain food, though all is not food that enters the mouth, and it has to supply water and oxygen to the gills. But its power as a grasping organ depends upon the rate of swimming and the kind of animals pursued. Fast swimmers have a large gape, sharp teeth, and a wide gill-opening; slow swimmers have all sorts of mouths, even very large ones, but broad and not pointed. Most fishes are carnivorous, and in the predatory kind the teeth are not only pointed, they are very numerous and on other bones as well as the jaws, vomer, palatine, tongue, etc., as in the Salmon and Tarpon.

It is easy to imagine that the mouth and teeth are adapted to particular purposes, simply because they are used in this or that manner, but a general comparison of many cases often reduces the value of the adaptation. For example, it has been thought that the loose front teeth of the Angler, which fold into the mouth and are pulled up when the prey tries to escape, simply by the action of the latter, is an adaptation of special merit. But if the fixed teeth of a Hake or a Conger get hold of anything, there is not much chance of the prey escaping. To make the loose teeth an adaptation, we should have to imagine that the prey on entering the mouth had first of all loosened the teeth, for that is the peculiarity that requires explanation. We might in reality regard this as a fault or due to some fault in the development and then recognize that the fish does very well even with this faulty construction.

In general, indeed, the jaws and teeth are in a somewhat elementary condition among the fishes, that is, too much under the influence of other forces than the purposes they serve to be regarded as adaptations. But extreme cases may well be called so, for example, the suctorial apparatus of the parasitic Hag-fishes and Lamprey. Myxine has only a single tooth in the roof of its mouth ; it is not really a tooth in the same sense as the teeth of the Shark or bony fish, being only a horny cone hardened by much use, yet it is very efficient. Getting this tooth into its victim, preferably one already caught on lines or in a net and unable to escape, Myxine bores its way into the tissues by means of its rasping tongue and leaves very little for the fisherman. Regarding the destructiveness of the American Borer, Jordan and Evermann write, " Large fishes of even 30 pounds weight are often brought up without flesh and without viscera, and they certainly do not swim into a gill-net in this condition." Instead of a tooth the Lampreys have a large funnel with a sucking disc at the end, which serves the same purpose. And they do not readily let go ; the marine Lamprey has been carried by the Salmon long distances into fresh water. One American Lamprey seems to have acquired such a mastery of this apparatus that it plays with it. Bashford Dean has described how the males and females take part in a game of lifting stones, piling them up in one corner, removing them, and stealing them from one another, whilst endeavouring to make a nest. It is quite possible that this playing with a structure or organ the fish may have acquired, has had a good deal to do with the development of adaptations.

We can readily imagine this to have been the case with the tactile barbels or tentacles about the mouth seen in many fishes, especially those which have few or small teeth, like the Red Mullets, Sturgeon, and some of the Siluroids. Among the Gadoids we can follow its increase in size from a small knob in the Haddock (it is quite absent in the Pollack and Green Cod) to a little longer barbel in the Cod, still longer in the Lings, and then to a varying number up to five in the Rocklings. The Lings and Rocklings have rather smaller eyes than the others, like the

Siluroids, so that the barbels come to be very useful in finding the food. In the case of the Siluroids, as observed in aquaria, the fish seems quite dependent on the barbels or other taste organs of the body for the detection of its food. In the Rockling the vibratile first dorsal creates water currents along the dorsal fin groove, which is lined with taste buds and is a food-detecting organ (Thomson, 1912).

The long, spade-like snout of the Spoon-bill (Polyodon) is supposed to be of use in stirring up the mud when the fish is searching for its food. But the species that lives in the Chinese rivers is said not to use its snout in this way; in fact, the fish seems to be greatly inconvenienced by it. And we know, of course, that the Carps can stir up the mud without a snout.

One of the most peculiar adaptations connected with the head region is the enormous sucking-disc acquired by Echeneis and Remora, by means of which they attach themselves to boats, turtles, and more especially Sharks. The disc is supposed to be a modified first dorsal fin, but it may have arisen quite independently. In addition to the power of suction it gets a good hold by means of twenty transverse ridges and the small sharp spines with which these are provided. The spines point backwards, so that the object attacked, when moving forward, becomes more firmly fixed. And such is the character of the fish, that it will be torn in pieces rather than let go, and sometimes leaves its sucker, when roughly handled, on the object to which it was attached. Such a small fish "weighing but one and a half pounds has lifted a sixty-three pound turtle" (J. M. Mellen). How it fixes itself to Dog-fishes will be

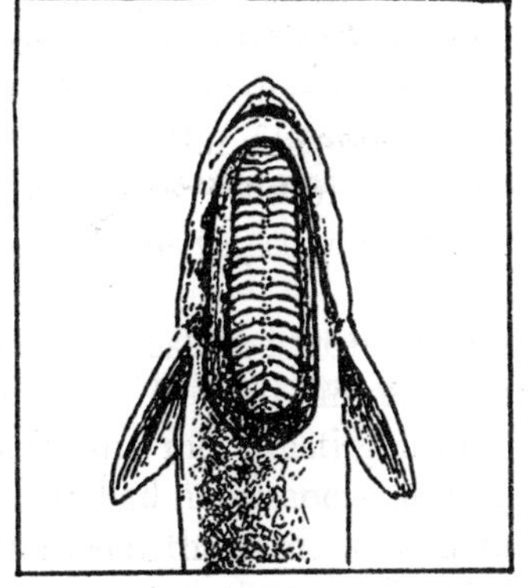

FIG. 67.—Sucker of Echeneis. (From New York Zoological Society Bulletin.)

seen from the excellent photograph by E. R. Sanborn, reproduced in Pl. XIII. As it is a strong swimmer and feeds chiefly on fishes, the reason why it should hold on to anything, especially by the head, is a mystery. It is a relative of the Pilot-fish, which also has a partiality for Sharks, and it may possibly have learnt the habit from playing a deadly game of hide-and-seek with the monster.

The Angler (Lophius) has learnt another variation of this game. Deformed by circumstances in its early days, with a terrier-like stump of a tail (Fig. 4), belly of elastic dimensions and very broad head and mouth, it makes quite a good use of its infirmities. Some parts of its flesh and bone have been pressed upwards above its head; on two of these, one on each side, are the smelling organs, directed upwards, and between these a long flap has developed into a ray-like tentacle with a greenish piece of skin at the end, which may be "phosphorescent"; close by are the watchful, squinting eyes. Further back it has a few more rays or tentacles, branched and looking like sea-plants or zoophytes. With this queer shape it has naturally a sluggish disposition and rests for the most part on the bottom with its mouth wide open and the green leaf just above. This is the lure and the bait is probably the tongue within, or it may be idle curiosity that leads other fishes to enter this well-conceived trap.

The Angler is not particular in its choice of victims; Herrings and Mackerel by the score, many other kinds of fish, Plaice, Rays, Gurnards, etc., and even diving birds have been found in its stomach. But it is also able to move on occasion; it has been seen to capture birds on the surface and has often been taken on attempting to swallow Cod and Conger when these were being hauled up on the long-lines. Its fishing habits seem so deliberate that we must credit it with a considerable amount of intelligence.

Some deep-sea fishes of the same group (Pediculati) have gone a stage further with the lure, as it seems to be. In Ceratias the leaf is changed into a luminous organ, like a pocket-lamp, which is supposed to play on the susceptibilities of the other creatures of the deep. Many other fishes, for example, the Blennies and Cottids, have loose flaps of skin above their heads, but they are not known to use them as a lure. Possibly they are rendered less conspicuous among the seaweed where they live, so that the unsuspecting prey approaches them and are then captured by a short but quick effort.

Noteworthy examples of the adaptation of various structures to one and the same end are given by the fishes which attach themselves to solid objects for some reason or another. In the rapid streams of the Himalayas the Anabantoids use their movable spines to anchor themselves to stones, but the Loricariidæ of Central and South America, which have no spines, use their mouth as a sucker and keep such firm hold they cannot be pulled off without difficulty. It is by this means that they manage to reach up to great heights, over 11,000 feet, among the Andes. Still more remarkable is the behaviour of the Globe-fish (Tetrodon) of East India. It is a voracious and predatory fish in spite of its diminutive size (about 3 inches), and is able to inflate its body like Diodon. To avoid being carried away by the stream, it creeps through narrow openings between stones, or behind empty shells of snails which it has piled up, and swells out its body like a balloon. It is then so jammed in between the stones that it remains just where it wants to without effort (Thilo).

Another form of attachment is the adhesive disc on the belly of the fish, and this would seem to be the most natural apparatus for the purpose of stemming a current. And if structures arose and developed on account of their utility we should expect to find it in the freshwater forms, but it is only found in marine forms. The nearest approach to the condition in fresh water is Gastromyzon of North Borneo in which the whole ventral surface of the belly, with the pectorals at the sides in front and the ventrals behind, forms a large sucker and is used by the fish to adhere to stones in the mountain torrents. Another freshwater form, Pseudecheneis, lives in the rapids of the Himalayas and is provided with a transversely plaited disc between the pectoral fins. But apart from these examples, freshwater fishes manage to keep their places without sucking-discs.

The origin of these discs must be referred, therefore, simply to the habit of resting on the bottom, and the necessary condition for their development appears to be the suitable position of either the ventral or pectoral fins. The sucker is only developed on the anterior half of the belly, that is, when the ventral fins are anterior. The absence of suckers in the freshwater fishes is thus accounted for by the fact that in most of them the ventrals are posterior.

In many marine forms, for example the Anglers (Pediculati) and the Rocklings, the belly is expanded laterally, and soft and pulpy as if the fish literally clung to the bottom. But a more definite sucker is formed by the ventrals in the Cyclopteridæ (Lumpsuckers) and the Gobies, some of which enter fresh water. In these the fins meet in the middle line and a fold of skin behind completes the cup. In the case of the Gobies which live in deeper water with little or no current, the sucker is but feebly developed and is perhaps more a tactile organ than a sucker.

A still greater development of this apparatus is seen in the Cling-fishes (Gobiesocidæ). Here the ventrals are still further forward and more to the side; between them lie the expanded ends of the clavicles and behind are the large post-clavicles. The skin covering these bones is pulpy and glandular and thus in itself adhesive. When the fish settles down, however, the suction is increased by a slight withdrawal of the central part of the disc, and it is not easy to move such a fish. A Lumpsucker has been placed in a bucket of sea water and allowed to fix itself on the bottom, then the bucket, water and fish have been lifted simply by pulling up the fish, so firmly did it cling.

The pectoral and ventral rays in some fishes are also used just like the barbels on the mouth and throat as tactile organs, most probably in the search of food. It is probable that this habit arose in some cases from using the rays as supports or stilts. In the Gurnards the three lower rays of the pectorals are used in this way, and Williamson and others have shown that they are richly supplied with nerves and end-buds, which indicates a sensory function. The Blennies use their ventrals in a similar manner. The

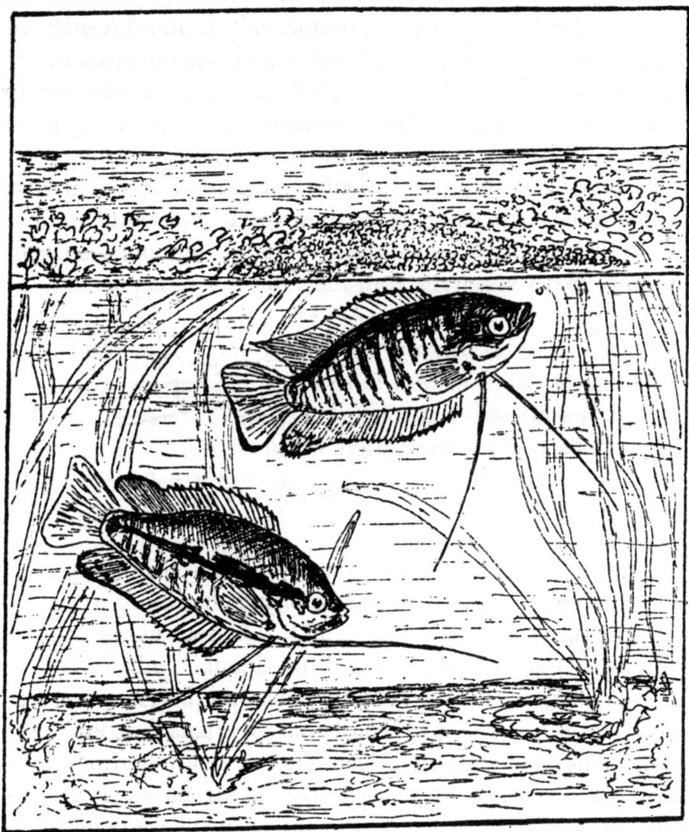

FIG. 68.—The thick-lipped Gourami (*Trichogaster labiosus*). The male above building the nest.

Gouramies (Trichogaster) are, however, the great masters in the art of using the fins. Here the ventral is as long as the fish itself, and it is one of the quaintest sights in an aquarium to watch the little fish send out the fins forwards in front of the head, or downwards or backwards, either both together or one forwards, the other backwards, as if seeking for something—and all with the greatest serenity, remaining quite still the while, as if it were the easiest thing in the world (Fig. 68). We have but to practise with our arms to discover that this art takes some learning.

Among the cartilaginous fishes (Elasmobranchs) the ventral fins, or rather their base, have been modified in the males to serve as claspers. During sexual congress these are inserted through the cloaca of the female into the oviducts and the seminal fluid flows along them into the upper parts of the latter, where impregnation of the eggs takes place. In some of the bony fishes or Teleosts it is the anal fin that is so modified. This is the case especially among the viviparous fishes of the Cyprinodont family, where in some forms the anal fin serves no other purpose but as a kind of intromittent organ (Fig. 73). The great majority of fishes do not have these adaptations, so that it is clearly impossible to regard them as necessary or indispensable for the preservation of any species. Since they only occur in the males their origin and development have to be associated with the nature of the latter. In other cases, especially among the Cottoids, Lumpsuckers, Cling-fishes, Gobies, and other fishes that live amongst the seaweed in shallow water, the urinogenital papilla swells out into a large organ at the breeding season and is probably used as an intromittent organ in the viviparous fishes (some Blennies, Embiotocids, etc.). But in the non-viviparous fishes, where it is equally well-developed, the eggs are fertilised after extrusion. Here as with other structures the origin and development of the organ can clearly be distinguished from its later use.

In the sedentary forms, or those that do not move about to any great extent, various other structures of interest are developed at the breeding season. In some of the Pipe-fishes (Syngnathidæ) lateral folds of skin expand outwards under the belly to meet and form a closed pouch, in which the males receive the eggs and hatch out the young. According to Duncker the eggs become attached to the skin of the male and by this means the embryo receives nourishment as through a placenta. In Aspredo of the Guianas it is the females that undergo this change and nourish the eggs.

The origin of the skin fold in the Pipe-fishes may have come about in the following manner. These fishes have narrow bodies and have been derived from others with deeper bodies (Clupeids) and longer ribs. Here the ribs are short or absent and the superfluous skin remains as a lateral fold. We find it again in other fishes, such as the Sand-eel (Ammodytes), which apparently make no use of it.

The male Sea Scorpion (Cottus), when it takes charge of the eggs, sitting on them like a hen, develops hollows and ridges along the inner margins of the pectoral rays. In this way it is able to bring the eggs together and keep them in a compact mass. The phenomenon is none the less wonderful because we ascribe it to the regulative law of action and reaction. What is perhaps still more wonderful is that these hollows and ridges now develop with the other secondary changes in the male before the eggs are laid—thus in anticipation of their use. We can understand this phenomenon from the view that the hormones stimulate these and other parts of the body to develop as the sexual organs ripen.

3. Adaptations connected with the Respiration

The necessary elements of a respiratory organ consist simply of a separating membrane between blood on the one side and a medium containing oxygen on the other. It is not essential that the organ should take the form of gills. Hence we find that in fishes almost any part may on occasion serve the purpose. For example, Ryder (1893) has shown that the fins of the young fry of the Embiotocids are richly supplied with blood capillaries and clearly serve as respiratory organs. In many fishes (Loaches, Loricariidæ and others) parts of the alimentary canal are used for the purpose. Hickson (1889) has described how the Walking Goby (Periophthalmus) often sits on land with its tail only in the water (Plate XIII, p. 296). The tail is serving as a breathing organ. All breathing membranes, whether gills or lungs, require to be moistened; given the moisture in sufficient quantity many fishes can live out of the water for quite a long time. Whether the skin of any fish, like that of the frogs, can serve as a respiratory organ, we do not know with certainty, but it is quite possible that this is the case in the Lung-fishes and Eel-like fishes (Amphipnous) which can live away from the water.

In addition to the principal breathing apparatus or gills, certain subsidiary and accessory parts are frequently developed. If a fish like the Wrasse or Cod is watched in an aquarium, particularly when at rest, it will be observed that just inside the mouth two flaps or folds of skin are automatically beating backwards and forwards. These are the so-called breathing-valves, withdrawing as the oral cavity widens and water is drawn in, returning when the cavity contracts and the gill-arches and cover open to let the water pass behind through the clefts and over the gills. These folds of skin have probably arisen through the constant regurgitation of the water, and there is no doubt that they must save the fish a great deal in muscular energy. The Wrasse breathes only at the rate of fifteen respirations in the minute, but the Stickleback needs almost ten times as many.

Many fishes, especially the freshwater fishes, on account of their organization and higher metabolism, have greater need for oxygen than others. When the Minnows come up to the surface, it is not only to catch flies; there is more oxygen in the surface water, and if they are kept down below, by a net or other apparatus, they perish. The Dipnoi, though provided with gills, must also have free access to the air. Other fishes come regularly to the surface and take in gulps of air, and from this habit several important organs have developed.

In the simplest cases (some Clupeids and the Ophiocephalidæ) the air settles in a cavity or fold above the ordinary gills and the lining of the cavity becomes highly vascular or a part of a gill projects into it. In others (Anabas (Fig. 69), Osphromenidæ, and some Siluroids) a bony support is developed in the cavity from one or more of the branchial arches, and the vascular membrane lining the support receives the ordinary venous blood of the gills and, after cleansing it, returns it to the dorsal aorta. With this extra supply of clean blood to the head it is not surprising that these fishes are remarkable for their intelligence.

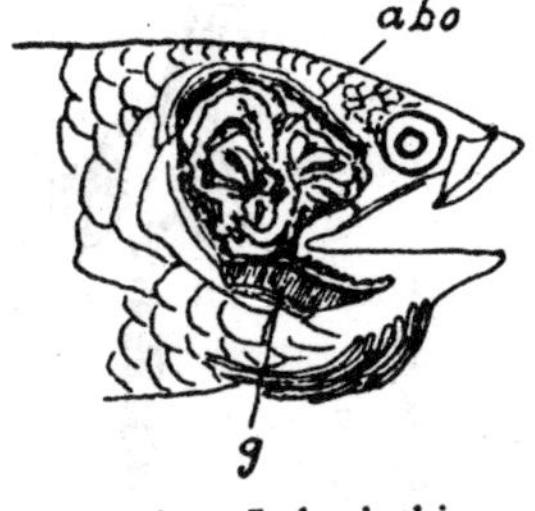

FIG. 69.—Labyrinthine or accessory breathing organ of Anabas. (From C. et V.)

In other cases the air passes into lung-like pockets or sacs, which may even extend backwards to near the tail (Saccobranchus). In the amphibious Cuchia of India (Amphipnous), there is a distensible air-sac on each side behind the head, communicating with the gill-cavity, and the fish fills it with air derived immediately from the atmosphere.

The so-called swim or air-bladder has almost certainly arisen in the same way. Even at the present time it is noticeable that the larval fishes, which from some cause or another are prevented from coming up to the surface (Elasmobranchs and Blennies), do not attempt to form an air-bladder, whereas those which are hatched out at the surface almost always have it. The Herring which hatches out at the bottom has the beginnings of an air-bladder before it manages to reach up to the surface, but the principal development takes place at the surface. It is quite possible, therefore, that the modern fishes still take in a gulp of air, as it were, to start the air-bladder.

However it may have arisen, the organ possesses several important functions, and these, again, have reacted on the structure. It lies in the abdominal cavity below the vertebral column and thus acts as a balancing and hydrostatic organ as well as a reservoir for gases. These three functions are quite distinct and have clearly come into conflict at various times with one another. Seeing that many fishes can balance themselves and rise and descend in the water without it, we may take the air-bladder as another example of the organs and structures which have arisen without any reference to their utility; in fact, we may doubt if it has been a useful organ in these two respects. Its failure as a balancing organ has been exemplified by the Flat-fishes and Pipe-fishes. As a hydrostatic organ, it compels the fish to keep to a certain level of pressure or bathymetric range, even though an open duct may be present. This duct or communication from the alimentary canal to the air-bladder allows air to enter, but not to leave apparently. Owing to this difficulty probably, many of the Clupeids have made a new opening or "safety-valve" to the exterior, and it is to the latter that they owe their great versatility. Even though provided with a duct the Pollan (Coregonus) of the Swiss lakes, when brought up from deep water, have the air-bladder swollen to such an extent that they are unable to move, and soon perish unless the air-bladder is pricked to relieve the pressure (Semper).

The same phenomenon is more frequently seen in marine fishes, the majority of which have no open duct. A curious thing is that the gaseous contents of the air-bladder (oxygen and nitrogen with a small proportion of carbonic acid) vary in the opposite direction to what one would expect. The proportion of oxygen is least in fresh-water fishes, greatest in deep-sea fishes. The latter have 80 per cent. more oxygen in their air-bladder. At a depth of 1500 metres this means a partial pressure of 90 atmospheres, whereas in the surrounding water the partial pressure of oxygen is only about one-fifth of an atmosphere. The activity of the cells must produce this difference and it is under nervous control. If the air-bladder is emptied, it soon fills again with gas, mostly oxygen (Pütter, 1911).

This would indicate that the fish is able to exercise some selection of the gases, and the "red glands" or retia mirabilia of the Physoclists are evidently used for this purpose. They consist of a dense complex of glandular cells surrounding the capillaries or small blood vessels in the lining membrane of the sac, and enable the fish to increase or decrease the amount of the gases in the reservoir, but the reabsorption is so very slow that it does not permit of rapid changes of level. Experiments have shown also that a fish may be asphyxiated, though the air-bladder still contains a considerable amount of oxygen.

In some cases, particularly among the Siluroids and Sciænoids—freshwater and marine fishes respectively, the

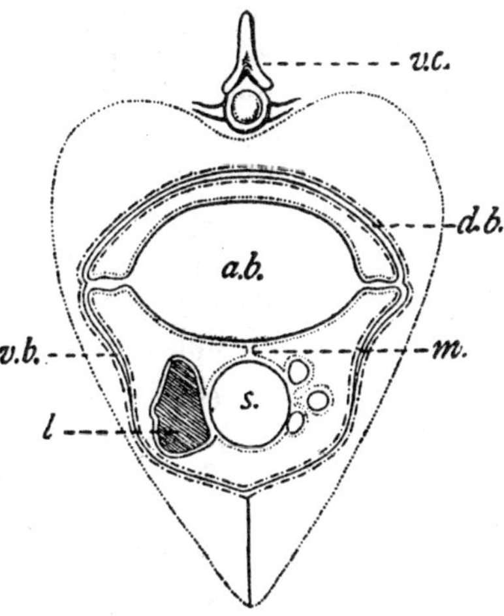

FIG. 70.—Air-bladder of the Sciænoid Collichthys. (From Günther.) *a.b*, *d.b.*, *v.b.*, air-bladder with dorsal and ventral extensions; *v.c.*, vertebral column; *l.*, liver; *s.*, stomach; *m.*, mesentery.

former physostome, the latter physoclist—the amount of gas taken or secreted into the reservoir is so great that side passages or galleries, as it were, are developed in all directions, between the muscles of the body and even round the abdominal cavity (Fig. 70). In many ways these are just as remarkable as the large air-sacs of birds. What use they may be to the fishes we can only conjecture.

Through its tendency to encroach upon other parts of the body, the air-bladder has developed various communications to the exterior. The most frequent of these is found just above the base of the pectoral fin, behind the shoulder girdle. Here a superficial, bony plate serves as a drum, apparently, and conveys differences in the external pressure to the air-bladder. At any rate, fishes like the Loaches, which are provided with this apparatus, are very sensitive to changes in the atmospheric pressure and have been called "weather-fishes." But the most remarkable of the air-bladder's activities is that which has brought it into communication with the statocyst or "ear."

In many fishes, like the Gadoids, the anterior end of the bladder is continued by two solid cords which reach forward to the head and are fixed to the prootic, just outside the labyrinth of the ear. Originally these were probably filled with fluid, and even though solid they are not merely attaching structures but probably convey some impression of the condition of the bladder forward to the prootic, which may serve as a drum. In the Clupeids these anterior prolongations penetrate into the semicircular canals and expand into bullæ, with the lining membrane in contact with the endolymph of the canals. Thus, differences of pressure in the air-bladder are conveyed to the endolymph and thus to the nerve ampullæ within. It was formerly thought that these prolongations were direct continuations of the air-bladder and filled with gas, but Thilo has shown recently that they contain a liquid, probably serous fluid, and are shut off from the air-bladder by a membrane. In this way they serve as a manometer or barometer.

A further stage of this remarkable apparatus is found in the Ostariophysi (Carps, Siluroids, etc.). The growth of the first two or three ribs in these forms has been interfered with in some way, and the air-bladder, by means of the rudiments and perhaps also of the transverse processes of the corresponding vertebræ, has obtained a most elaborate mechanism connecting it with the statocyst. The accompanying figures will show how the apparatus works. When the air-bladder is loose and contracted, as in the first figure, the ossicles are not in contact with the statocyst; but when the air-bladder is distended, the apparatus becomes taut and the front ossicle (*c*) closes like a cap on a fenestra in the extended wall of the endolymph chamber which is part of the statocyst. In this way changes of pressure or vibrations within the air-bladder are communicated directly to the sense-organ. This apparatus is only developed in certain freshwater fishes, the most numerous and "dominant families of freshwater Teleosts at the present day" (Bridge).

This wonderful mechanism resembles a good deal the arrangement of the ear ossicles in higher animals and serves the same purpose, the conveyance of pressure waves

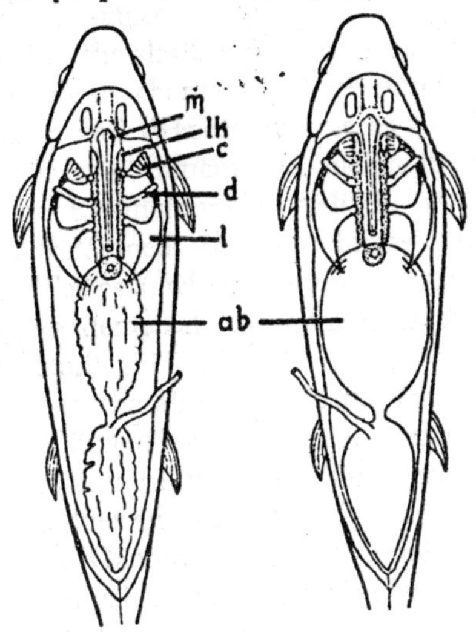

FIG. 71.—Small bones connecting the air-bladder with the "ear" in the Carp. (From Thilo.) *ab.*, air-bladder; *l.*, lever; *d.*, director; *c.*, cap; *lk.*, lock; *m.*, medulla.

to the central organ. The ear ossicles have probably arisen in the same way, but from a different set of bones. From what forces of pull and stress they have developed we do not know, but they are in some way connected with the superior pharyngeal retractor muscles.

The "lungs" of certain fishes have probably no connection with the air-bladder, though they may have arisen from the same cause. They are found in those forms which have the sac beginning near the mouth or pharynx and have a short but wide open duct. Such air-sacs are found more particularly among the Dipnoi or double-breathing animals, and as they are highly vascular, with numbers of small pockets or sacculi, they approach very closely to the lungs of Higher Vertebrates. In the dry season of the year the South American Lepidosiren and the African Protopterus bury themselves in the mud and can remain there for nearly half a year. Even in the water they come up to the surface from time to time, like porpoises, to empty their lungs and take in a fresh supply of air. The Queensland Neoceratodus does the same thing, but does not seem to bury itself at any time. The adaptation, therefore, has arisen in the water, like the air-sacs of other fishes, or the accessory breathing organs of the Anabantoids. The two species of Dipnoi mentioned, like the Higher Vertebrates, have simply made a more special use of an organ or adaptation that had already been acquired.

It should be noted further, that a sac of some kind or other has been found by Starks (1911) posteriorly in the abdominal cavity of an Ophiocephalid. As in the rest of this family the air-bladder is exceptionally large, but this extra sac, though closely apposed to it, has no communication with the air-bladder and opens to the exterior through the genital pore. A similar condition of things seems to be present in other fishes also, and it is possible that this may be a regulating mechanism for the entrance of water, not of air. In the Herring, at any rate, the air-bladder has frequently been found with a considerable quantity of water in it, which could only have entered by way of the abdominal opening.

Whilst the air taken in by the fish has led to the development of air-sacs and lungs, the water to the development of breathing-valves, the food and solid objects entering by the mouth have led to other adaptations. The gill-filaments are arranged along the posterior or hind margins of the arches (in Teleosts); along the anterior margins we find a parallel series of rough knobs or slender fibres, the so-called gill-rakers. They do not rake the gills, as a matter of fact, but they are certainly of great importance to the latter. We can see from various things that they have arisen from the irritation caused by foreign bodies on the delicate lining membrane of the arches and they increase in number with the growth. Where the fish lives mostly on other fish or the hard-shelled Molluscs, the rakers are solid and hard, like tubercles, as in the Turbot (Fig. 72, 1), but where it feeds mainly on plankton—the floating or swimming organisms of a smaller kind—the rakers are long and bristle-like, as in the Herring or Shad (Fig. 72, 2). In both cases the rakers may be branched and so densely packed on the gill-arches as to form a perfect sieve or filter. The most perfect example of a sieve is found in the Basking Shark, a large fish that lives near the surface and feeds mainly on the small fishes and marine Invertebrates.

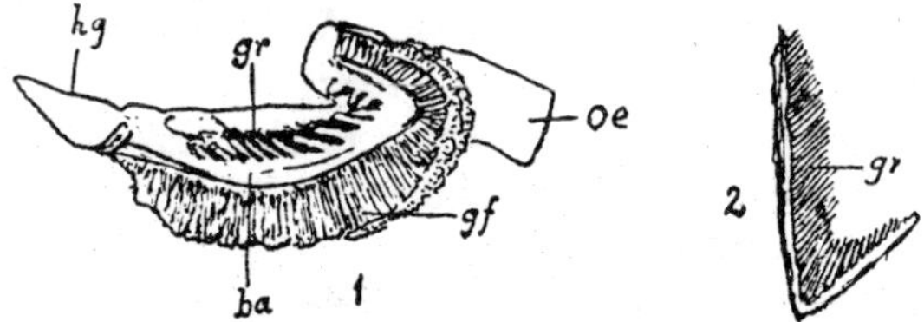

FIG. 72.—Gills and gill-rakers. (From Zander.)
gr., gill-rakers; *gf.*, gill-filaments; *ba.*, branchial arch; *hg.*, glossohyal (tongue); *oe.*, œsophagus.

It would be wrong, however, to conclude that the mode of life of the *adult* fishes has called forth this remarkable apparatus. It is the last of the permanent structures to be developed in the Teleosts, and its development is connected with the conditions in the early stages, during the first few months of the larval life. Several fishes, including the Eel, have no gill-rakers. One may infer from this that the young Leptocephali, during their three years' sojourn in the North Atlantic, do not encounter irritating organisms or do not permit them to enter the mouth. Hence the so-called grasping teeth, which are conspicuous objects in many deep-sea forms, may serve as a filtering apparatus. These are " milk teeth," formed on the edge of the jaws, and are probably not altogether useful. Other deep-sea fishes, Sternoptyx, Macrurids, and the like, do not have them and develop gill-rakers. It may be also that the great difference in the size and development of the Leptocephali comes in part from the difference in the amount of food they can get into their mouths.

The principal function of the gill-rakers is to protect the delicate gill-filaments—as one might conclude from the law of action and reaction. It is only secondarily that they are connected with the food. The " selection " of the latter depends on the senses, and a fish, we may be sure, does not trouble to consult the gill-rakers when it is hungry. The Herring has a very good " filter," which keeps the irritating Copepods from the gills, but it feeds very largely on young fishes, especially Sand-eels (Hardy). Forms that live in clear water have a comparatively small apparatus, those that grub in the mud have a fine sieve. And it has been pointed out, especially by Schiemenz (1905), that fishes take to different kinds of food according to the season and their needs, irrespective of the nature of the apparatus. The Carps may be called omnivorous, feeding on insects, shell-fish and even plants, yet they have such an apparatus, derived from their habit of stirring up the mud.

In the Teleosts the last branchial arch, where the oral cavity narrows to the gullet, is usually modified to serve as a grinding or tearing apparatus. As fishes have no salivary glands we can hardly say that they masticate their food, and the teeth which develop on the last branchial arch are rather to be regarded as specialised gill-rakers. They have probably arisen in the same way and become specialised in various directions according to the nature of the food and solid particles taken into the mouth. The lower part of the arch (ceratobranchial) is the one that most frequently bears teeth, and in some groups the two halves of each side are fused together to form a large and solid plate (Scombresocids, Labroids, Cichlids). In others the upper parts (pharyngobranchials and epibranchials) may also be fused together (Scaridæ).

This fusion of the pharyngeal bones, like the reduction of the gill-arches in the Pipe-fishes, has no definite relation to the food or mode of life. It has arisen from the general co-ordination of the different parts during growth, and we can hardly say that it is an adaptation necessary to the well-being of the species. The food of the Wrasses, small fish, Molluscs, and shrimps, are taken by other fishes also; but in working the booty into the gullet it is conceivable that the hard shells would lead to a stronger development of the teeth. The Carps have no teeth on the jaws and only a few on the lower pharyngeals, none at all on the upper pharyngeals, which form a fused pad against which the lower teeth work. Yet they can take any kind of food, though they prefer the soft bodies of the insect larvæ.

The Grey Mullets also have no front teeth, but the skin of the tongue (glossohyal) and palate is covered with tubercles and even small teeth. Round the pharyngeal bones are thick, fleshy swellings which obstruct the passage into the gullet. Thus a kind of chewing or rubbing of the food goes on in the mouth before it enters the latter. This is ribbed with filaments and coated with thick mucus, and the pyloric portion of the stomach is provided with strong muscles, which seem to act like the gizzard of a bird. From the nature of the food, mostly Bivalves or Univalves (Mussels and Rissoa) with some Crustacea, it would appear that the whole apparatus is a good example of the fact that fishes are not restricted to one or other particular structure. If they lose their teeth, they can use their tongues, and if the pharynx does not grind or tear their food, they can adapt their stomachs to the purpose.

CHAPTER XII

FISHES AND THE WEB OF LIFE

A SINGLE individual is a rogue or an outcast; the phenomena of life require at least two, and around these are the links which bind them to others of the same kind. Further away, but still pressing upon them, are the relations, friendly or hostile, with an infinite number of other forms of life. Apart from the sex problem, it is mainly a matter of food that brings one form of life into touch with another; perhaps "hunger" would express the relations in either case. Then, too, the pursued seek to escape the attention of their pursuers, and this leads to other curious relationships. How these express themselves in the case of fishes may be gathered from the following.

1. SEX, COURTSHIP AND REPRODUCTION

The question of sex-determination is one of the most difficult problems the biologist has to deal with. In general, we may accept the distinction so clearly drawn by Geddes and Thomson (1899), that the female is a relatively anabolic organism produced under conditions favourable to the storing-up of materials, whereas the male is a relatively katabolic organism produced under conditions that make for a preponderance of waste over repair. The female is physiologically more passive, the male more active. The sexual dimorphism is thus an expression of the conditions occurring during the maturation of the sex-cells and should be treated as a variation, like any other variable character. Males and females, we might say, are the extremes of a physiological condition of balance or two different balancings of the same materials, and the difference also expresses itself in the physical balance or proportions. As a rule, the difference persists throughout life, yet the essential identity of the materials is shown frequently in strange ways, by various forms of hermaphroditism and the change from one sex to another.

This theory of the physiological distinction between the sexes, with the wealth of evidence in its support brought forward by the authors mentioned, has been pushed somewhat into the background by the interest taken in Mendel's laws of heredity and the chromosome theory of sex-determination. Although these have led to fruitful investigations in other branches, they have not been taken up to any extent in the study of fishes. Here the natural conditions are all against such exact investigations, and where artificial cultures have been undertaken, as in Johs. Schmidt's experiments referred to in a previous chapter, the results have been rather against the Mendelian theory.

In addition to the evidence in favour of the physiological theory given in the "Evolution of Sex," some further experiments have been made of late years on the factors determining the sex in frogs. These, of course, are not far removed from the fishes and the conclusions should be even more applicable to the latter.

Normally, on any theory of heredity, there should be an equal proportion at birth of males and females. Yung, in his classical experiments, found that tadpoles left to themselves tended to give such proportions, but if fed with nourishing materials, beef, fish, and frog-flesh, the proportion of females was greatly increased. This result was explained from the basis that tadpoles are really hermaphrodites to begin with or pass through a hermaphrodite stage; and this is possibly true of all animals. But Yung did not pay sufficient attention to the mortality among his tadpoles, for this may have been differential, *e.g.* more males than females.

R. Hertwig (1912) has carried out the converse experiments. By fertilising unripe and over-ripe frogs' eggs he obtained a very large percentage of males; in some cases, indeed, all were males. Consequently, the condition of the sex-cells at the time of fertilisation is of the greatest importance. If the manager of a trout or salmon hatchery keeps the females back until all are ripe at the same time, the chances are that he will obtain too many males in the offspring.

In other experiments Hertwig placed two sets of eggs, from the same parents, under different conditions of cold and warmth. In the former case he obtained about three males to one female, in the latter only a slight excess of males. And he found also that in the frogs developed from the cold cultures transitional stages from female to male sexual organs were unusually abundant; that is to say, individuals which normally would have been females had turned into males.

Other experiments showing the influence of the chemical conditions on sex-determination have been made by Helen King (1911). In faintly acid solutions the percentage of males was increased, whilst in hypertonic solutions (equal percentages of salt and sugar), which reduced the amount of water in the egg, the great majority became females.

These results clearly indicate that sex is a variable character, so far as the little specialised frogs and fishes are concerned. Sex chromosomes may be present, but it is probably the inherited balance rather than the substance that is of chief importance. As with other characters, whatever upsets this balance changes the course of development and alters the structures.

The experience of the practical breeder of aquarium fishes tells in the same direction. In a brief but interesting notice Thumm (1908) narrates how the proportion of the sexes becomes exceedingly variable in many tropical fishes (Cyprinodonts) which have been introduced into Europe. Sometimes the broods contain only males (Jenynsia), at other times almost entirely females (Mollienisia). He points out that the result depends a great deal on the size and vigour of the male parents. On pairing a female of three years old with a young male one year old, only 50 females were obtained in 800 offspring, but in the next brood from the same male, then two years old, he obtained 300 females in 400. On pairing large, late-ripe males with large females (Pœcilia) 76 per cent. of the offspring were females; but middle large females with the same kind of male gave 92 per cent. of females. Small males always gave a larger proportion of males. In general, the percentage of males is greater in the spring, very low in the autumn (viviparous Carplings); which, of course, is also the experience of poultry breeders.

From an examination of well over a hundred thousand spawning fishes of various species (Trout, Salmon, Coregonus and Salvelinus) Surbeck found that the proportions of the sexes varied considerably in the different lakes of Switzerland, from almost equality in the case of the Sea-trout up to nearly 700 males to 100 females in the case of Coregonus (1913, 1914). Following the result of Thumm's experiments, this disproportion is referred in the main to the practice of using the milt of the smaller males with the roe of larger females. But this is admittedly only a partial explanation. In the Finnish lakes Järvi (1920) has found similar disproportions for the Coregonus there, 271 males to 100 females, and from other data it seems the rule that the males of most freshwater species in Europe preponderate over the females, even under natural conditions.

A curious thing is, that the males mature earlier and die younger, in some cases at least. Järvi found only a small proportion of males in their fifth year and thereafter only females, but the numbers were small. In the remarkable little Salmonoid Plecoglossus of Ayu, which normally lives only one year, some manage to survive the second winter in a warm pool, and these are always females (Nomura, 1922).

What the proportions of the sexes are in warmer climates is not so well known. According to a note by E. G. Boulenger (1924) the females of Xiphophorus (a Mexican Cyprinodont) outnumber the males by four to one. If this refers to the natural conditions in Mexico, the temperature experiments of R. Hertwig are recalled to mind. Behind the variability lie the inherited balance and the differences of temperature. It has to be remembered that these European freshwater families came originally from warmer zones, even though some like the Salmonoids have adapted themselves to more northerly climates. The tendency of the males to mature one year earlier, which is a fairly well-established fact for northern species, both freshwater and marine, may have come from this change of climate or be simply an attribute of "maleness," according to the theory of Geddes and Thomson. Among the fish represented on the spawning grounds, therefore, the males have an extra year's group, whilst the oldest year-groups, mainly females, contain fewer individuals. Part of the disproportion in the numbers of the sexes can be accounted for in this way and the disproportion will go on increasing, as in the Swiss lakes, unless some means are taken of securing proper mating. The experiments of Thumm are of special interest in this connection.

With regard to the marine fishes of northern waters the evidence for a disproportion under natural conditions is still obscure, though much has been written on the subject. When we take the natural law into consideration, that the males mature earlier, it seems that on the whole the females are in the majority. In the first years of growth the numbers are approximately equal, but the males die off more rapidly whilst there seems no natural limit to the age and growth of the female.

Returning once more to the interesting Cyprinodonts further reference should be made to the fact that the females are able to produce several broods, up to seven, from a single pairing with the male. It has been found that the males preponderate in the later broods, which is in remarkable agreement with Geddes and Thomson's theory and goes far to explain the great variability in the proportions of the sexes in this group of fishes. Another curious thing is, that the females sometimes assume the secondary sexual characters of the male. This phenomenon was first discovered in 1891 and has been noted by several authors, more especially by Phillipi (1904, 1908). One female even while still carrying young gradually assumed some of the male characteristics. On examining another which had assumed these characters, he found that the female sexual organ was quite absent though the duct remained; in another testes had developed, one containing two large eggs, and the gonoduct still remained.

These curious phenomena may be explained in two ways. On the one hand, the months-long retention of the sperms in the gonoduct may react on the characters of the female (hormonic stimulus); on the other, as in the case of the frogs, it may be that the balance of these fishes approaches more nearly to the hermaphrodite condition than in other fishes. The viviparous Norway Haddock (Sebastes) is also said to change its sex, and the Hag (Myxine) is first of all a male, then later a female. Many cases of distinct hermaphroditism have also been found in the Teleosts, especially among the Gadoids. And Grassi explains many puzzling phenomena connected with the Eel from the standpoint that for several years the young Eels are hermaphrodite or non-sexual (1919).

However sexual dimorphism may be explained, it appears from these observations that the outer conditions and habits react on the inner constitution. The Cyprinodonts have an extraordinary power to adapt themselves to all sorts of conditions, from deep-sea, salt, and fresh water, up to heights of 14,000 feet. Whether as cause or effect, perhaps both, this adaptability has been accompanied by a loosening of the bonds of heredity. Not only is the balance of sex unstable, but even the old rule that different species cannot yield fertile crosses is broken. Quite a number of crosses between Xiphophorus, Platypœcilus, and Mollienisia, different genera of the viviparous Cyprinodonts, have been produced, one cross has even been taken for a new species, and the crosses have remained fertile and constant (Milewski, 1920).

The possibility of one sex changing into another brings us to a consideration of the meaning of secondary sexual characters. These are the outcome of sexuality; quite apart from the special organs the males frequently have a different appearance and different structures from the females. In other groups of Vertebrates these differences are usually of a constant character; the males are larger with distinctive markings, the females smaller and plainer; but in fishes the range of variation is very great.

Among the marine Teleosts external sexual differences are seldom apparent, the sexes of the Gadoids, Clupeids, Scombroids, etc., are precisely alike. But among the inshore forms, particularly those that live among the seaweeds like the Labroids, the males show a more brilliant coloration than the females during the spawning season. In the Dragonet Callionymus, the male is much the larger, with longer snout and dorsal fins, whilst the body during the breeding season becomes ornamented with yellow and blue bands, the female remaining of a dull greyish-yellow. In an Indian form, however, it is the female that is the more brightly coloured. In the Elasmobranchs the colour is not distinctive of the sex, but the basal part of the ventral or pelvic fins in the males is used as a clasper to keep hold of the female during pairing; in the Chimæras there is a second clasper in front of the ventral fin and still a third on the top of the head. During pairing the male twines his body around and lengthwise along the body of the female so that the top of his head comes to rest on the back of the latter. The male dog-fish envelops the female in the same way but has no head clasper.

It is amongst the freshwater Teleosts, however, that the secondary sexual characters are most developed, and every part of the male fish from the mouth to the tail is affected, naturally, different parts in different groups. Usually it is the pectoral fins that are longer in the male, sometimes, however, the ventrals or the dorsal and even the caudal. The coloration of the male may be remarkably brilliant, with red and black dominating, whilst the female usually

PLATE XII

Courtship of the Dog-fish.

(From a photograph taken by A. Schensky in Helgoland Aquarium.)

remains of a drab appearance, but in some cases, for example, the freshwater Sticklebacks, the female also assumes a bridal dress. As a rule the female is larger than the male.

Formerly, it was thought that these differences in the sexes had been developed by some process of selection, the females choosing the brightest partners whilst their own drab colours were a protection to them. As a better explanation of the phenomena, however, it is now more generally believed that the colours and other characters are simply attributes of "maleness." The hormones of the sexual organs are conveyed by the blood and lymph to the different parts, pectoral or ventral fins or mouth, which the different groups have developed as special organs in connection with the pairing. That the males usually have a more brilliant coloration than the females is due to their greater metabolism and activity.

The males of some forms have structures lacking in the females which seem to have nothing to do with the act of reproduction. In some species of the flat-fish Arnoglossus the anterior rays of the dorsal fin are more elongated in the male than the female, whilst in the allied genus Bothus the migrating eye of the male travels further back on the head than in the female, whilst at the same time parts of the mouth bones project as tubercles through the skin. In these cases the differences can be ascribed to a difference in the balance of the whole body between the males and females. The old males of the Salmon also develop a peculiar mouth. The maxillaries above separate, forming a groove in the middle into which fits a very peculiar growth of the lower jaw. It used to be thought that this peculiar hook was a kind of spade used in working holes to receive the eggs. But, so far as known, the female is more active without the hook in this operation. Other suggestions have been, that it was used by the male to hold fast to the female or that it was a protective organ of some sort, but these explanations do not fit the facts. "The least violent explanation appears to be that the hook is an essentially pathological appearance, characteristic of age, but produced by the irritation caused by blows on the snout, both during combats and in leaping over obstacles " (Smitt).

This author gives an interesting story which illustrates how impartial, even indifferent, the females are to the nature of their partner. A female had prepared the bed as usual and then conducted a male to it; this male was caught and removed. She searched for and obtained a second mate; this was also removed, and so on until she had conducted in all nine males to her spawning bed; " and when even the last male was caught, she returned with a large Trout in her train."

The courtship of fishes, in so far as it throws light on their mental processes, will be dealt with in the following chapter, but the physical side of the matter may be discussed here. It is very infrequent for the female to show any activity in the choice of a mate, as in the case of the Salmon. As a rule she keeps out of the way, if possible, and lets the male do all the work. In the case of the Cyprinodonts it has been remarked that the female shows a decided aversion to having anything to do with any male, no matter how brilliantly coloured he may be. This is not coyness, it is simply their nature to be passive and anabolic, whilst the male is using up his stores of energy in activity. The resulting phenomena can be understood from this basis.

Among the marine Teleosts there is no courtship, so far as one can determine. The females assemble on suitable grounds, shed their eggs at intervals and then depart as quickly as they can; the males remain for a longer period and serve a number of females indiscriminately. But it seems to be the rule that the male keeps in close attendance on the female and even helps in the extrusion of the eggs by rubbing his body along her side. The only example of a definite sexual intercourse among Teleosts with pelagic ova is provided by the Dragonet, as described by Holt (1898). The Pipe-fishes seem also to be monogamous, for a season at any rate, and the male and female play with one another for quite a long time before the eggs are laid. When this occurs the pair are in close embrace and it is the male that takes charge of the eggs, receiving them either into a groove under the tail or abdomen or into a closed pouch, where they are fertilised.

Among the Dog-fishes and Rays there is little pretence at courtship, and the masterly Stickleback, who is most peremptory in the management of his mate (or mates), simply shuts her into the nest he has made and stands on guard until the eggs are deposited. Then he drives her out and takes charge over the future proceedings.

Among many of the freshwater Teleosts the male shows undoubted eagerness in his pursuit of the female, but we can hardly call it courtship. If the eggs are laid and fertilised externally, as among the Gouramies, he urges her to perform her part even before she is ready, by pushing her about, rubbing his snout against her body, and even seizing

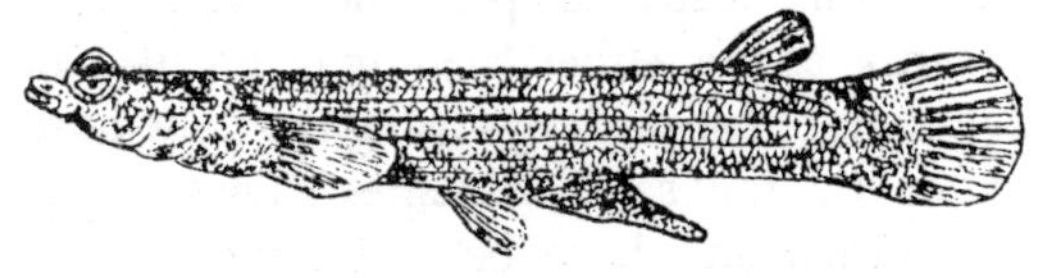

FIG. 73.—Anableps tetrophthalmus from Central and South America. Male with modified anal fin, about ½ natural size. (From Boulenger.)

her with his teeth. In the viviparous forms (many of the Cyprinodonts) this eagerness is carried to an extreme. The males seem constantly in chase of the females, even when quite immature, and also attack their own sex. Among these Cyprinodonts also several of the rays of the anal fin are developed into a large gonopodium or spermatopodium, which is fixed into the genital opening of the female, and the sperm packets are then shot into the duct, not through the gonopodium but along the front of it. This gonopodium is most developed in the curious Anableps, which swims on the top of the water (Fig. 73).

It is probable that in these phenomena we see the origin and cause of viviparity. Many authors have taken this to be a development of parental care, but this explanation seems too anthropomorphic. It is more probable that the over-eager attentions of the male have led to the female receiving the sperms into the oviduct before the eggs were ripe and these were then fertilised and developed before extrusion. The female has been in all cases apparently an unwilling party to the transaction and revenges herself later by eating up the young when they escape, if the male leaves her any.

2. Commensalists and Parasites

In moving about from place to place the fishes meet with many forms of life other than their own kind and enter into various relationships with other animals and even plants. They do this sometimes in the search of food, or protection from fishes bigger than themselves, or, it may be, simply to have a quiet corner where they can spend most of their time without troubling about food or enemies. In many cases this leads to permanent relations between different forms of life, either friendly or hostile, the successive generations of fishes always using the same host or being sought after by the same parasite. These relations are many and varied and have been classified under a large number of names. Some examples may be given here, beginning with the friendly or it may be quite neutral relations.

When a Blenny creeps into a hole between the rocks or a Goby into an empty shell, it chooses a house for itself where it can settle down and rest and lay its eggs in peace. Fierasfer, also a sort of Blenny, has chosen a more select kind of house (Fig. 74). It inserts itself, either tail or head first, into the cloacal chamber of a Holothurian, and from this refuge it projects its head or whole body from time to time in search of food. If Holothurians are not handy it chooses a starfish or a large bivalve Mollusc. In either case it is simply a lodger, but a related form Semper describes as a parasite, in that it feeds on the viscera of the Holothurian. On the other hand, Fierasfer is sometimes not so fortunate in its choice of lodgings. A species that occurs in the Bay of Panama has occasionally been found in pearly oysters, and these have a drastic way of dealing with foreign bodies, surrounding them with a pearly covering which puts an end to their activities and renders them harmless.

If the lodger obtains food with the assistance of his host, he is called a commensalist. The young of many species, Gadoids, Horse-Mackerel, Nomeus, etc., commonly called scads and mackerel midges (Rocklings), live under the umbrellas of large jelly-fishes of various kinds. They are protected in this way, it is supposed, from larger fishes

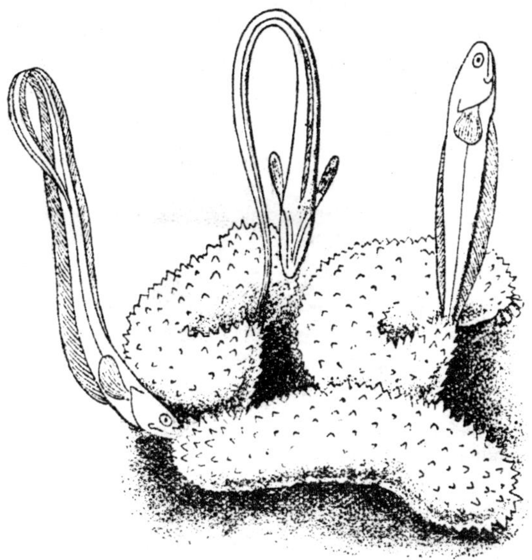

FIG. 74.—Fierasfer lodging with a Holothurian. (From Emery.)

that may not like the look of the jelly-fishes, whilst a quantity of smaller animals paralysed by the stinging-cells of the medusæ fall to their lot. In the tropics many fishes of different kinds (Labroids, Chætodonts, etc.) live a similar sort of life, playing hide-and-seek among the corals. One small Percoid (Amphiprion) even lives inside a large Actinian and seems to share its food with the latter, which reciprocates by giving it shelter. In such a case, where mutual benefit accrues to both host and lodger, the commensalism turns to symbiosis.

A remarkable instance of partnership for mutual benefit, one may call it commensalism or symbiosis as one likes, is that existing between the Pilot fish (Naucrates) and a Shark. It seems well-established from many observations that the small fish, usually in pairs, lead the big fish about. They probably have sharper eyes for large booty than the Shark and then obtain their reward in the crumbs that fall from the latter's mouth. If the Shark is caught from a boat, the Pilot then follows the boat even for some days until it finds another partner. Echeneis, the fish with a sucker on its head, which may have developed from the Pilot fish, has given up leading the Shark and become a semi-parasite (Pl. XIII*b*). It is supposed to obtain its

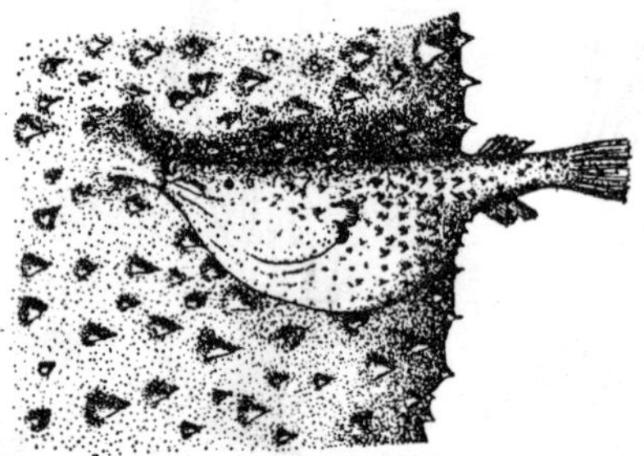

Fig. 75.—Small Ceratias parasitic on body of large female Ceratias. (From Sæmundsson.)

food more easily in this way, but it is obviously no longer of any use to the Shark. Its usefulness in another way will be noted in the last chapter.

A relation of a different and stronger kind is displayed by the Ceratiids. Sæmundsson (1922) describes how the large female Ceratias carries about with it a number of small fry of its own kind (Fig. 75), which are so minute that they seem like loose pieces of skin. On closer examination Regan (1925) has found that these pigmies are really the males, and it seems that the same phenomenon occurs in other genera of the Pediculati Ceratioidea. The connection is not merely that of an external parasite. The snout and chin of the tiny male are produced forward into outgrowths,

PLATE XIII

MUD-SPRINGER OR WALKING GOBY.
(From Hickson.)

SHARK-SUCKER AND DOG-FISH.
(From *Bulletin* of the New York Zoological Society.)

which unite in front of the mouth. The swollen ring of tissue thus formed surrounds and is united to a thick papilla projecting from the skin of the female. And the union of the tissues is so complete that the vascular system of the female becomes continuous with that of the male.

Though some water may pass into the mouth at the corners and the male thus retains in some measure its own respiration, in other ways it is quite dependent on the female. It takes in no food by the mouth; the digestive tract remains rudimentary and undeveloped; the liver is apparently absent, and Regan is doubtful whether even the kidneys are present. The dwarf male, we may say, has become part and parcel of the female, retaining only the sexual organ, a single testis, as evidence of its separate individuality. Since the males are thus undeveloped, Regan thinks it probable that they succeed in attaching themselves to the female whilst still in the larval condition or shortly after this stage.

A curious instance of symbiosis is given by the small Bitterling (Rhodeus) of Central Europe. The female possesses a long genital papilla during the breeding season, and by means of this organ introduces her eggs into the gill-cavity of pond mussels (Anodonta), where the young fry develop until the yolk-sac is absorbed, about a month later. In return, the Mollusc sheds its eggs about the same time and the young larvæ (Glochidium stage) fasten on to the skin of the Rhodeus. Here they become encysted and develop into young mussels. It seems rather a case of "tit for tat" than true symbiosis.

From these stages to parasitism is but a short step. For example, a small Cat-fish or Siluroid, Stegophilus, inserts itself into the gill-cavity of a larger Siluroid (Platystoma) and there sucks the blood of its host. As already mentioned in a previous chapter, the Hag-fishes and Lampreys have developed this habit to a remarkable extent. They manage to get hold of their prey, Cod, Flounders, or Salmon, when these are resting or caught on lines, and bore their way inwards, eating as they go until nothing is left but the skin and bones.

Not many fishes, however, take to this mode of life; as a rule, they are the sufferers from all kinds of external and internal parasites (Crustacea and worms). But a remarkable instance of what seems to be friendly parasitism was discovered by Max Weber on the Siboga Expedition to the East Indies. In freshwater ponds or craters he found some small fishes (Anomalops and Photoblepharon) with what appeared to be a luminous organ under the eye. This organ was later examined by Buchner, who has shown that it is composed of a compact mass of bacteria. From this discovery, as already mentioned, Buchner has come to the conclusion that all the luminescent organs in fishes have developed under the influence of bacteria, these being the actual source of the luminescence (1924).

It has been supposed by many, that these luminous organs were of some special use to their possessors, either as attracting friends and food or warning enemies, and so on. These small fishes, however, swim about in shallow water, and Steche (Brehm's *Tierleben*) found that the light was "burning" during the day as well as the night. Once he came upon quite a colony of them and compares their illumination to a number of small electric lamps, which faintly lit up the coral blocks among which the fishes live. But the fishes are able to turn out the light by rotating the whole organ inwards as well as by pulling up an outer fold of skin like a shutter. Steche thinks that in this way the fishes are able to attract the small plankton animals, like shrimps, which are attracted by light. As most fishes manage to capture these without any lights, this explanation seems a little strained; indeed, no explanation is needed. The fishermen of Banda (Malay), where these fishes live, make good use of the phenomenon, however; they cut out the luminescent organ and attach it to lines just above the bait on the hooks. With this strange fishing apparatus they catch the larger fish which live in deep water outside the crater.

3. Diseases and Enemies of Fishes

Although fishes have displayed a wonderful power of adapting themselves to all sorts of conditions, this is a general rather than a particular power. The time required to effect adaptive changes has been and is considerable. When these have been accomplished the fish becomes specialised to a definite range of temperature, for example, and thus more sensitive to conditions outside the range of the customary. It is probably for this reason that the freshwater forms seem more subject to illness and diseases than the marine forms.

The great attention devoted to the rearing of freshwater fishes, with the introduction of many tropical and subtropical forms as aquarium fishes, has revealed that fishes are not so immune to the ills of flesh as was formerly thought. They seem even to catch cold in the usual unaccountable way. Sometimes whole populations are swept away by epidemics as mysterious as any that trouble man and just as much seemingly beyond control. Possibly, inflammation of the gills—the equivalent of cold on the lungs—produced by changes of temperature or chemical influences, is the first cause of the epidemics, followed by loss of appetite and sluggishness, just as among human beings. Such an epidemic has also occurred on a large scale out at sea. The American Tile-fish (Lopholatilus), which lives off the coast of New England in the warm waters of the Gulf Stream, was almost wiped out in the year 1882. Many millions of the fish were killed, and it is said that "their dead bodies literally covered the surface of the sea for hundreds of square miles." It was believed that ice-cold water, coming in the train of severe gales, had displaced the warm water over a large area and caught the fish unawares. For many years not a single Tile-fish could be found, but it has reappeared again in its wonted haunts and is now as abundant as formerly. Similar phenomena have been noted by Hjort in the case of the Arctic Capelan or Mallotus.

Similarly, the extension of factories and industrial works

with the outpouring of chemical waste-products into the water has had a serious effect on the fish population of the rivers. The Thames was once a Salmon river in days long gone by, and some stragglers have occasionally been seen at the mouth trying to enter, but none are now caught in the river. The Salmon has almost disappeared also from the Elbe and Weser.

As the result of change from warm to cold water or from soft to hard water, fishes may suffer just as we do. The skin begins to peel in various places and serious wounds may develop. These may heal after a few days, but in the meantime some deadly enemy may seize on the exposed flesh, in the form of either plant or animal parasite. So long as the fish remains strong and healthy, and even if the parasite is placed deliberately on the exposed part (Edington, 1889), it is able to recover from such wounds. But when the Salmon, for example, has become exhausted by spawning and fighting with its neighbours, the deadly fungus (Saprolegnia) makes its appearance and rapidly covers the fish with soft thread-like outgrowths. This was formerly thought to be the cause of heavy and frequent mortality among the Salmon in Scottish (and Irish) rivers, but it has since been discovered that an even more deadly enemy has previously been at work. This pest is a bacterium, Bacillus salmonis pestis, which grows best in cold water, thus in the winter months when the Salmon are spawning. The disease is very contagious, and whenever fish infected with the fungus are seen, the endeavour is made to remove them from the water as soon as possible. The bacillus has no influence on warm-blooded animals, not even on frogs.

The fungus disease is also very common among Carps and other Cyprinids, in fact on all fish exposed for a short time in unclean air, but here also it is thought that other bacterial parasites begin the trouble, for example, in the kidneys, and weaken the fish to such an extent that it cannot resist the attacks of the Saprolegnia. It is probable, further, that bacteria lie behind the lymphocyst disease, which shows itself as rough, tuberculous swellings on the skin and fins of many fishes both marine and freshwater. All one can see in the swelling is a number of large connective tissue cells, somewhat like cartilage cells, and these giant cells have sometimes been taken for eggs. In other cases, though not often, malignant growths have been found in fishes (Johnstone, 1924), and the proliferation of these bodies has a similarity to the growth of cancerous tumours, but there is no reason to suppose that this disease in human beings comes in any way from fishes.

Other bacteria getting into the blood and lymph (B. pestis astaci, Hofer) cause inflammation of the internal organs as well as of the skin (lymphocytosis). The scales stand out and matter can be pressed from under them by rubbing the hand along the side. In some cases the eyes bulge out from an excess of fluid within or behind the eye-ball, and it is probable that the huge excrescences seen on the top of the head in some of the artificially produced Japanese varieties of the Carp (see frontispiece) are pathological growths of this nature. In these the disease (or growths) appears as the fish grow older and they are then unable to maintain their balance and sluggishly move about head downwards.

Another remarkable disease, supposed to be due to a sporocyst, affects the labyrinth of the ear to such an extent that the fish becomes, as it were, dazed and turns round and round on trying to move. This is frequently a cause of much trouble and anxiety in the Trout hatcheries and leads to serious loss. Sometimes the posterior portion of the tail loses its colour and may even turn black, so that it looks as if the profundus branch of the great lateral nerve, which serves the cutaneous sense organs and skin of the caudal region, becomes affected as or before it emerges from the skull.

Even the single-celled plant organisms are able in some way to penetrate the outer skin of fishes and grow within the tissues. A small green alga seen as fine green points through the skin and others that infest the gills may cause serious inflammation and thus weakening of the fish, predisposing it to the attacks of the ever-present Saprolegnia. One of the most deadly of these fungoid growths is Branchiomyces, which somehow gets into the capillaries of the gill-filaments and its hyphæ spread there to such an extent that the fish (Carp) is soon suffocated. This occurs mostly during very hot days in summer and is more common in some waters than others. In fact, it is a peculiarity of all these diseases that some rivers, for example, the Tweed, are worse than others. A deficiency of oxygen or air in slowly moving or stagnant water, when greatly heated in the summer, also has a similar effect on the gills.

Of the parasites that infest fishes there is almost no end ; various Copepods without and many worms within. There are over fifty species of Trematods alone. The latter probably live in the sexual condition in gulls. The larva (Diplostomum) develops from the egg in the fish, mainly freshwater forms, but also some marine, and makes its way in the blood to the eye-ball, where it settles at first in the lens and causes cataract, the eye becoming quite opaque. Although quite blinded by the parasite, the fishes seem able to find their food by means of their other senses. In the Elbe this disease has been well known for a hundred years and is exceedingly common, yet it does not seem to have affected the stock of fish to any extent and the blinded fish are quite good to eat (Ehrenbaum, 1917).

A more deadly parasite is one of the Band-worms (Diphyllobothrium or Bothriocephalus), which begins its life-cycle in a small freshwater Copepod (Cyclopus or Diaptomus). This is eaten by the second host, Pike chiefly, but also Perch, Trout, Whitefish, etc. Here the parasite penetrates from the gut into other organs or the muscles, and when again the fish is eaten by cat, dog, or man, the full-sized tapeworm develops and may become a serious danger. It is said to be the commonest parasite of human beings in Turkestan and Japan, but it also occurs in Europe, Switzerland, North Russia, and the Baltic provinces of Germany (Plehn, 1924). Unlike some other worm parasites, which sometimes cause a serious mortality among fishes, the Band-worm does not seem to give the fish any great concern and, if the fish is well-cooked, it does no harm to man. The disease is common in the above-mentioned countries simply because the natives mostly eat the fish raw.

Whilst the fishes are thus subject to many diseases and illnesses just like human beings, they have also to suffer from other dangers now almost unknown to the latter. That fishes are cannibals and partake of their own kind, especially when they are younger or weaker, there is little doubt. Naturally also the larger prey upon the smaller, and the smaller may combine to attack the larger. The Salmon is one of the most powerful fishes in fresh water, yet the much smaller Eels have been known to work together to devour it even in the pride of its strength before spawning. Before it enters the rivers also it has to run the gauntlet of the marauding seals. Upstream the bold Trout follows it to the nests and takes a heavy toll of the eggs. The Trout itself, like other freshwater fishes, has to keep a wary eye on the pike, the otter, and other enemies from above. No wonder he is clever and prefers to seek his food by night or when the water is disturbed.

At sea a shoal of fish, Mackerel, Whiting, Sprats, or Herring, is invariably indicated by the presence of gulls, which seem to know that fishes cannot hear, and keep up a ceaseless chatter and screaming as they fill themselves with the rich booty. The diver and the cormorant (Black Shag) do not keep to the surface however ; they follow the fishes down to the bottom, where they are sometimes caught in the fisherman's net or the wide gape of the Angler.

We can but faintly picture what goes on at the bottom of the sea, but the stomachs of bigger fishes keep a record. It is a rarity to find a Cod without remains of some other fish inside it, and often it is over-full with Sand-eels (Fig. 76), Herring, and the eggs of the Herring. The latter, like other Clupeids, is relentlessly pursued by all other animals that can swim or use a net, the Mackerels, Tunnies, Porpoises and man. We can form some idea of the numbers taken by man, and some think, indeed, that man is the chief enemy, but we do not know. The Gadoids, Tunny, whales and the gulls must be formidable competitors for the honour, if added together.

Another species that is relentlessly pursued by all kinds of fishes is the Sand-eel. On the bottom, it falls a prey to the Rays, the larger Sand-eel, and the Turbot; if it emerges a little, the latter and the Cod or other Gadoid follows it, and higher up are the Green Cod, the Salmon, Bass, and whatever fish may be in the neighbourhood or

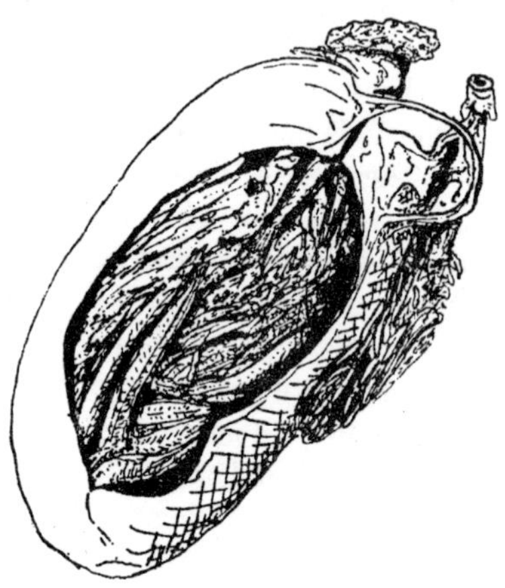

FIG. 76.—A good catch of Sand-eels in the stomach of a Cod. (From Ehrenbaum.)

locality. The best bait that can be used for the Bass (Labrax), as he comes along the coast in the summer and autumn, is the Sand-eel. Here again, in spite of this merciless persecution, the Sand-eel is one of the most abundant of all fishes, not even the Herring yielding larger numbers of young fry to the plankton nets.

We can form but little notion of the quantities of young fry destroyed or consumed in the open sea. This is probably the most dangerous period of their career. The eggs which float on or near the surface are perhaps immune, except from jelly-fishes, and the loss in the egg stage is due to storms or simply to the fact that a large percentage is never fertilised. But the young fry as they sink down to the bottom or play about near the surface are exposed to the attacks of one another, jelly-fishes, Sagittæ, Mackerel, or even the Herring, if they should be in the neighbourhood.

Whatever may be thought of this ceaseless pursuit and destruction, and it is difficult to refrain from calling it a struggle for existence, we must remember that fishes, like ourselves, must eat to live. How the balance of nature could be maintained otherwise, one cannot imagine. And if we say that the pursuit has strengthened the muscle and bone of the Dolphins as well as the fins of the Flying Fishes or increased the wariness of the Trout with the cunning of its pursuers, we are simply repeating a truism. But we cannot deduce from this that the struggle alone has made the species what they are. It is still as true now as of old, that "the race is not to the swift, nor the battle to the strong."

The Herring, as defenceless as any fish, not at all swift by comparison with many others, and certainly not strong or clever or protected from enemies, is descended from the longest direct line we know of and still maintains its superiority in numbers over its pursuers. We have no reason to believe that they have influenced or affected its characters or habits in any way. The most deadly enemies to any species are, not its nearest rivals or pursuers, but the physical and chemical conditions, and after these come the lowly organisms that are ever present and ever ready to destroy any weakened fish. Those who overcome these difficulties, if any, owe their survival not to any particular character, but to their great vitality, their ability to keep moving and to adapt themselves to the altered circumstances. If there is any moral to be derived from the struggle for existence it is, that the clean and the healthy are the most likely to survive and make progress.

CHAPTER XIII

THE FOOD QUESTION

THE relations of fishes to one another and to the other forms of life which serve them as food constitute an important part of their biology. Strange as it may seem, one of the most interesting facts brought out by recent investigations is that fishes may not get enough to eat and their growth is accordingly seriously affected. Owing to the natural law of increase there seem too many of them, and from various signs we may conclude that man is able to influence their abundance, in some cases at any rate. How fishes obtain their food, its nature, and the part taken by man in regulating their numbers will be the subject of the present chapter.

1. THE FOOD OF FISHES

Taking them altogether, the diet of fishes, like every other character, is exceedingly varied. It is easier to say what they pass by, the medusæ with their stinging cells, the sponges, and most of the star-fishes. Presumably they have learnt by experience that these should not be touched, and their taste-buds now warn them that acids and too much chalk or silica are not nutritive. Apart from these, anything from a micro-organism about the size of a needle point to a whale is good booty. In general, the larger fishes take the larger prey, whilst the small and the young fishes live on the tiny organisms, but there are exceptions and vegetarians occur as well as flesh-eaters.

If we accept the theory of Pütter, we must go even further back. According to the views prevailing hitherto, and, it may be said, still maintained by most biologists, fishes are dependent for their food on organised bodies, that is, on forms that already contain protoplasm or living matter. Pütter, however, is of the opinion that the tissues, gills and alimentary canal, even the skin in young forms, are able to absorb organic and inorganic substances dissolved in the water. If living organisms could make complete use of the food-stuffs they take into themselves, the end-products would be simply water and carbonic acid, but in most cases the assimilation is incomplete and large quantities of other metabolic products are set free in the water. The most abundant of these are the carbohydrates (sugars), but nitrogenous compounds (free or in albuminoids) and phosphates or phosphoric acid also occur. It has been found also, that the latter are present in the water of the sea and that their quantity varies in different places and even in the same water at different levels and different seasons of the year (Matthews, 1916; Raben, 1916). If Pütter's views prove to be correct, this variation may be of importance to the young fishes.

It is known that some organised bodies, bacteria and parasites without digestive tract, are able to absorb dissolved organic substances into their bodies, and the question is how far this may occur in fishes. The possibility is not to be excluded without careful investigation. The Leptocephali of the Eel, as noted in Chap. III (Fig. 12), spend three years out in the Atlantic, and very little food has been found in their digestive tract, especially in the third year. It has been assumed that they live on the nannoplankton, microscopic organisms that escape through the meshes of the finest plankton nets, but we know as little about these as about the importance of the dissolved organic substances. The fact that the whole internal structure of the Leptocephalus is bathed in a watery fluid, which may have free access from and to the exterior, suggests that it serves the general purposes of respiration, digestion and excretion. Again the young of viviparous fishes (Zoarces, Embiotocids, and others) spend some time within the body of the mother-fish and emerge with all their structures already formed. They have no organic connection with the mother-fish and must obtain their nourishment from the ovarial fluid, which is rich in dissolved organic stuffs. The distance between these and the broken-down ova or tissue cells, on which they are supposed to live, is not very great. Lastly, many Teleostean larvæ pass through a difficult stage when the mouth is being formed (Chap. IV) and do not appear to feed during this period.

These examples indicate that Pütter's theory may be quite correct, so far as the young post-larval stages of the Teleosts are concerned. In the older stages, when the tissues have become more differentiated, it is open to doubt whether any assimilation of food-stuffs can take place except through the digestive tract. So far as experiments have gone, the living membranes of the gills prevent the diffusion of anything but water and gases, but it has to be noted that these observations are still few in number and insufficient to decide the question.

Apart from the above possibility it may be said, as it is said in general, that all flesh is grass. The animals feed upon one another, but when we go low enough or far enough back, the smaller animals are found to be feeding on plant life. Along the shores are the immense quantities of sea-weed, and out at sea the teeming algal populations which are of importance more directly to the fishes. By the aid of the energy derived from the sunlight these plants are able to build up organic substances simply from the water, carbonic acid and a small percentage of salts. There are several communities of plankton algæ, but the most important appears to be that of the diatoms, very small microscopic bodies with a brownish colouring matter which enables them to assimilate the carbonic acid, just like the chlorophyll of land plants.

During his intensive study of the plankton in the Irish Sea, Herdman found that the diatoms flourished best when the temperature of the water was rising. The monthly average catch increased from 65,000 in January to 26½ millions in May, then followed a rapid decline to the minimum of over 13,000 in August, with a second smaller maximum of 319,000 in October. In an interesting manner he correlates this variation with the changing condition of the water. In spring, when the diatoms are absorbing the carbonic acid, the water is alkaline; in the summer the alkalinity declines because the animal plankton is then eating up the diatoms and giving off carbonic acid. The last increases to a maximum in the winter, when the diatoms begin again, and so on.

A curious thing is that although diatoms are present in such enormous quantities, the number of species or genera is relatively small and the genera reach their maxima in different months. Thus Biddulphia is most abundant in April, Chætoceras in May and Rhizosolenia usually in June. Some idea of the quantities present may be gained from the fact that in one haul about 58 millions of Chætoceras cells were obtained, and of these 51 millions belonged to one species.

Allen (1917) has made some noteworthy experiments in order to determine the conditions and substances necessary for the growth of diatoms. Artificial sea-water was built up by dissolving in distilled water the chemical compounds known to occur in sea-water, after seeing that the distilled water was perfectly pure. Then minute quantities of potassium nitrate, sodium phosphate, and iron were put in and a small quantity of a diatom culture, Thalassiosira, was added. At first the results were not favourable, the alga did not grow; but on adding a minute quantity (1 per cent.) of natural sea-water, good cultures were obtained. This unexpected result led to a search for the substances or body which must exist in the natural sea-water but not in the artificial, and which is obviously necessary to give a stimulus to the diatom growth.

What the substance is could not be definitely determined, but it has the nature of an organic compound—that is, something complex derived from animal or plant organisms. It appears, therefore, that the lowly diatom is dependent for its growth on other and higher forms of life. The compound or bacteria, as it might be, is in some way connected with the waste products of the latter, and it is significant that whilst the coastal waters, with their proximity to the drainage from the rivers, are rich in plankton, the open ocean by comparison appears to be a desert. We see again how closely the many varied forms of life are linked together.

The plankton algæ are eaten by the larval fishes and the young stages of the innumerable forms of Invertebrates, Gastropods, Bivalves, etc., that swarm in the water during the spring and summer. It is also believed that they constitute the principal diet of the planktonic Crustacea, Copepods, Mysis, and the like, and various observers have recorded their occurrence in the digestive tract of these forms. But a doubt on the matter has been raised by Pütter (1924) just with regard to the Copepods, and it may be useful to note his arguments.

In the first place, he is very sceptical with regard to the previous records of Diatoms occurring in the stomachs of Copepods, but a knowledge of the literature on this particular subject is not a strong point with him. Apart from the earlier records one may refer him to the recent work of S. Marshall (1924) on the food of the common Copepod, Calanus finmarchicus, which was found to be living mainly on Diatoms in the summer and Radiolarians in the winter. However, he indicates that the Copepods and Diatoms do not occur in the same layers of water at the same time; thus in Herdman's lists of the plankton the Diatoms attain their maxima long before the Copepods, and out in the open sea the Diatoms are near the surface whilst the Copepods live mainly in a deeper layer. Again, he doubts whether the Copepods are furnished with the necessary acids to digest the Diatoms, and in one form (Haloptilus) he points out that the stomach is short and ends blindly, yet this Copepod is able to grow and increase its volume by 9 to 15 times. Even if the Copepods do eat the algæ, the amount is only about 2–3 per cent. of their requirements, according to his calculations.

Pütter comes to the conclusion, then, that by far the greater quantity of the food of Copepods must come from the dissolved organic substances in the water. This is not taken into the digestive tract but penetrates through the surface of the body, just as the oxygen does. Since vitamines (A) have been found in these Crustacea, he suggests that the Diatoms may be useful in supplying these accessory substances. As already remarked, these views of Pütter cannot be lightly rejected, but we must await further investigation before giving up the old belief that the Diatoms and other algæ form the " grass " from which the flesh of the Copepods and other animals arises.

With regard to the food of very young fishes, the information given by M. Lebour (1921, 1924) is very instructive. Dealing mainly with the earliest stages of the Herring, this author shows that the diet is varied. Up to a length of 12 mm. the larval Herring prefers the more satisfying larval Molluscs, especially Gastropods, if it can get them, but it certainly takes quantities of the unicellular organisms, Diatoms and Infusoria. As it grows, however, it passes to an exclusively Copepod diet and is here fairly catholic in its tastes. The size of the mouth is probably the determining factor in its choice of food, and it is of interest to note that as the young post-larval Herring sink down from the surface they seem to take less food. " Thus more than half those from near the bottom and from midwater were empty, and only a little more than a third were empty from the surface." This is just the stage when the jaws are ossifying and when the little fish has difficulty in taking in food (see Chap. V).

The adult Herring is also not very particular ; it is not restricted to any particular form. It is not above eating young Sand-eels and other young fish, even its own fry, but to satisfy its voracious appetite it finds the thick " soup " of Copepods more suitable. Herdman has found that the summer Herring off the Isle of Man feed mainly, if not wholly, on Temora. This Copepod may be so abundant as to cause large patches of a red colour on the surface of the sea, known to the fishermen as "fish-food, or spawn."

For the winter Herring of the Channel M. Lebour found that Euphausids (Nyctiphanes) formed the bulk of the food, partly on account of their large size, but the smaller Copepods, Temora, Calanus, etc., were also abundantly represented. As it is generally thought that fishes do not eat when spawning, it is of interest to note that the Herring is not troubled in this way. Running females sometimes were full of food, "one with newly hatched Herring, Herring eggs, and Sagitta."

Other pelagic fishes, like the Mackerel, feed in the same way on these plankton Crustacea, at least when they come near the coasts to spawn, but whether they direct their movements according to the presence and abundance of the Copepods, as the Herring seems to do, is not yet certain. During and after spawning they pursue the young fish, "Mackerel midges," young Whiting, Green Cod and Scads, and a bent piece of white metal with green on the one side forms an attractive bait to catch them in the summer months. When they retire to deep water, they feed mainly on Schizopods, also Amphipods, occasionally Copepods and a Sand-eel or two (Ehrenbaum, 1923).

The bottom fishes, like the Haddock and Flat-fishes, feed on the bottom animals, worms, star-fishes (Ophiuroids), other fishes and particularly various kinds of Molluscs. Whilst the diet is for the most part mixed, it is noticeable that each species has its own preferences. The Plaice when living near the coast in the summer and autumn lives mainly on the Molluscs (Mactra and Cardium, etc.), but offshore, when it has spawned, it seeks out the worms. The Dab, on the other hand, prefers the worms at all times and is, perhaps in consequence, a more delicate and tastier fish to eat, when fresh. The Sole is also partial to the worms, siphons of Solen and Cumaceæ (E. Mohr).

Other fishes, like the Mullets which have no teeth, feed on the mud or detritus, the remains of plant and animal bodies which sink to the bottom. Very few of the marine fishes are strictly vegetarians, but some of the Sparoids of the Mediterranean and tropics are said to live only on the seaweed, Fucus and Zostera. The Wrasses (Labroids), which live among the seaweed, feed chiefly on Molluscs and Crayfish, whilst their relatives of the tropics grub on the polyps of the corals.

The freshwater fishes have not such a wide range of diet as the marine, and that is probably the reason why most of them take to grubbing in the mud on the bottom. Some are vegetarians, but the idea that the Carps keep to a vegetable diet has proved not well-founded. The common Carp of European waters (Cyprinus carpio) is really omnivorous, though apparently it must eat plant stuff to obtain the indispensable vitamines. Insects, ants, bees, flies and their grubs, form the chief diet of freshwater fishes and they help to keep down the numbers of these pests. In Central America and the West Indies the Cyprinodonts feed on the larvæ of mosquitos and have been found exceedingly useful in combating tropical fevers. These illnesses come from the bite of the mosquitos, by which various Flagellates (Plasmodia) get into the blood, and formerly they caused heavy mortality among human beings. The introduction of various American Cyprinodonts, especially Girardinus (Lebistes, the millions fish), into the Barbadoes and the Panama zone has done a great deal to clear those regions of malaria.

2. The Valuation of the Sea

Since the fishes depend for their food on the lower forms of life in the waters, the quantities of the latter must determine the quantity of the fish. If now we wish to ascertain how many fish live within a given area, the simplest procedure would seem to be to determine, if possible, the amount of food present. Where the conditions are fully under control also, as in the rearing of Trout or Carp in ponds, pisciculturists can tell to a nicety the amount of manure and food required for a given size of pond and weight of fish. The question is, whether the same can be done for the sea, whether, for example, the North Sea can be turned into a sort of " farm," with sowing and reaping and rotation of crops, or whether such a comparison is quite misleading.

The primary conditions necessary for such an exploit is some measure of the quantity and quality of the organisms available in the sea, and it is to Victor Hensen, the Kiel physiologist, that we owe the first expression of this vast idea. How vast it is we can only realise when we come down to details. Whether such a measure is obtainable or not, the influence it has exerted on the investigations of the past forty years is one of the most signal examples of the power of an idea in scientific work. When Hensen put forward his plan in the 'seventies of the past century almost nothing was known of the forms living in the sea, they could hardly be distinguished from one another at that time, and, more especially, very little was known regarding the eggs and biology of the fishes. The efforts to realise the plan, however, or to refute its practicability, have lain behind most of the intensive investigations which have been carried out during recent years.

Hensen's principal aim in the beginning was to determine the quantities of the plankton in the sea, that is, the small organisms, plant and animal, which can be taken in a net of convenient size. Knowing this size and drawing the net vertically from the bottom to the surface, the volume of the water column could be calculated and hence, from the contents, the quantity of the organisms in the column. From many similar columns and the total volume of an area, let us say, the North Sea, the total quantities of the plankton could be calculated.

The technical difficulties alone in the way of carrying out this apparently simple scheme are very great. Nets are of many different kinds and each gives a different picture of the plankton. Again, if the smallest organisms like the Diatoms are present in large quantities, they soon clog the meshes of the nets and only part of a column may really be fished through. On the other hand, the distribution of the plankton is by no means uniform, which is a necessary condition for the successful estimate of the quantities over a large area. The plant organisms flourish best under certain conditions of light, temperature, and chemical composition of the water. They alone have the power of reconverting the end-products of animal metabolism, carbonic acid and ammonia compounds, into useful food for animals. Consequently they are more abundant in the littoral waters and the currents of the sea distribute them unevenly, in " streaks ", over a wide area. Lastly, it has been found that even the finest nets do not capture the smallest kinds of plankton (nannoplankton), which may perhaps be considered as the basis of all the rest. And, if we take Pütter's theory into account, we must go still further back and estimate the dissolved organic substances in the water, phosphates and carbohydrates, which are not at all represented in the plankton nets, or the filtrates obtained by means of a centrifugal pump.

These difficulties have only become apparent of recent years. It may appear to some, therefore, that Hensen began at the wrong end, that the first thing to do was to determine exactly the different forms of plankton life and the limits of accuracy in his methods. But it has to be remembered that knowledge on these points would probably not have been obtained but for the necessity of critically examining every point in the practical application of the method.

In one respect, however, it is possible that Hensen's conception might be put to good use. An absolute determination of the plankton may be beyond reach, and even the relative value—that is, comparison of one area or zone with another—may be misleading; but if we take the larger objects, such as the fish eggs and early larvæ of fishes, a rough measure is obtainable of the fish wealth of the sea—for example, in a restricted region like the Kattegat or Southern North Sea. Here again, however, the method has to be used with care. Fish do not spawn uniformly over an area, nor do they get rid of their eggs all at one time, and the currents soon carry them away from the actual spawning ground. Hence the different stages of development have to be taken into account, and it is probably the earliest stages only, if samples are taken constantly during the whole spawning period, that can be considered in the estimate of the actual quantities of fish present.

Worked out in details, therefore, we see that Hensen's method remains more or less an ideal to be aimed at in one or two particular cases, but unrealisable with any great degree of accuracy. To put the matter bluntly, we have to recognise our own limitations. Even if our knowledge of details were adequate, we cannot be everywhere at once, and the multiplication of vessels and investigators brings in another series of practical questions, personal equations and expenditure, which experience has shown to be not the least of the many difficulties to be encountered in the study of the sea.

The "farming" of the sea is a similar idea. Man neither sows nor does he manure or clear away the weeds, he only reaps. The comparison is therefore three parts fallacious, and the regulation of the fourth part, the reaping, involves a number of grave questions which have nothing to do with farming.

Less ambitious in aspect, but more realisable in practice are the conceptions put forward by the experienced Danish naturalist, C. G. Joh. Petersen. Instead of trying to break the whole "sheaf of arrows" at one time, Petersen has emphasised the importance of making certain beforehand whether an aim is practicable and attainable. If we wish to know, for example, how many Plaice are living within an area, the fishing tells us how many are taken and marking experiments tell us what proportions escape. Thus we have learnt that more escape during the year than are taken. Again, if we wish to know whether one area is better than another for a species, we have to determine the kinds of food selected by that species, for example, the Plaice, and then the quantities of the particular food on the grounds in question.

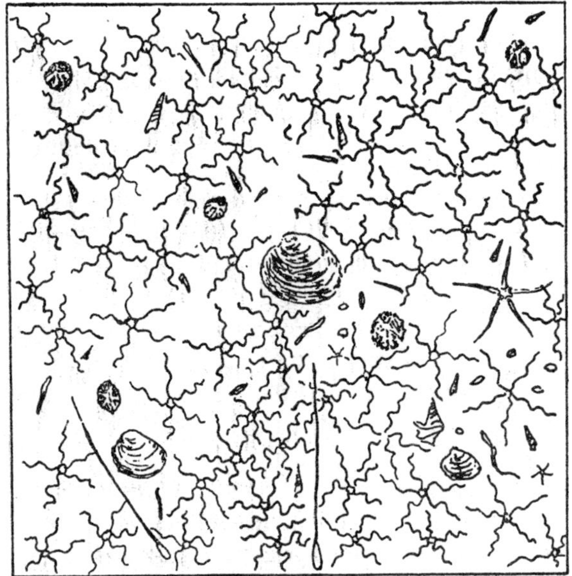

A. Echinocardium-Filiformis community.

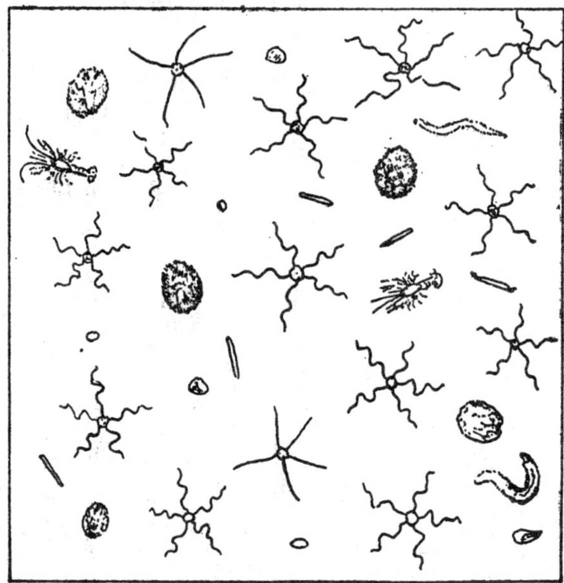

FIG. 77.—B. Brissopsis-Chiajei community. Bottom-samples from the Kattegat. (From Petersen.)

For the latter purpose he has devised an apparatus to take large samples of the bottom-soil. It descends through the water open and the heavy wings bite deep enough into the soil, if not too hard, to collect a known quantity. By means of a falling weight the wings are then allowed to close and the apparatus brings up a sample of one square foot, or even more according to size, from which the mud and sand are passed through sieves of various grades, leaving the animals behind. These are then counted and identified and their food value determined by chemical analysis.

With the aid of this bottom-sampler Petersen has made a quantitative comparison of the various Danish waters, and the method has also been employed by others in the North Sea. An ordinary dredge is also able, of course, to give an approximate estimate of the richness of any given ground in bottom-animals, but this method is much more accurate and reliable. The accompanying figures represent the quantities taken at two stations in the Kattegat. As a rule the majority of the animals in the bottom-soil are useless as food to the Plaice, which lives chiefly on the smaller Molluscs and Worms; hence the quality rather than the quantity of the ground is of importance, and we should not expect to find the same unvarying quantity or quality from one year or even season to another.

3. Resources of the Sea

Regarding man simply as one of the agents influencing the ways and life of fishes, we come now to a number of problems of considerable practical and scientific interest. Man takes a heavy toll of the fishes yearly and the necessity of deciding questions arising between different classes of fishermen and different countries has led to intensive investigations and the accumulation of a weighty literature. The freshwater fishes, under conditions to a great extent of man's own making and somewhat similar to those of plant crops and poultry breeding, need not be considered here. As yet, the fishes of the sea are under natural conditions, with a much greater fecundity than freshwater fishes and requiring no assistance in maintaining the numbers, at least of the young.

The magnitude of the world's sea-fisheries can hardly be stated with any degree of accuracy. As the fishing is prosecuted almost entirely in the coastal plateau down to 100 fathoms, the most extensive fisheries are in the northern hemisphere with its greater coast-line. Here the annual value of the catches landed, including those of Japan, Canada, and the United States, but excluding other Asiatic countries and the countries bordering on the Mediterranean, has amounted during recent years to over £100,000,000. The American authors have estimated the annual value of the whole world's fisheries at over £200,000,000. If we take any figure in between we obtain a sufficiently clear idea of the value of the fisheries to the fishermen. The value of the fish when it comes to the consumer, whether fresh, pickled, or preserved, is of course about three times as much. The capital employed in reaping this harvest—boats, gear, canneries, etc.—can only be imagined.

The number of men engaged in the fishing differs greatly in the different countries according to the nature of the fishing and the size of the boats. Great Britain, which has specialised in the offshore fishing and contributes a very large proportion to the above sum, has only about 80,000 fishermen; the United States has about 180,000 and contributes somewhat less; but Japan, with its coastal fisheries, has 1,400,000 regular fishermen and about half as many again occasionally employed.

With regard to the value of the fisheries, it should perhaps be stated that the price of fish is very much greater now than formerly. Thus in England the average value of all fish landed in 1913 was about 12*s*. per hundredweight, in 1923 it was twice as much. The probability is that it will remain about the latter figure, since the costs of fishing have increased to a greater extent.

It is still more difficult to state the total quantities of fish landed yearly over the world. In 1920 the North

European countries, excluding the Mediterranean, took about 5,600,000,000 lbs. of fish from the sea; Great Britain and Ireland alone about 2,400,000,000 lbs., Norway a little more than half that amount, Germany and France each about one-sixth. The quantities landed in the United States are about the same as those landed in Great Britain and Ireland, and in Japan also the same, whilst the catch of fish in Canada is estimated at about 900,000,000 lbs. Adding these together we have already over 11,000,000,000 lbs. as the product of the fisheries in the northern hemisphere excluding the Mediterranean countries, the islands of the Atlantic, Central America, India and China. If we could add these and also the fisheries in the southern hemisphere, the quantities of fish taken annually from the sea all the world over would probably be not far short of 15,000,000,000 lbs. Since the Clupeids, especially the Herring, make up more than half the total amount, we may estimate the average weight of the fish to be about one and a half pounds, which gives as a rough estimate the number of 10,000,000,000 fish taken annually from the sea. The estimate is of no scientific value, but it gives some impression of what the sea yields in food to man.

In addition, the shell-fish, oysters, mussels, shrimps, lobsters, etc., provide employment and food for many thousands of people. The corals, sponges and pearls, though not exactly food, come from the animal life of the sea and also the many shells of a pearly lustre, like Haliotis, more beautiful even than pearls. The cuttle-fishes are not to be despised as food and still less parts of the Tunicates and sea-cucumbers (Trepang, " bêche de mer "), whilst the turtle has only two faults—it lives far away in the tropics and is not sufficiently abundant to do away with substitutes. The whales and seals provide sport and livelihood to a separate little world of fishermen.

The plant-life of the sea is probably just as great as that on land and, like the latter, has its annual and perennial growths. Many seaweeds are edible (dulse), and the Japanese especially make good use of them, both as food and in the preparation of glue, agar-agar, etc. Lastly, it has to be remembered that valuable mineral products, such as iodine, potash and salts, are obtained from the sea, where some believe that even gold is to be found. Our ancient derivation from the sea is also indicated by our need of salt. If we were obliged to do without this simple but essential compound, we should turn, that is if we lived and used other kinds, into peculiar monstrosities according to our present notion of æsthetics.

Returning to the fishes, however, we may examine more particularly the produce of the North Sea. For some years now the countries bordering on this rich, littoral region have paid special attention to the collection of data as an aid to the solution of practical problems, and the statistics now available are probably as reliable as we can ever hope to obtain. That is to say, they contain a smaller probable error, but it would be absurd, as some have done, to attach too much importance to a thousand kilograms or so; the probable error is always much greater.

The following table gives the quantities of the more important species landed from the North Sea during the years 1920, 1913, and five preceding years. The figures given represent thousands of kilograms; 2,200 English lbs. is a near approximation to 1,000 kg. The fisheries of Norway and Sweden are not included here.

	Herring	Cod	Haddock	Plaice	Grand Total.
1920	461,021	97,243	208,464	53,982	958,307
1913	674,307	87,976	94,230	49,878	1,059,181
1912	606,896	92,683	126,681	52,193	1,016,527
1911	609,686	91,060	135,676	51,458	1,007,285
1910	597,199	89,950	122,051	48,008	976,164
1909	539,321	87,468	136,265	48,008	925,169
1908	554,630	70,304	158,908	47,320	939,900

In 1920 the fishing had not become what may be called quite normal, yet all the fish quantities are well up to the average with exception of those of the Herring. We have to go back to the year 1903 to get a smaller amount of Herring than that yielded in 1920, but this did not mean any decrease in the actual quantities. On the other hand, the quantities of the ground fishes, Cod, Haddock and Plaice, are well above the average, perhaps as the result of the, for them, comparative peace during the years 1915–1918. But this cannot be taken as the whole explanation, since the Plaice, the most ground-loving of them all and the one supposed to be the most subject to the attacks of the trawlers, is the one that shows the least difference from the average. Previous to 1908 the quantities of every species showed a great variation, from 10 to over 60 per cent., in the annual catch, but here again the Plaice showed the least variation. Since the beginning of these statistics in the year 1903, the Plaice has shown remarkable constancy in the quantities landed year by year, the yearly fluctuation being only about 10 per cent. over ten years.

The outstanding feature of the fisheries in all countries is the prominent part taken by the Herring and other Clupeids. Here in the North Sea region we see that over 60 per cent. of the total quantities landed consists of Herring. In spite of the most rigorous pursuit by all kinds of enemies, there is no sign of diminution. By means of the scale-studies it has been found recently that as many as twenty year-groups may be represented in the catches; thus a particularly good year like 1904 may continue to be prominent in the annual catches right to 1922. In spite of the persecution, therefore, a very large proportion of Herring must escape and contribute to the quantities of the following years.

The Haddock is the most variable species, so far as the quantities taken are concerned. When the quantities have decreased, the variation has frequently been credited to the trawlers, who have been accused of removing too many small fish. Here also, however, we must recognise that the natural causes are far more influential than the fishing by man. For example, it was pointed out in the Statistical Bulletin for the year 1908, that the years 1909 and 1910 would be good years for the Cod fisheries of the North Sea, owing simply to the large amounts of the "small" Cod landed in the years 1906, 1907 and 1908, whilst the same years would be lean years for the Haddock for the reverse reason; and so it proved. The importance of this simple deduction lies in the fact that although immense quantities of "small" fish are caught by the trawlers, yet sufficient escape and remain in the sea to yield a heavier catch of "large" in the following years.

This natural fluctuation in the quantities of fish, quite independently of man's action, is a point that cannot be too strongly emphasised. The assumption that man is the only or chief factor in producing fluctuations in the supply of fish has led in the past to very serious errors. We have now the experience of nearly a century of trawling in the North Sea, and if the predictions of alarmists had been justified, the North Sea should long ago have been exhausted. The figures quoted above for recent years furnish the best contradiction of these groundless fears. We have more justification for saying that another century of fishing in the same way will still see the North Sea far from exhausted. This leads to a brief consideration of the various attitudes which have been adopted from time to time with regard to the fisheries.

Among the old complaints against the trawlers it was said that they destroyed the spawn and fry of fishes. It may appear strange to us at the present time that such complaints ever received serious attention, but forty years ago people did not know that the spawn of fishes—that is, those referred to in the complaints—floated up in the water whilst the fry lived near the shore. Hence it was thought advisable, as a practical measure and as a means of soothing obdurate complainers, to try and increase the stock of fish in the sea—on the assumption that the trawlers removed so many fish that nature could not maintain the stock. In various countries extensive hatcheries were established where, following the experience gained with the freshwater fishes, the spawn of various kinds, Cod and Plaice particularly, was specially obtained and kept until the fry hatched out. The comparison with the freshwater fishes broke down completely, however, when it was found that the young larvæ could not live through the artificial conditions and had to be set free in the open waters before the yolk-sac was absorbed, thus before the most difficult stage in their career began. As a practical fisherman once put it, the whole operation looked like "blowing tiny bubbles into the sea," and judged from the standpoint of the present time the simile was quite justified. Apart from the improbability of these artificially reared larvæ surviving, it has long been known that the sea does not suffer from a lack of young fish; quite the reverse is the case.

It was thought also that the trawl was most destructive to the ground-feeding fishes, Gadoids and Flat-fishes, because they were unable to get out of the way of the apparatus. This assumption to some extent underrated the intelligence of fishes and their ability to learn by experience. So far as the Plaice and Sole are concerned, every fisherman knows that of half a dozen trawlers fishing after one another through a narrow way between rough ground—a favourite practice of trawlers in spite of natural and artificial restrictions—it is not the first coming that is best served. Trawling is an art that requires more study than lining, and the fisherman that knows how to change his gear to suit the fish and the time is the one that succeeds best. The Flat-fishes in particular seem to have become as wary as the Trout in its element. A change in the gear or a new fishing apparatus catches them for a time in greater quantities than before, on grounds supposed to be fished out; they are caught unprepared, as it were, but they learn to overcome the difficulties. This factor has to be taken into account, for several reasons.

When Plaice have been marked and set free again, the percentage of recaptures within a year thereafter has not exceeded 50 per cent. of the liberated specimens, except on one or two occasions in the Kattegat. We may conclude, therefore, that at least one-half of the stock escape from one year into another. And it has to be noted that even if the Plaice could be swept clean from the open grounds, an unwarranted assumption, about 13 per cent. of the North Sea region is not suitable for trawling and thus offers protection to the species.

There is always the feeling, however, that something should or might be done by man to increase or improve the stock of fish in the sea, as agriculturalists have improved the yield of the land. And from time to time glowing accounts have appeared of growing Plaice " like cabbages in a back garden." This, indeed, has to some extent been accomplished in the Limfjord, where the clever Danish naturalist, Dr. C. Joh. Petersen, has developed a system of helping the young Plaice over barriers at the mouth into the rich feeding grounds within. In imitation of this, experiments have been made in the transportation of Plaice from the inshore areas of Denmark and Holland out to the Dogger Bank, where they have thrived very well, in small quantities. But the North Sea is not like the Limfjord. In the latter there is no competitive fishing and no Plaice except the transplanted, and a small number makes a big difference. To make any appreciable difference in the forty-five million kilograms of Plaice landed annually from the North Sea would necessitate the transportation of more than the known fluctuation, that is, more than five million kilograms, or more than forty millions of young Plaice. And then they would have to be " protected " there, as in the Limfjord.

It has been thought also that the decrease in the quantities of large Plaice is a cause of anxiety. This phenomenon has now been fairly well studied in a number of regions, on Norwegian banks, in the Kattegat as well as in the Barents Sea. It means that the old, slow-growing " accumulated stock " on a new fishing ground is more or less rapidly fished up and " current stock " growing more rapidly takes its place. For example, Heincke has found that the large Plaice of the Barents Sea reach an age of sixty or more years, but the largest Plaice of the North Sea are only about twenty-five to thirty years old. The largest Plaice are not of much use as food, and there can be no doubt that fishing improves the quality and rate of growth of the fish. The question is, therefore, whether nature can provide an optimum quality and quantity at the same time or whether man's assistance is necessary to regulate the complicated matter. Petersen has recently found in the Belts of Denmark that an intensive fishery has led for a time to an increase in the annual catch and rate of growth of the larger Plaice.

One of the strange paradoxes frequently met with in nature has arisen from the intensive study of these questions during the past twenty years. As mentioned above, it was formerly thought that the removal of the large fish, by diminishing the number of spawners, would diminish the supply of young fry. But it has been found that the young fry are actually too abundant and crowd one another on the nursery grounds to such an extent that there is not enough food for all. Possibly the amount of food varies from year to year. Hence we have to recognise that the removal of large quantities of young Plaice—however wasteful it may seem—can be compared, for the present at least, to the gardener's thinning out of his young carrots and onions, etc., in the spring-time or, as Petersen has it (1921), to "a lawn which is cut many times per year, in lieu of once every second year."

Another example of the complications to be faced by investigators of these problems may just be mentioned. In theory the proportion of the sexes of the Plaice should be equal and the samples taken by the investigation steamers in the North Sea have yielded comparative equality. But we do not know that theory and practice agree. From the proportions among the freshwater fishes of North Europe it is possible that the males outnumber the females in the proportion of 2 or 3 to 1. An analysis of the catches made by Lowestoft trawlers during the winter months, the season when the Plaice are mostly sought after offshore, has given the proportion of about 7 males to 1 female. Which result are we to believe? But, it should be noted, the males and females do not behave in the same way; the males mature earlier and migrate outwards from the nursery grounds sooner than the females ; further, they are more active and do not always frequent the same grounds as the females. Hence different areas and different seasons give different results. Who can tell the normal proportion of the sexes in the sea ?

A picture of the immense quantities of young Plaice which crowd yearly into some of the shallow waters of the Eastern North Sea has been given by Lübbert (1925). In a small area at the mouth of the Elbe, fishing with a shrimp-trawl of 27-foot head-line, he obtained from 300 up to 1,000 Plaice in hauls of 15 minutes' duration in the months of May to October 1922. These were all less than two years old, thus born in the years 1921 and 1922. In September and October the great majority belonged to the 1922-group, thus to the time when the accumulated war-stock had already for the most part been fished up.

The destruction of young Plaice in certain parts of the North Sea has been going on for a long time, but every new investigator is impressed by the great waste and would like to find some means of stopping it. How old the practice is may be judged from the words written by Holt in 1893 : " My information goes to the effect that the wholesale destruction of small fish on these grounds had been going on for many years (ever since about 1830, when British trawling vessels seem first to have taken to fishing the Dutch grounds) before it attracted public attention. It escaped notice for this reason, that as long as the supply of fine fish held out, the small ones, which were at least as numerous as they are now, were shovelled overboard, and thus never made their appearance in the market. The same thing is going on at the present day, only the items of the catch which are too small for the present market are indeed minute."

These examples will serve to show how complicated the problems of the fisheries are. One may doubt, indeed, whether they are to be solved by the microscopic methods necessary for the exact determination of details. It is advisable to take a broad view, and this can be done by the fishing industry as well as, probably better than, any others. Mountains have been in labour now for many years, trying to find something practical to show, but the only result has been a very tiny mouse.

CHAPTER XIV

THE MENTAL LIFE OF FISHES

We have seen in Chapter VIII that the fishes may be in league with elemental forces, powers of darkness, which man is only beginning to discover and which he is still far from understanding. We turn now to the lighter side and seek to discover whether they do not also display the more humane forms of radiant energy, intelligence and feeling. Formerly, it was the custom according to one school of thought to place a wide gap between man and the lower animals, the former being credited with reasoning powers, the latter only with instinct. Other schools of thought, however, including most naturalists, have emphasised the essential unity of structure and activities, and some, indeed, have gone rather far in ascribing human powers and tendencies to lower animals. The differences in these views are, perhaps, more wordy than real, but it is possible to avoid discussion by steering a middle course of scepticism towards extremes.

1. Tropisms and Reflex Actions

From the operation of our own minds we infer what is going on in the minds of animals ; there is no other method open to us. To say that the actions of animals are automatic and instinctive simply leads us in all fairness to inquire whether our own actions and those of our neighbours are not just of the same kind. We have, then, to admit that automatic actions play a dominant part in our daily life. In fact, the more automatic our operations become, the better they are adapted to satisfying the end in view. This makes just the difference between the skilled or trained worker and the unskilled. The man who uses his brains more than his hands must "train" in his own way, and this leads to instinctive associations and ways of thinking.

In interpreting the phenomena of life, therefore, we must regard all actions as potentially instinctive. We sometimes speak, for example, of the social and moral instincts which bind human beings into a more or less well-organised community, and classify actions according as they agree or disagree with these instincts. But the study of history has shown that these have developed from cruder forms and have to be unlearnt from time to time and transformed into new ones by the light of intelligence and reason, and then again sink into the subconscious world of instinct. We have no reason to believe that the lower animals learn by a different process.

On the whole, fishes do not belong to the class of sociable animals; self-help is the dominant instinct. Yet examples are not wanting where individuals combine to carry out a common work. The spawning habits of the Brook-Lamprey (Petromyzon), so well described by Bashford Dean and Sumner, provide an instance of this kind. The Lampreys come together when making a nest and move the stones about by means of their suctorial mouths. But these authors are careful to point out that there seems no method in all this work. It is more a kind of play, the instinctive use of an apparatus they have acquired, just as we employ our limbs in sport.

Though not strictly speaking sociable, fishes are certainly gregarious. The single individual living by himself, a shark for example, is the exception. They live together in schools or shoals, and it is on this instinct that the fishing for them depends. That the Trout and other freshwater fishes do not do so may be taken as a sign of intelligence, a lesson learnt, a subject to be discussed later. In most cases we can ascribe a purpose to this gregariousness and say that they assemble on certain grounds to spawn or to feed, but behind these phenomena is a more general instinct which we cannot fail to recognise. Perhaps the best example is that of the little Sticklebacks, especially the freshwater forms. In the summer they come into the shallow water and play about in the sunshine. In the autumn and winter they retire into deeper water and roam about in large companies; solitary individuals are seldom met with. The same is true of the marine fishes, and it would almost seem, in some cases, as if they recognised a leader or posted sentinels.

Instinctive or subconscious actions may be resolved into two elements, the reaction of the whole organism to an external stimulus and the reaction of a part only. The former are generally known as tropisms, the latter as reflex actions. These grade into one another, but on the whole can be readily distinguished.

The tropisms constitute the most important part of the activities of all animals and plants, from the lowest to the highest. In the simplest forms of life, unicellular organisms, directed activity is apparent in the presence of diverse stimuli. Some move towards light, for example, others away from it, and other forms of energy, gravity and chemical composition of the medium, lead to similar reactions. Within the cell itself it would seem that movement comes from local differences in the protoplasm (Pantin, 1923), partly chemical and partly physical in their nature, which are connected with, though not altogether determined by, the varying stimuli. In the digestion of food and the movements of leucocytes in general the many-celled animals, including fishes and man, are absolutely dependent on these instinctive activities of simple cells.

It is not only the single cells, however, that display tropism; the whole body of a highly specialised organism may act as one piece in the same way. We know how insects and birds are attracted to a light in the darkness. Many animals, like the Earthworms, remain in their burrows by day and creep abroad at night; they respond to a certain intensity of light-waves. Similarly, the Eels migrate by night, and Petersen has shown that the light of a powerful lantern will induce them to deviate from their course and collect in a narrow passage beyond the rays. They shun the light; other fishes are attracted. The only method of fishing for Sticklebacks, a valuable fish in some ways, is by means of an artificial light. Baron Cederström has described the behaviour of these fishes in the following manner: "When the light approaches them, they generally remain quite still at first for a few moments, apparently unconcerned; but suddenly a fish starts up, casts itself to one side, and joins some comrade disturbed in the same manner, or takes its place in an army already formed and advancing in the immediate neighbourhood, an army which like a long, moving wall roves along the bottom, at first, as a rule, in a circle round the light. Gradually the advancing wall increases in height, length, and breadth, while it circles hither and thither, as if to collect more and more stragglers, in curves of greater or less extent, until at last, when the army appears to be sufficiently strong, it suddenly dashes up and assembles beneath the light. Here the crush that now follows is tremendous, and the movements of the fish culminate in a strange confusion, exactly as though they intended with their immense numbers to overpower and vanquish the fire. In spite of the hand-nets now plied, they still rush on, undaunted as before. When their numbers are so few that the fisherman does not think it worth while to use the net there any longer, he moves to another spot between 50 and 200 yards off, or sometimes even nearer his former station; and here the same occurrences are repeated" (Smitt).

Other fishes are affected in the same way. In the Mediterranean the fishermen make regular use of lights for the capture of certain migratory fishes (Gast). The adults of the Tunnies, Auxis, and Pelamys do not seem to be attracted, but the young are, as also Sardines, Anchovies, Sprats—even a Cuttle-fish (Loligo), Rays and Dog-fish. Some are attracted as by a magic craft, others, like the Hake and Mackerel, come after the food, Crustacea, Worms and young fish which collect in the circle of light. The Sardines, Anchovies and Sprats, however, seem at first to try and get away from the light, but it draws them nearer and nearer. And when the light is strong and kept going for some time even the Mackerels, which have been making lightning darts at the food out of the darkness and away again, come at length under its influence. In fishing the boat has to be kept quite still, without movement; on the other hand, the fishermen can talk or shout as much as they please without disturbing the fish.

This powerful effect of light, however it may act on the organism, resembles greatly the phenomena produced by hypnotism. The regulatory system and central control of the fish are completely mastered and the fish turns and moves towards the source of the radiant energy. The "curiosity" which leads animals to their doom at times is not quite the same thing. The latter may be a testing or trial movement, the former is a fascination, as uncontrollable as the attraction of iron filings to a magnet.

There are many degrees of radiant energy as well as different kinds, and fishes are adapted accordingly. During certain conditions, as at the spawning period, these influences are more dominant than at others. The tendency to seek certain optima of pressure, temperature, light or darkness affects the whole body and is quite beyond the control of the animal. In the ordinary life of the fishes tropisms play a great part. When we say that they are adapted to salt or fresh water, to tropical or temperate zones, etc., we mean simply that their whole body is tuned in definite ways, so that if a change in the conditions occurs, externally or internally, the fishes are urged to change their abode. This sometimes leads them into strange places. A few years ago, when an old house in Ireland was being pulled down, a number of Eels were found in the rain-pipe of the roof. In their efforts to get to a region of lower pressure they must have wriggled up the drain-pipe—no mean performance.

During experiments made in Sweden some years ago

(Schäffer) Eels were carried to a distance away from water and then set free. They invariably found their way to the nearest stream, though miles away, and that with hardly any hesitation. Is this tropism or a water-sense ?

Reflex actions or motor responses to particular stimuli embrace a wide range of activities. Here we may think of the brain and spinal cord as a sort of mirror. From without a stimulus passes along the sensory channels to this mirror, whence it is reflected on to the muscles without any deliberation or consciousness of what is happening. We instinctively close the eyelid when anything approaches too near, the hand closes on seizing an object, the mouth opens and shuts with the entrance of food, and so on. The nerve paths from the sense organ receiving the stimulus to the muscles responding form a continuous course.

Many of these simple reflex actions can be studied in the fishes of an aquarium. When a fish is resting, we can see the breathing valves opening and closing to take water into the mouth, a slight rippling along the vertical fins or a beating of the pectorals shows how the fish is balancing itself, and when it moves we can see how the swing of the tail is sending it forward. It does not matter what sense is stimulated, the balance, smell, or eyesight, the response seems to come automatically.

The reflexes are very complicated when more than one set of muscles come into action. To get at its food the fish may have to turn and twist about or change its direction up and down in the water. The central organisation in the brain here plays a part in preventing the inappropriate muscles from acting. Using the simile above, we may say that several mirrors are in the path of the impression, but the correct one is chosen at the right time. But the simile has to be changed when we come to still higher levels of action, when the fish, for example, has several courses open to it, as when offered different foods, and definitely selects only one of them. In his experiments on Eels Schiemenz pressed some food with the fingers of one hand, then placed both hands, no food in either, into the water. The Eels came and rubbed their mouths on the hand with the smell of food on it, but paid no attention to the other.

As a matter of fact, the simile is only useful for the simplest kind of reflex action, but its attempted use helps us to appreciate the wonderful powers of co-ordination between sense and muscles, even in a fish. How this co-ordination may have arisen is a question that should be considered, even if briefly.

According to Darwin, instincts are to be regarded in the same light as other characters; they depend upon some inherited modification of the brain. Their usefulness is a sufficient explanation of their existence. Since instincts vary, their origin may be referred to a natural selection of the better with an elimination of the worse. Darwin did not deny the possibility of intelligence taking a part in forming an instinct or habit which could be inherited—such as the fear acquired by birds on uninhabited islands after man had come amongst them—but he considered that the great majority came from the natural selection of small variations of unknown origin or cause. Weismann carried this view still further and cut out the intelligence and acquired inheritance altogether; all instincts arise from variations in the germplasm.

The difficulties in the way of accepting the selection theory as an explanation of the phenomena of differentiation in fishes have been discussed in previous chapters. The same difficulties are met with here. Some of the Three-spined Sticklebacks, for example, do not build nests, others do; who can say which is the more useful instinct? Weismann's endeavour to derive characters useful to the adult from variations in the germplasm—though a logical conclusion from Darwin's position—shows, indeed, the inadequacy of the doctrine of utility.

Both Darwin and Weismann have taken a very restricted view of the inheritance of acquired characters, including thereunder only the morphological features and instincts. From the past history of fishes, however, we know that whole populations have been wiped out at various times. The species that at present occupy the fresh waters of Britain, for example, have arrived since the Glacial Age. Some of them are relicts of a more northern fauna, which have become so specialised and differentiated in separate waters that each lake contains its own peculiar race or species. We must conclude from this, as from variations in general, that the environment cannot be left out of account.

Taking this into consideration, the simplest explanation of instincts is, that intelligence was present at their origin and that improvements came from the acquirement or learning of the co-ordinations at increasingly early stages under the same environment. The modifications in the nervous system have arisen, in fact, in the same way as the modifications in the external characters.

Even the inheritance of instincts acquired or learnt by the individual during his lifetime is a possibility that cannot be excluded. One or two or half a dozen generations do not provide a fair test. We have to take the time factor into account; in the case of the above-mentioned fishes at least ten thousand years have passed since they became isolated in British waters. And we know that among human beings faculties have been handed down from parent to offspring until after a few generations they seem to be inherited instinctively. Apart from the inheritance of the same environment, it is possible that the regulation of the nervous system in particular ways may have a definite effect on the germplasm.

Recent physiological investigations into the nature of the central nervous system are able to throw a good deal of light on these complicated problems of reflex actions and instincts. Certain paths or connections between different parts of the organism may be marked off as definite and fixed; for example, the connection between the eye, optic lobes and the chromatophores of the skin; similarly, the connection between the lateral line organs, the tuber acusticum and the muscles of the fins. These may be called definite reflexes. But the outstanding characteristic of the brain-cell is its plasticity, its power of amœboid movement. To this power is ascribed the ability to control opposing tendencies, so that, for example, the closing muscles of the mouth do not operate at the same time as the opening muscles.

It is evident that this power or condition of the brain-cells brings forward the question whether instincts in the old sense really exist. It implies a constant watchfulness on the part of the organism, and the disturbance of this watchfulness by shock, poisons, or abnormal conditions is perhaps the best external indication of its existence. Instinct, therefore, if the word is to be retained, has to be kept distinct from reflex actions.

2. Intelligence and Adaptations

It is often said that the instincts of animals are replete with intelligence and we may take this to be literally true in theory and in fact. In addition to their presumed utility or purposefulness they display a power of selection on the part of the organism which forms their essential quality. It does not matter whether this power is determined, in a metaphysical sense, or not. If it is, the activities of all forms of life from the highest to the lowest are brought within the same category. For all practical purposes, however, the instinctive power of selection may be said to come from intelligence, to which, as shown above, we can ascribe a physical or biological basis.

As the result of the intensive investigations of fishes in recent years, it has become known that each species has its own particular food, or rather, that it prefers a certain kind if it can get it. The Sea-Trout when young will eat almost anything from a leech to a frog, but as it grows older it shows a marked partiality for the larvæ of insects. If other insect-loving fishes are living in the same waters, however, it suffers from want of food even though it might eat the fishes.

There is often, too, a seasonal variation in their likings. The female Plaice, after they have spawned, seek out the grounds where they can live mostly on a worm diet, the main diet of the Dab, though before that they prefer certain Bivalves. The Herring lives a great deal on the Crustacean plankton, but at certain seasons it largely prefers this food already prepared for it in the form of Sand-eel (Ammodytes).

Even the young larval fishes exercise a choice when different foods are present. If the food is too large for their mouths, they do not touch it, but if they prefer a small species they will leave a larger species alone though quite able to swallow it. When young Whiting of 16–19 mm. in length were placed in a tank in an aquarium, they were fed on Calanus and ate them greedily. But when Pseudocalanus and Temora were put into the tank along with Calanus, the Whiting took Pseudocalanus in preference. "It would never eat a dead Copepod and always refused dead food of any kind or chopped worms. It is also very fond of Acartia and prefers it to Calanus" (M. Lebour).

Regarding the mode of feeding of young fishes this author also writes: "Some deliberately stalk certain selected food, others apparently eat the first thing that comes. Some feeding at all times, others only at night. . . . The Whiting, of all the young fish observed, is the greediest, and very clever at recognising the different foods and selecting the best first. The Pollack comes second. These stalk their food, usually Copepods, coming up behind and giving a curious sideways dart so that the Copepod is swallowed head first."

As soon as they are old enough to make their way against currents—and this they are able to do at a remarkably early age—the little fishes separate and make for different grounds or waters. Johs. Schmidt has provided a classical example of this form of selection in the case of the Eels. The Leptocephali of the American and European Eels keep company together in the Gulf Stream as far as off the Bermudas, then the American Eel makes for the west, the European Eel for the east. The young of the Flat-fishes distribute themselves in a similar manner; the Flounder into brackish and fresh water, the Plaice along the coast, the Lemon-sole further out in deep water, never inshore, and the Pole-dab or Witch in still deeper water. Again, the parents of many species have the bad habit of eating their young ; the young seem to know this instinctively and hide themselves as soon as possible among the weeds or wherever they can find shelter.

This is all instinct or ancestral memory, one may say, but it is likewise all intelligence, of the same order as when we try to avoid the traffic in a crowded street or take to the life that suits us best. How long it has taken the different species to learn in this way, it is hardly necessary to inquire, but we may be certain that in the learning the balance and structures have been effected under the guidance of the intelligence.

In the case of mutations the first step towards differentiation was beyond the control of the individuals ; yet the main point is, that to survive at all has required a great deal of intelligence. If we could imagine ourselves in the position of a Flute-mouth or a Pipe-fish, we should have to picture the difficulty of earning a living, in the midst of plenty, whilst suffering from lockjaw or something like it. It is just among such deformed fishes that we find most signs of intelligence.

People are apt to think that fishes are stupid The Gadoids, Herring and especially the Sand-eels are probably stupid by comparison with the Flat-fishes, the latter by comparison with a Stickleback or a Trout. It is all a matter of comparison and no fisherman would agree that any fish is stupid. It is said that Carp placed in a running stream elude all traps and nets in a most knowing manner and the hook has to be cunningly baited and placed to capture them.

The Common Grey Mullet has long been known as one of the most cunning of fishes and how singularly the aptitudes are inherited is noted by Couch. " Even Mullets of extremely small size have been seen to throw themselves, head or tail foremost, over the head line of a net, where it would have seemed much easier for them to have passed through a mesh." In another place he writes: " In the port of Looe, in Cornwall, there is a saltwater mill pool of thirteen acres that is enclosed on the side of the river by an embankment, and into which the tide flows through the flood-gates that afford a ready passage for fish to the space within. When the tide begins to ebb the gates close of themselves, but even before this has happened the Mullets which have entered have been known to pass along the enclosed circuit within the bank, as if seeking the means of deliverance, and, finding no outlet, they have thrown themselves on the bank on the side to their own destruction."

Day has given us a similar picture. "At Mevagissey a shoal entered the harbour, and having been perceived, the entrance was at once barred by nets. The fish first tried to jump over, but a net was raised so as to bar that route. The water was very clear, and the fish were seen to swim round and round, to try to find an exit. Next they attempted to get under the foot rope; at last one made a push, but became meshed. When this was done, another came and lay down beside it, and nothing could drive it away. In short, all escaped but these two." Couch also states, that the ancient Romans (quoting Pliny) used to employ one of the Mullets as a decoy to catch others, such was their fidelity to one another.

Among the more obvious examples of intelligence the habits of the Pipe-fishes must be placed in the front rank. They can move about on excursions and are not always found among seaweed. Couch describes the Aequoreal Pipe-fish as swarming on the surface over deep water during the summer evenings, and other species have been observed skimming over the water and even leaping into the air. But when they do come among seaweed or inshore, they take to surroundings that in a marked degree resemble themselves.

Heincke writes as follows: "The power which the Syngnathi possess of adapting their colour to their environments, is the most perfect instance of the kind that we know among fishes. If we place some Deep-nosed Pipe-fishes in a large aquarium, together with a quantity of seaweed

(Zostera marina) such as that which grows in the native haunts of the fishes, after some time we may make an interesting observation. The leaves of the seaweed have partly risen vertically or obliquely through the water, and stand motionless, slowly swaying to and fro if the aquarium be slightly shaken. Among them, motionless or swaying with the leaves, the slender Pipe-fishes have chosen their positions, and we can only just see how the gill-covers expand and contract, or how the perfectly transparent dorsal fin ceaselessly continues its vibrating and undulating motion. The colour of the fishes, often down to the most delicate shades, is exactly like that of the seaweed. Often we imagine that we are gazing at a blade of seaweed, and only on closer inspection do we discover that it is a Pipe-fish. . . . Among the green, still living seaweed there lie here and there a number of partly or entirely dead leaves, in all shades of colour from green to dirty-brown and brownish-black. At these spots the Pipe-fishes assume a different hue, their colour passing gradually, according to their different surroundings, into brown or brownish-black, until, whether their position be vertical or horizontal, they are scarcely to be distinguished from a dead blade of sea-weed" (Smitt).

The colour is, however, only one of the resemblances. Those species which attach themselves to seaweed select the various kinds that suit them best; the Sea-horse (Hippocampus) and Phyllopteryx prefer Fucus, the Worm Pipe-fish is brown in colour and lives among the brown algæ and among stones, the large Nerophis lives over sand and in the Laminarian zone, but the Straight-nosed Pipe-fish (N. ophidion) attaches itself to the floating seaweed called Chorda filum and the Deep-nosed Pipe-fish (Syphonostoma typhle) to Zostera.

Behind the many changes that have taken place in the structure and habits of these and other fishes, intelligence must always have been present, not causing any change itself but directing the movements and choice of environments that led to the changes. Since intelligence is weak by comparison with the forces ranged against it and has to reconcile opposing interests in the different stages of a short life, we can well understand why adaptations are continually imperfect. A gain in one stage or direction is counterbalanced by a loss in another. If the Pipe-fishes, like the coral fishes, can see their own images in their surroundings and alter themselves accordingly, they have lost the power of swimming boldly through the water retained by the Herring and Mackerel families, which have not developed an æsthetic sense.

The æsthetic sense may be useful to fishes—why should it not be?—but for many reasons we cannot suppose it to have had a determining influence on the survival of the species. We may imagine, as many have done, that it gives them some sort of protection when they are not playing about freely in the water, but we do not know that they require any protection at that time in the inshore waters. Such a hypothetical explanation obscures the main interest in their proceedings, namely, that they select their own surroundings.

Almost all fishes, apart from the brutal Sharks and Rays, are timorous, particularly the smaller kinds and those that live in shallow waters. We may admit that this timidity is not merely a matter of nerves but is an acquired habit like any other and is directed by intelligence. But there are exceptions. The little Stickleback fears nothing and the not much larger Shanny and Butterfly Blenny are bold as brass, whilst the Wrasses are more or less indifferent. The explanation of the difference seems to be that the chief enemy of the shallow-water fishes is not its neighbours or its own kind, but the sun. The inshore fishes, like the freshwater fishes, avoid the heat and glare of the sun, and their pigment is believed to be a protection against its rays. The differences in the behaviour of these fishes may thus depend simply on the thickness of their skins.

Another sign of intelligence is the patience and persistence which some fishes display in their pursuits. When a Shark takes up his position in a particular bay, it might be said to live on hopes of future favours, which implies memory of good things in the past. The little Pilot fishes attach themselves to a big wandering Shark and conduct it to food by means known only to themselves. It is not merely to share in the crumbs, for they are quite able to catch small fishes for themselves ; it seems more a feeling of protection or perhaps an attraction to something bigger than themselves. What sort of language they use in communicating their wishes and discoveries to their comrade is not known, but many witnesses have testified to the fact of the communications. The Frog-fish (Lophius) which lies in wait and lures on its prey, must have learnt to exercise as much patience and having learnt it, retained it in memory, like the most experienced angler for Trout.

The cleverness of the common Eel is almost uncanny. Not only can it climb slippery posts and wander over fields without the aid of ventral fins or legs, it shows a remarkable sense of knowing what it wants. Recently, it was found that an Eel placed in a tank of the aquarium in the Zoological Gardens, London, had managed to escape. It was discovered in a tank some distance away and replaced in the original one. A few days later it was again missing and again found in the same tank ; being evidently so decided in its preferences, it was allowed to remain there. As it had passed another tank on the way, it must have had some idea in its head. Still more wonderful is the way it gets out of a tank. If it cannot squirm up the side head first, it turns round and throws its tail over the ledge, gets a hold and swings its body up and over.

It was a common belief among the country people of Sweden, that the Eel took to green food, peas and the like, when wandering over the fields, and Smitt describes how special efforts based on this habit were made to capture it. Most authors, of course, are sceptical about such stories, but Smitt, usually very careful to distinguish between fact and fancy, comes to the conclusion that this particular one cannot be rejected. There is no doubt that the Eel does eat vegetable food and its partiality for peas—or rather the moisture and slugs connected with peas—is quite intelligible.

Many other fishes besides the Eel leave the water and wander about overland, some from necessity, others from choice. A Pike when trapped in a drying-up pool has been known to wriggle out and cross over to a stream near by. The small Cyprinodonts of Mexico and Central America are said to hop about on land like fleas until they reach water again. The Climbing Perch (Anabas), whose abilities have erroneously been called in question by sceptics, has often been observed making its way from drying-up pools to other water basins. Bolau has described how some specimens kept in the Zoological Gardens at Hamburg persistently leapt out of a small tank, over a partition two feet high, and scrambled about on the sand of the reptile house. In a larger tank they behaved better, but they always leapt out of the small one—evidently they objected to narrowing restrictions and preferred a wide perspective.

The mud-springing Gobies (Periophthalmus) of the tropical waters of Africa, Asia, and Australia, prefer to seek their food out of the water. They live like frogs on marshy or muddy ground among the mangrove swamps or along the shore between tide marks. Sometimes, with the aid of their tail and ventral fins, and clever bending of the body, they clamber up stems and slide along branches. When discovered or disturbed they drop back from the branches, several feet above, into the mud below, and are said to remain out of the water for hours together. Hickson, who studied the habits of these fishes in Celebes, has given us the study from life shown in Pl. XIII*a*, and the American authors Jordan and Seale have also drawn it as they found it along the shores of Samoa (see cover). "This extraordinary little fish abounds especially in muddy bayous, freely leaving the water to climb bushes, to skip through the grass, or to lurk under piles of stones to await the returning tide. It is exceedingly quick of movement and tenacious of life. Specimens placed in a pail of formalin escaped when the lid was raised."

As the name implies, Periophthalmus is able to turn its eyes in all directions. It may not be a beautiful fish to look at but its half-comical, half-serious and wholly intelligent appearance is perhaps the best expression one can find of the mental life of fishes. Some specimens have been brought to Europe and, placed in tanks with suitable opportunities to exercise their proclivities, they refrained from leaping out and soon adapted themselves to the conditions. They proved themselves neither shy nor stupid.

From the biological standpoint the most interesting feature about these fishes that wander about on land, is that some have organs specially adapted to this mode of life, others not—a difference that the natural selection theory is quite unable to explain. The climbing fishes (Anabantoids) have developed an accessory breathing organ (Fig. 68), which is believed to be of assistance in assimilating the oxygen of the air, but the others do not have it. The ventral fins of Periophthalmus are no doubt of use in sliding along slippery stones and branches, but the Eel has no ventral fins at all. The spines of the gill-cover and fins of the Climbing Perches enable them to get a hold as they leap over the ground, but the Carplings (Cyprinodonts) have no spines. Eigenmann has described how he once, accidentally, when crossing a river creek, caught a Rivulus in his hand. Surprised for the moment he let it go and the little fish sprang on to the vertical face of a rock, where it held on by its tail. Instead of then dropping into the water, however, it sprang upwards to a higher spot on the same rock.

Flight over the land is not the only resource of fishes. When a muddy pond known to be populated by Eels has been drained, the Eels have apparently disappeared and few are to be found. As the pond dries up they bury themselves deeper into the mud, even to a depth of two or three feet. Here surrounded by their own slime they wait until the water returns; in the winter months they also bury themselves in the mud. Many other freshwater and brackish-water fishes have learnt the same habit, the classical examples being the Dipnoi. Protopterus and Lepidosiren bury themselves in a sort of burrow, lined by the mucus secreted from the skin, when the dry season approaches. They do not wait until the ground has become too hard.

Of the fishes so far mentioned, the Eel, Pilot-fish, Angler and the others, it would be difficult to say which is the cleverest, but, if all stories are true, the palm has to be awarded to another. One has heard of birds being trained to catch fish and other birds, few would believe it possible that a fish could be used for the same purpose. This is the famous Sucker-fish, Echeneis, a not very distant relative of the Pilot-fish and one already noted as also keeping company with a Shark. Many observers have described its performances; Haddon on the Cambridge Anthropological Expeditions to the Torres Straits, Gill and Semon from the same region, Bullen from China and Lady Brassey of the *Sunbeam* from Venezuela. More recently, Gudger has collected these references with additional observations of his own into an interesting account of this wonderful fish. It appears that in the year 1494, when Christopher Columbus discovered Cuba, he found the natives there using this method of fishing. The sporting tendencies of Echeneis have thus been exploited in all parts of the world; did the natives teach the fish, or the fish the natives?

The method of fishing is essentially the same everywhere. The fish is readily caught and a line is fastened round its tail; the fishermen take it with them, either inside or attached outside by its sucker, when they row out in search of turtles—the chief object of the fishing, though in Australia even the Dugong has been taken in this way. When they have approached a turtle, this naturally disappears from the surface, but the Sucker-fish is then loosened and goes in pursuit. How it finds the turtle is known only to itself, but it does and fastens on to its back. The fishermen know from the line when it has struck and approach as near as possible. There is no fear of the turtle escaping, for the Sucker-fish will be torn in pieces rather than let go until the right moment. Its lifting capacity, however, is but small, as Gudger found by experiment and the natives apparently by experience. One of them dives down and fastens another line on the turtle. And the most remarkable thing, vouched for by Haddon, is that when the diver arrives and gets hold, the Remora shifts its position from the upper to the lower surface. No wonder the natives have a great respect for the Sucker-fish : " Gafm savey all same man. I think him half devil." And they seem to have, or imagine they have, mysterious communications from the fish. If it is disinclined to work, they say it is a bad day for them and do not work either. If it goes in an opposite direction to their wishes, they do not seek to hinder it. Where it points, they follow—the fish is master of the proceedings.

3. Reason and Parental Care

When the glowing light of intelligence displays itself in a series of actions implying memory and association of ways and means to a definite end, we may say that reason is present in the mind of fishes. This may not be reason as some philosophers would have it and perhaps we should insist upon ideas being also present. They may be ; can we credit ideas to the fishermen and none to the Sucker-fish, in the case just described ? It would be more difficult to say when reason is not present, however we may define it. But avoiding formal definitions, we have simply to compare the actions of fishes with those of our neighbours, and each one is at liberty to decide whether reason is present in either case or not.

One of the tests of reasoning powers is whether an animal is able to learn from experience. We know that a burnt child avoids the fire, but how long would it take an ordinary man to give up smoking, when he learned it was not good for him ? We have information that fishes undoubtedly learn and profit by their learning. Their behaviour under natural conditions, the avoidance of nets by the Flat-fishes and Carps, already referred to, may be taken as evidence. It was known already to the Romans that Carp were not to be caught by ordinary means. They may be taken on hooks carefully baited; they play with the bait, even taking it into their mouths, but if the hook is felt or at all in evidence, they spit it out bait and all. The Bass and the "podlies" along the piers have learnt the same trick. Anglers can tell many stories of the big ones that escaped.

All fishes probably learn by experience. That hunger drives them to learn, is perhaps not a fair test of reasoning. Yet it must be noted that when in captivity fishes are very punctual at meal-times. The Chinese have a gong or bell to summon the Golden Carp to their food, but it is more probably the sight of their keeper that collects the fish together from all parts, as fishes have no apparatus to appreciate the wave-lengths we hear. On the other hand, we know from experience that an incautious footstep near the bank will send an impression through the water, that the Trout does not wait to "hear" repeated.

The little Carplings (Cyprinodonts) get so excited when their keeper approaches, they spring out of the water in their haste to get at the food, just as chickens leap in the air when the corn comes in sight. And who would think of training them to come one at a time in proper order, as children have to be taught?

Eels kept in a large pond or lake assemble at the proper time, but again we cannot tell whether it is the sight of the keeper, his whistling or bell-ringing, the smell of the food in the water, or their own persistent hunger that brings them together. But they are certainly always on the watch and know the precise spot where they are usually fed.

More direct evidence is available from experience. Möbius trained the rapacious Pike not to injure their noses against a glass partition, even though their choice food was on the other side. Triplett repeated this experiment with the Perch. In a large tank he placed some Perch on the one side and Minnows on the other side of a glass partition. To begin with, naturally, the Perch strongly objected to the partition and damaged themselves not once but often. In about a month, however—the Perch is not to be compared with the Carp or Eel in quickness of learning—they had gained so much experience as to pay no attention to the Minnows and the partition was removed. Even then the Perch did not venture into the Minnows' preserve. How long the absent partition remained in their memory is not told, but when one of the Minnows accidentally got amongst them and passed rapidly by, it promptly disappeared. On another occasion, after the partition was removed, one of the Minnows trespassed, but so long as it swam quietly about it was not molested. Then it took fright apparently, made a quick movement, and paid the penalty of its transgression. There is probably a psychological explanation of the latter phenomenon. Our nerves, for example, have become inured to the rapid movements of trains and motors, so long as they are at a safe distance, but if anything small, or let us say not too large, passes quickly in front of us, there is an urge within to seize it or say something. In her interesting notes on the behaviour of fishes in the New York Aquarium, Ida M. Mellen has emphasised this peculiarity of living nature.

In the Aquarium just mentioned an experiment was made to bring about an association between the Shark-sucker and a small Dog-shark (Mustelus canis). "When the former attached itself to the Dog-shark, the latter struggled for hours in the attempt to dislodge it, but finally accepted the inevitable. After some months of their intimate association, both species remain in good condition" (C. H. Townsend). The photograph (Pl. XIII) shows the nature of the association, and we can admire therein the reasoning submission of the rapacious Dog-shark to the master hand or rather head of the Echeneis.

In the exercise of their reasoning powers many fishes have developed strange habits and aptitudes, quite beyond our powers, and thus often treated as fancies by unbelievers. Could any one think it possible for a fish to shoot down its prey? Yet the small Toxotes jaculator (Pl. IV, p. 40), from the wonderland of Siam, has received its name just from this proclivity. Its methods are described in Brehm's Tierleben in the following manner: "The shooter posts itself in a horizontal position near the surface, but so that no part of the body reaches above. Remaining for a moment quite still, it fixes its eyes on an insect—taking aim—and then through the closed mouth, which is directed upwards naturally, it spits a small quantity, just a drop, of water in a direct line and with considerable force and rapidity right at the object, and seldom misses."

An observer, Meissen, has watched the proceedings of the Jaculator in the open. Like an experienced hunter it never gets too near its booty but " measures " the distance. When an unsuspecting fly alights on a water-plant, perhaps a foot above the water, the Shooter glides quietly along into position, aims for a moment then fires—a spring forward like a greyhound's and the insect is gone.

The same observer, according to Brehm, has kept the Jaculator in an aquarium. At first, the fish was exceedingly shy and hid itself among the vegetation near the surface; but after a few days it got over this fear, and then Meissen made the interesting observation, that it was able to distinguish between him, its keeper, and strangers. One day an ant crept up outside the glass tank and two Jaculators fired at it in turns—naturally without result. But they seemed to gather very soon how useless the proceeding was and turned to their keeper, as if to indicate they were hungry. He captured a fly and placed it about half a foot above the water on a piece of grass and retired a little. The fly was at once attacked from two sides and soon disappeared.

In the course of a few weeks, the Jaculators had become so tame that they would take anything from the hand of their keeper, when held four or five inches above the water. Then they also became accustomed to strangers and developed a game with them. The first case was perhaps accidental; an observer was shot right in the eye. But afterwards they practised on the nose, ears and lips, and seemed to do so intentionally from a sheer sporting love of the thing. " With what certainty and celerity the fish had learnt to shoot can be judged from the fact, that even when

PLATE XIV

Male Lumpsucker (*Cyclopterus*) guarding the Eggs; Female above.

(From photograph by A. Schensky, Helgoland Aquarium.)

one knew the shot was coming, at three feet away, one had no time to close the eye" (Brehm).

As already noted in describing the habits of the Sucker-fishes, Echeneis and Remora, it is quite possible that these aptitudes arose originally as a kind of game or sport without any reference to their usefulness. As the structures, head and mouth, gradually responded to their peculiar use, after many generations, the sport became their principal means of earning a living.

The Characin fish Pyrrhulina filamentosa from South America has also been thought to use its mouth in the same manner as the Jaculator, but with a different purpose. It appears that the male and female at the breeding time, in close embrace, climb out of the water and the eggs are laid and fertilised on land. Then the pair drop back and the male takes up his position near the eggs. One account has it, that it then deliberately squirts water over the eggs from time to time to keep them moist; but according to the more precise description in Brehm, the tail is the organ used for the purpose. The male does not remain constantly by the side of the eggs but returns about every half-hour and beats up the water over the eggs with its tail. The young fry escape from the eggs in about three days and slip into the water, where the parent pays no further attention to them. How does the fish know that the eggs require moisture?

This brings us to the extraordinary habits displayed by many fishes of taking care of their eggs and even building nests for them. Aristotle had noted and described an example of this kind in the ways of a Siluroid fish (Glanis) in the rivers of Greece, and though ignored and treated as fable, his account has been proved correct. Care of the eggs in one form or another is now known to be almost as common as the reverse. The different stages may be briefly enumerated. In the simplest cases the spawning fish select a retired spot, clear it of stones or make a hole in the sand, lay and fertilise the eggs, then cover them over with stones (Queensland Siluroid) or the male remains on guard (Blennies, Gobies, Cottoids, Lumpfishes), sometimes both (in some Cichlids). Again, definite nests may be constructed by the ingenious use of mouth, fins, and special secretions (Labroids, Sticklebacks, Macropods), or the male takes special charge of the eggs either in a pouch (Pipe-fishes) or in his mouth (some Siluroids), and sometimes it is the female that undertakes the latter performance (African Cichlids). Lastly, we have the viviparous fishes (many Teleosts and Elasmobranchs) which may be regarded also as showing parental care, though the origin of this habit has probably been of a different kind, as explained in a previous chapter.

It should be noted, however, that the habits are varied in closely related species. One Cottus takes care of the eggs, another pays no attention to them. Some Siluroids do not build nests, others do, and some take the eggs in their mouths, whilst a distant relative, Aspredo, broods over them to such an extent that they become attached to her body. This diversity is of some importance in helping us to arrive at an understanding of the way in which the habits arose.

The significance of this parental care, to the fishes, may be illustrated by some extreme examples. In a recent description by Gudger of the American Gaff-Topsail, a marine Catfish or Siluroid, we find that the male carries about 55 eggs in his mouth for a period of nearly 70 days. Each egg is 19–20 mm. (nearly an inch) in diameter, the largest fish eggs known among the bony fishes, and it would appear impossible for the fish to take food during the whole of this period. What can be the meaning of this self-sacrifice ?

Prof. McIntosh has given the following account of the paternal solicitude of the Lump-sucker for its eggs : " About the middle of May a male was found at St. Andrews, a short distance from low-water mark, in a broad runlet, with its head close to a mass of ova placed on the seaward edge of a stone. The stream of sea-water was so shallow as to leave the stone partly exposed at ebb-tide, and was quite insufficient to float the fish, which was 11½ inches in length. Accordingly, for a considerable period, twice daily, the devoted male had to lie in the runlet on his side, a portion of his body, including the region of the upper gill-cover, being above water. From the situation of the eggs on the stone just described, the current of the runlet flowed into the mouth of the fish, which, in the warm sun of June, must have been less comfortable than under ordinary circumstances (showing that the performance was deliberate). . . . For some weeks this faithful male was found at low tide in this position, sometimes on one side, sometimes on the other. In order to test the case still further, Dr. Scharff removed the fish a couple of yards from the eggs and placed it on a stone. It wriggled actively into the water, at once rushed to the eggs, and assumed its former position with the snout almost touching them. . . . The solicitude of the males for the eggs which they have under charge was further illustrated by the occurrence early in May of a heavy sea which swept masses of eggs from their positions all along the rocks. As soon as the sea became calm, numerous anxious males, like pilgrims, were seen seeking for their charges. While on guard, the males are frequently attacked by rooks and carrion crows—and yield their lives to the faithful discharge of their duty."

If it is not a sense of duty, what can be the meaning of these performances? But the word implies the presence of an altruistic instinct in fishes, and we are perhaps not yet prepared to admit so much. Observations on other cases seem to lead to precisely the reverse conclusion. We shall see presently, how these views can be reconciled.

One of the most famous nest-builders among the fishes is the small Stickleback of the coast and fresh waters. The marine species (spinachia) does not assume a bridal dress like the freshwater forms, but otherwise their proceedings are similar. The nests are constructed during the months of May and June and consist mainly of seaweeds, though any odd thing lying about, even a piece of rope, may be used for the purpose. In the marine form Fucus is the basis of the structure with added fragments of Ulva, Corallina and even Zoophytes. The whole are woven together by a thread-like, mucus material, like a spider's web, which is secreted and manipulated by the male.

From the investigations of Prince (1885) it appears that this mucus secretion is formed by the kidneys, either by the cells of the urinary tubules or the cells themselves degenerate and serve as mucus. In either case it is the excretory organ that is used for the purpose, and the male appears to select a suitable mass of seaweed and to attach to it by mere contact the viscid mass of mucus protruding from the urinary aperture. Then passing and repassing over and under the growing nest he binds all firmly together. When all is ready he goes in search of a female, usually not very far away, and induces her by various means to enter the nest, where she sheds the eggs. Sometimes she escapes and he has to obtain another or even a third, but when he has collected a sufficient number of eggs and fertilised them, he gets rid of the females in a summary fashion and takes full charge.

The nest building in the freshwater form aculeatus has been carefully described by several observers, and Smitt in the "Scandinavian Fishes" gives the following account (from Brehm) of the wonderful performance. "At times the male shook the building and then pressed it together again; at times he kept swimming over it. With his fins, which he kept in continual and rapid motion, he produced a current and thus washed away from the nest the pieces that were too light and the loose stalks, which he then took up again and tried to fit in more durably. It took about four hours to procure the various building materials; but at the expiration of this time the outlines of the nest were ready. Its completion, the removal of the parts that are too light, the arrangement of the separate stalks, the plaiting of their ends, and the addition of the sand to weigh them down, require several days. While the Stickleback is building his nest, he thinks only of his work and endeavours merely to provide against any interruption in its progress or hindrance to its completion. He labours indefatigably and watches with suspicion every creature that approaches the nest with or without evil intentions, whether it be another Stickleback, a newt, a water-beetle or a larva. A water-scorpion (Nepa) in one of Evers' aquaria was seized by the cautious builder thirty times or more, and carried in his mouth over to the opposite side of the aquarium.

"When the Stickleback has finished his building operations, he endeavours to attract a female to the nest. . . . The male distinctly shows his delight at the arrival of the female, swims round her in all directions, enters the nest, sweeps it out, returns in a moment, and tries to drive the female in by thrusting at her from behind with his snout. If she will not obey of her own accord, he also employs his spines, or at least his caudal fin, to overcome her reluctance; but in case of need another female is fetched."

If a man had to build a house for himself, could he do any better? We may call it instinct, but in the fitting of the different parts to one another, the selection of the materials and the fearless care taken of his house and property, the Stickleback surely displays reasoning powers of a high order. Some one has remarked, that we should be poor creatures to stand up against the Sticklebacks if they were as big as ourselves.

Whilst the house built by the Stickleback represents the highest level of craftmanship reached by the fishes, a more interesting method of adapting simple means to a definite end is shown by the Bubble-nest builders. These are freshwater fishes, mostly very small, belonging to India and the Malay Archipelago, with only one species in Africa. They have been introduced and their habits studied in northern countries, particularly Germany and the United States. Like other related fishes (Anabantoids) they possess an accessory breathing organ in addition to the gills, and the extra supply of oxygen thus obtained probably accounts for their high spirits and mode of building nests. The remaining Labyrinthici, as they have been called, the Ophiocephalidæ and Anabantidæ, have not learnt this habit, however, so that the Bubble-nest builders (Osphromenidæ) are specialised offshoots. It is the practice among them all to rise to the surface and take in gulps of air, and in all, especially the males, the lips are thick and swollen.

The nest building of the Rainbow Fish (Trichogaster, Fig. 68, p. 271) was carefully studied and described by Carbonnier as long ago as 1875. "As the spawning time approaches, the male, spreading his brilliant fins, plays round the female, showing her his bright colours; with his long ventral filaments he pats and touches her in all directions, until over-excited by his caresses, she takes to flight. I believe that all these graceful movements of the male fish, all these amorous proceedings, influence the physical condition of the female and aid the maturation of the ova.

"The male fish then commences the preparations for oviposition. Seizing a little Conferva in his mouth, he carries it to the surface of the water. The plant, from its greater density, would fall back very rapidly to the bottom; but our little workman sucks in a few bubbles of air, which he divides and places immediately beneath the plants so as to prevent them from descending. He repeats this process several times, and thus in the first day forms a floating island 8 centimetres in diameter. The bubbles of air are not coated with a greasy liquid as in the case of the fish of paradise, Macropodus viridi-auratus; all those which approach sufficiently to touch, unite together and fuse into one.

"The next day the male continues his provision of air, which he now accumulates towards the central point. These bubbles exert a pressure from below upwards, the consequence of which is the elevation of the vegetable disk, which, issuing from the water, becomes converted into a sort of dome floating on the surface.

"The nest being completed outwardly, the fish busies himself with giving it a firmness which may protect it from shipwreck. With this view, he creeps upon it in all directions and glides over its walls to smooth the surfaces; he forcibly presses this felt with his muzzle and his chest. If one of the twigs is too prominent, he seizes it and removes it

PLATE XV

THE GROWLING GOURAMI (*Ctenops vittatus*) AND ITS BUBBLE-NEST. MALE ABOVE BLOWING BUBBLES UP TO THE SURFACE.

(From a water-colour by W. Schreitmüller.)

or, by means of successive pushes with his head, forces it into the interior. It is by turning and pressing the wall from all sides that he succeeds in rounding it nicely."

He continues in this way until he has formed a regular umbrella or mushroom-like cover on the top of the water. Then like the Stickleback he informs the female in some way, that it is now her turn to play her part. If she is unready or unwilling, he may pursue in a vicious manner, biting her fins and eyes and even killing her. But if she submits to his embraces, they coil their bodies round each other and the male presses the side of the female. The eggs are then fertilised as they emerge and float up to the surface. Needless to say, the male has induced the female to come under the umbrella and the eggs thus float up into the bubble-nest prepared for them. In one form (Betta) the eggs are said to be heavier than water and sink to the bottom, but the male follows after, takes them one by one in his mouth and blows them up into the nest.

The eggs are now safe under the protecting canopy. Whether the rays of the sun would hurt them without this shelter is uncertain, but the nest is so strong that the water-snails cannot get at the eggs without some difficulty, and before they do any damage the watchful male has spied them and driven them away. But if the nest gets broken in any way, the snail gets among them and devours the young. The male continues on guard until the young hatch out, repairing holes, blowing new bubbles, and picking up any eggs, which grow heavier as they develop, that may drop out and blowing them back again into shelter—as it seems. In some cases he may even continue the practice when the helpless fry hatch out, and he regularly encloses them in the nest by blowing up bubbles underneath.

These fishes take kindly to captivity and their bubble-nests form one of the great attractions of aquaria. Whether any end is served, for the good of the young, that is, has been much debated. But the proceedings that follow, when the young are able to move about for themselves, quite negative the idea that this so-called parental care is for the good of the species. Unless both the males and females are removed, they pursue and eat up the young until none are left. This is the case in almost all species, the exceptions being where the young are sluggish and do not move about to any great extent. Very few species take any care of the young. Speaking of the male Paradise Fish (Macropodus) Ida M. Mellen very aptly describes the position: "He suddenly shifts his unstable instinct for the preservation of his species to a shameless appetite for his own babies."

The simplest explanation of the parental care of fishes is that the fish has some notion of possession and acts accordingly. The male, being usually the stronger, keeps the female away, and when she manages to get possession, as in the case of the African Cichlids, it is said that he shows every sign of resentment. It is quite probable that he has no intention to eat his babies, but if the female is present or another, it is a case of first come, first serve. Then there is the irresistible urge to chase and capture something smaller than itself. In aquaria it has been found necessary to remove the parents before the young hatch out, or place a fine-meshed partition in the tank with holes in it through which the small fry can escape.

This interpretation of the parental care of fishes, and other animals, has been admirably expressed by Theodore Gill, one of the most thoughtful and careful of naturalists. "The attribute of parental care must," he writes, "be regarded as an outcome of selfishness, or, if you will, self-love, a result of the sense of proprietorship. The eggs are the fish's own, and therefore they and the resulting larvæ are to be cared for as such. Perhaps it may be urged that the attention of the parental fishes is of the same nature as that of the hen to her young. We are not prepared to deny it. It may even be conceded, and yet the claim that the sentiment is the offspring of self-love can still be maintained. In fact, there is a regular gradation of self-love into the ennobled sentiment which impels the human mother to sacrifice her life cheerfully for *her* child and the degraded passion which emboldens the miser to suffer death rather than lose *his* gold. It is the basis of the courage of the farseeing martyr for his religion, for he is willing to sacrifice the present for an illimitable future.

Wonder may be entertained that one and the same method of care should have originated independently many times, but this will diminish on reflection. When the sense of proprietorship in the eggs has been established, protection by hiding them or clearing away of foreign substances that would interfere with them would not unnaturally follow. The mouth is used by many fishes for carrying, and the instinct to take up the eggs into the mouth for protection would be a natural consequence which might be, and repeatedly has been, developed into a habit. These and other provisions for the care of the eggs do not make excessive demands on our receptive capacity or imagination. It is only when we consider the case of the Sticklebacks that the combination of aptitudes for nest making impresses us. In them complexity is carried to an extreme. There is a sympathetic development of the kidneys and the testes ; there is the synchronous response of both to external stimuli ; there is the reaction of both on the brain and of the brain to external conditions ; there is the elaboration of the wonderful thread which is used to bind the nest materials ; there is the instinct to use the thread ; there are, finally, the regular aptitudes and impulses which are shared with the majority of fishes. Such an accumulation and convergence of structural, physiological, and psychological characters almost force upon us a rejection, as explanation, of natural selection or sexual selection. The development manifested in the Gasterosteids is, indeed, one of the greatest wonders of the evolution of animal life."

According to the theory of sexual selection, the males that are strongest in the fight or most successful in attracting the choice of the females, either by display or by brilliant coloration, are the ones most likely to have progeny and pass on their characteristics to their descendants. Whether this process is at all operative among the fishes, we have no evidence. It does not appear that the males anywhere wait to be selected by the females. Where their coloration is most brilliant, as in the Sticklebacks and Cyprinodonts, it is quite clear that the females have not the slightest choice in the mating. And if a male Stickleback deludes or overcomes a marauder on his property, that does not necessarily mean that he is the stronger. Even a small dog can chase away a large dog, when the latter is trespassing.

Again, if the sombre colour of females has been developed by natural selection as an aid to their protection when caring for their eggs and young, as among birds, we are at a loss to understand why it is just the bright and conspicuous males that perform these duties among the fishes. In fact, the theory of selection from without, whether of one sex for the other or of a mysterious combination of incalculable circumstances, seems to take us away from the kernel of the matter, which is, that the intelligence of the animal is a power of a higher order than the contradictory and chaotic elements of an external selection. The idea of possession undoubtedly gives a better picture of the stage of evolution reached by the fishes and the question of greatest interest is simply, how and when this instinct lost its selfish character and became of service in other ways.

4. The Feelings of Fishes

Whilst the reason of man or fish is mostly engaged in obtaining or looking after possessions, the feelings are more directly concerned with the social relationships. They are the driving power behind the application of means to an end implied in reasoning, and come into play when others are concerned in one way or another. An individual alone in the world would have no urge to amass wealth and little use for his reasoning powers. Place another beside him and rivalry or partnership begins with expression of feelings.

People are accustomed to think of the fishes as cold-blooded and without feelings. In most cases certainly they must be of a very different kind from those we know, yet the curious thing is, that few of the feelings we recognise in human beings are absent from fishes. This has already been indicated in the previous section and some more striking cases may be mentioned here.

Allusion should first of all be made to the sounds made by fishes. These are of many and various kinds, though restricted, so far as we know, to a comparatively small number of species. When caught, and even when free in the water, some say, the Gurnards make a grunting sort of noise. One of the Osphromenidæ is called the Growling Gourami from the noise it makes. But the great sound-makers are the Sciænoids or Drummers, which have aroused the admiration and astonishment of many hearers. They occur mainly in subtropical waters, from America, Mediterranean and the waters of Asia, and some of the romance of a tropical evening enters into the descriptions of narrators. Bending low over the side of a boat, or better still with an ear to the water, one can hear from the depths the songs of sirens, rising sometimes into the tones of an organ or sinking to the low melody of a harp ; at other times the sound has been compared to the ringing of bells or the continuous roll of drums. In the Mediterranean the fishermen are guided to the whereabouts of the Eagle Fish or Maigre (Sciæna aquila) by the drumming it makes in the water.

The mechanism employed in producing the sounds is naturally various ; a smacking noise by the lips, a rasping sound by the rubbing of the spines together or in their sockets. A Cat-fish has been said to make a noise like a spitting cat when caught and the rush of water or air pressed out of the gill-covers makes a hissing sound. The drumming noises, however, mostly come from the air-bladder or the muscles connected with it. In the Sciænoids the air-bladder is developed to an extraordinary extent and with the movement of the gases inside from one compartment to another, it is thought that the wall of the bladder is set in vibration. The muscles surrounding the compartments may contribute to the vibration, by rapidly extending and contracting, so that the sounds may be deliberate or voluntary in this case.

As with everything else, supposed uses have been ascribed to this sound-making; that it frightens enemies and attracts friends or is used in courting; the male Drummer is said to make loud noises, the female quieter tones. It thus seems to be a secondary sexual character, like the coloration. But we really do not know what the fishes may or may not hear, except in the case of Amiurus, a North American Cat-fish, which seems undoubtedly to respond to a whistle. Most of them, however, do not possess any apparatus to be able to appreciate the wave-lengths we hear, but of course our range of hearing is very restricted and the fishes may detect in other ways what we call sound, when carried in the water. We can certainly conclude, however, that since the sounds made by them are louder during the breeding season, they are in some way an expression of the fish's feelings.

Courtship and the breeding time provide indeed the great test of the feelings. At other times fishes can display the commoner feelings; some are timid and some are courageous to brutal; when it rushes on ships or boats the Sword-fish is said to be deranged by fear, which makes its movements seem like blind rage. But, on the whole, it is the excitement of the breeding time that brings out the best expression of the feelings.

At this time the males become very combative. The little Stickleback challenges all comers; if an adversary of his own kind trespasses on his preserves, he drives him off or it is a fight to the death. It has been said that the victor then parades himself with an air of pride and satisfaction. Just as courageous, though a veritable pigmy in size, is the Betta pugnax. The natives of Siam and Cochin China, where it lives, rear this fish for the special purpose of indulging in a favourite pastime, just as game-cocks were formerly reared and trained in England. And the Kings of Siam are said to have obtained no small revenue from the licences granted for these fights. The natives bring

PLATE XVI

THE MALE STICKLEBACK (*Gastrosteus aculeatus*) GUARDING HIS NEST.

(From Smitt.)

their favourites in glass-bowls to the scene of combat and after the preliminary betting has been arranged, wives, children, and other possessions, it is said, being thrown into the balance, the glass-bowls are brought near to each other. The males at once storm at each other, spreading out their fins and displaying their gorgeous colours. The one able to show the most brilliant changes, is declared the winner of the first bout. Then the two are brought together in the same bowl and they forthwith begin to tear at each other with their mouths and sharp spines, until the one is overpowered. The victor seldom lives to enjoy his triumph.

But it is not only the males that fight together. Among the Cichlids, as already mentioned, the females believe in equality of rights and take rather more than their share. The result is that both are sturdy fighters in some cases (Haplochromis) and it has been found better to make them live apart; otherwise they kill each other. The equality of rights principle does not make for peace in the establishment of the Cichlids.

It is fairly certain that envy is at the root of most of the trouble. When the female Haplochromis picks up the eggs in her mouth, the male pursues and beats her. The female retains possession, however, and when the young hatch out, makes a meal of them unless prevented by a stronger hand.

That extreme jealousy can be displayed by fishes, is clearly shown by an experiment which Benecke made and which is described in Brehm's "Tierleben." He placed two females of the Paradise-fish (Macropodus viridi-auratus) in the same tank with a male. "After a short time they had arranged themselves in this way: the two females occupied opposite corners and the male visited them turn about. The two females behaved quite well, however, and even played with one another at times."

Then one of the females became ripe and the male began to build the bubble nest in company with her, paying her special attention all the while. They would play for a time together, greeting each other with the lips so heftily that small pieces of skin were torn off. "I could only regard this as kissing of a special inner warmth," reports the observer. After some time the attitude of the females towards each other became very cool, and this changed to deeds. Torn fins and tails made them unsightly, and to avoid extremities the observer placed a glass partition in the tank, shutting off the smaller female from the pair. "The two females rushed at each other, however, in such a temper, each completely oblivious of everything else in her efforts to get at the other, that I was obliged to hang a piece of cloth over the partition. But the female left in the company of the male soon quietened down and the cloth was removed. She now took up a position between the male and the partition, casting angry and threatening looks at the rival in the next compartment. I then placed a dulled piece of glass in front of the partition; but the mere shadow of the widow, when the sun's rays fell across the basin, so excited the pair, that I was obliged to fasten paper over the glass. And now I hoped that peace and order would reign. But what happened? One day I found the two females engaged in tumultuous combat; the widow had sprung over the partition, five inches high above the surface. Nothing was left but to place her in a separate tank."

It would appear also that a love of admiration may be present in the fish's breast. It has already been mentioned, how the male Stickleback celebrates his triumph over his adversary, "while the vanquished lays aside his brilliant dress as though overcome with shame." This statement is taken from Smitt, and he goes on to describe the behaviour of the females. "While the males disport themselves in these chivalrous tournaments, or rather fight for their nests, the females swim about in long troops of greater or less strength outside the battle-ground, and now and then a male selects his temporary mate from the company. The female that heads the troop, swims forward with rapid darts, followed by the others, suddenly stops, and assumes a vertical position, with head towards the bottom. The others assemble round her and range themselves in the same manner, as densely packed as possible. When she has thus collected the troop around her, she suddenly deals a blow that scatters the whole crowd in an instant. This sport is often repeated, but the rapidity with which they disperse, renders it impossible to observe whether it is always the same female that takes the lead, or whether they change places."

It may be that they perform in this way out of sheer sportiveness of spirit, but a deeper intent would be only human; they have no bright colours wherewith to attract the males.

It may be that the desire for possessions must always precede self-sacrifice or the sharing of goods and interests—otherwise, there might be nothing to share. But it is pleasant to record—and we may conclude on this note—that greed and rapacity are not the only attributes of fishes. Exceptional they may be, yet nevertheless striking are the examples of fidelity and affection even unto death shown by the Lumpsucker and Mullet, mentioned in preceding pages. The remembrance of their lost comrade which lingers in the mind of the Pilot Fishes seems also of a higher order than mere selfishness. But the crowning point of intelligence and feeling united in a good cause can be seen in some examples of true care for the young, as distinct from jealous watching over the eggs.

Both the male and female of Amiurus, the North American Cat-fish, take part in making a nest in the mud, sometimes spending two or three days in the work. When the young are hatched the male takes charge of them and leads them about, just as a hen looks after her chickens (Eycleshymer).

Many of the African Cichlids lay their eggs in the normal manner and both parents keep on the watch above them. When the young hatch out they are still very helpless, an unusual thing among fishes, and the parents take them in their mouths and place them in a nest in the sand, which they had previously prepared and cleaned. When the fry are able to move about freely, the male or sometimes both parents still watch over them and keep them together.

Finally, we have the story of another Cichlid (Tilapia)

carefully observed and described by the Rev. N. Abraham. This time it is the turn of the female to earn the honours. Her care of the eggs—she takes them in her mouth—may be ascribed to greed in the first place, but at any rate it represents some courage in withstanding the attacks of the male.

According to Abraham the mother in one case kept them in her mouth for about ten days after the young were hatched. Then one morning the observer saw a crowd of tiny fishes playing round the head of the parent. She herself was only about three inches long, yet there were more than sixty of the tiny fry. With the slightest disturbance of the water, the whole of them collected in front and the mother opened her mouth to let them enter. She kept them shut up for several hours, then two or three were blown out like smoke from a pipe, then several more until about thirty were free in the water, and the remainder were shaken out all at once by a rapid movement.

Five days thereafter she watched and cared for them. A few shrimps which had been placed in the tank were so harried and chased, that the observer had them removed; even later she attacked anything strange that appeared in the water. And every evening, when darkness came with the night, she withdrew into a corner and put them to sleep in her mouth.

PLATE XVII

HEMICHROMIS BIMACULATUS AND YOUNG.

(Drawn from life by W. Schröder.)

APPENDIX

SYNOPSIS OF THE FAMILIES OF FISHES

THE following synopsis continues in more detail the arrangement of fishes outlined in Chapter IX, p. 217. It gives one view of the relative positions of the principal families and genera mentioned in this book.

Class : PISCES

I. Subclass : OSTEICHTHYES

Phylum : *Dipnoi* (Lung-fishes) ; Neoceratodus, Protopterus, Lepidosiren.

Phylum : *Ganoidei*.
Subphylum : *Holostei ;* Amia, Lepidosteus.
Subphylum : *Chondrostei ;* Acipenser (Sturgeon).
Subphylum : *Crossopterygii ;* Polypterus.

Phylum : *Teleostei*.

Order.	Families.	Genera.
Clupeoidea	(Leptolepidæ)	Leptolepis (extinct)
	Clupeidae (Herring family)	Clupea
		Engraulis (Anchovy)
		Pellona
		Pristigaster
		Coilia
		Notopterus
	Albulidæ	Albula
	Elopidæ	Elops (Tarpon)
	Salmonidæ	Salmo (Salmon)
		Salvelinus

Order.	Families.	Genera.
Clupeoidea	Salmonidæ	Osmerus (Smelt)
		Coregonus
	Argentinidæ	Argentina
		Mallotus (Capelan)
	Cyprinodontidæ	Lebistes
		Pœcilia
		Fundulus
		Xiphophorus
	Esocidæ	Esox (Pike)
	Scopelidæ	Lampanyctus
		Scopelus
		Myctophum
		Ipnops
Ostariophysi	Cyprinidæ	Cyprinus (Carps)
	Characinidæ	Pyrrhulina
	Gymnotidæ	Gymnotus (Electric Eel)
	Mormyridæ	Mormyrus
	Loricariidæ	Plecostomus
	Siluridæ (Cat-fishes)	Silurus
		Malopterurus
Apodes	Anguillidæ (Eel family)	Anguilla
		Conger
	Gastrostomidæ	
Irregulares	Scombresocidæ	Scombresox (Garfish)
	Exocœtidæ	Exocœtus (Flying-fish)
	Fistulariidæ	Fistularia (Flute-mouths)
	Syngnathidæ (Pipe-fishes)	Syngnathus
		Siphostoma
		Nerophis
		Hippocampus
	Gastrosteidæ (Stickle-backs)	Gastrosteus
	Orthagoriscidæ (Sun-fishes)	Orthagoriscus (Mola)
		Ranzania
	Diodontidæ	Diodon (Porcupine-fish)
	Balistidæ	Balistes (File-fish or Trigger-fish)
	Ammodytidæ	Ammodytes (Sand-eels)

Order.	Families.	Genera.
Irregulares	Gadidæ (Cod family)	Gadus
		Molva (Ling)
		Merluccius (Hake)
	Macruridæ	Macrurus
	Stromateidæ	Stromateus
		Stromateoides
	Atherinidæ	Atherina
	Mugilidæ	Mugil (Grey Mullets)
	Sphyrænidæ	Sphyræna
	Ophiocephalidæ	Ophiocephalus
	Anabantidæ	Anabas (Climbing Perch)
	Osphromenidæ	Macropodus
		Ctenops
		Trichogaster
		Betta
Trachinoidea	Blenniidæ	Blennius
		Zoarces (Viviparous Blenny)
		Pholis (Shanny)
		Centronotus (Gunnel or Butter-fish)
	Anarrhichadidæ	Anarrhichas (Wolf-fish)
	Trachypteridæ	Trachypterus
		Regalecus
	Fierasferidæ	Fierasfer
	Batrachidæ	Batrachus
		Porichthys
		Opsanus
	Lophiidæ	Lophius (Frog-fish or Angler)
		Maltheus
	Ceratiidæ	Ceratias
	Symbranchii	Amphipnous
	Monopteridæ	Monopterus
	Gobiidæ	Gobius
		Aphya
		Mistichthys
	Soleidæ	Solea (Soles)
	Symphuridæ	Symphurus
	Callionymidæ	Callionymus (Dragonets)
	Trachinidæ	Trachinus (Weevers)

Order.	Families.	Genera.
Trachinoidea	Uranoscopidæ	Uranoscopus (Star-gazers)
	Pseudochromidæ	Lopholatilus (Tile-fish)
Carangoidea	Carangidæ	Caranx (Horse-mackerel)
		Argyreiosus
	Zeidæ	Zeus (John Dory)
	Toxotidæ	Toxotes
	Chætodontidæ	Chætodon
		Holocanthus
		Scatophagus
	Pleuronectidæ	Pleuronectes (Plaice, Flounder, etc.)
		Hippoglossus (Halibut)
	Rhombidæ	Rhombus (Turbot)
	Bothidæ	Bothus
		Arnoglossus
	Cyclopteridæ	Cyclopterus (Lump-sucker)
	Cottidæ	Cottus (Sea-scorpion)
	Triglidæ	Trigla (Gurnard)
		Dactylopterus (Flying Gurnard)
	Scorpænidæ	Scorpæna
		Sebastes
	Cichlidæ	Tilapia
		Haplochromis
		Hemichromis
	Labridæ	Labrus (Wrasses)
	Embiotocidæ	Ditrema
	Scaridæ	Scarus
Percoidea	Sparidæ	Sparus
		Chrysophrys
	Mullidæ	Mullus (Red Mullet)
	Serranidæ	Serranus
		Labrax (Morone)
	Sciænidæ	Sciæna
	Percidæ	Perca (Perches)
Scombroidea	Trichiuridæ	Trichiurus
	Coryphænidæ	Coryphæna ("Dolphins")

Order.	Families.	Genera.
Scombroidea	Echeneidæ	Echeneis, Remora (Sucker-fishes)
	Histiophoridæ	Histiophorus (Sail-fishes)
	Xiphiidæ	Xiphias (Sword-fish)
	Naucratidæ	Naucrates (Pilot-fish)
	Scombridæ	Scomber (Mackerel)
	Tunnidæ	Pelamys (Bonito)
		Orcynnus (Tunny)

II. Subclass : Elasmobranchii

Phylum : *Holocephali* (Chimæridæ) ; Chimæra, Harriotta.

Phylum : *Selachii* (Sharks and Dog-fishes) ; Scyllium, Mustelus, Acanthias, Alopecias, Scymnus, Carcharodon, Sphyrna, Læmargus. (Rhinobatis), Rhina.

Phylum : *Batoidei* (Skates and Rays) ; Raja, Myliobatis, Trygon, Torpedo, Manta.

Phylum : *Cyclostomata ;* Myxine, Bdellostoma (Hag-fishes) ; Petromyzon (Lampreys).

Incertæ sedis

Cephalochordata ; Amphioxus, Asymmetron.

BIBLIOGRAPHY

I. A Few General Books Dealing with Fishes or the Biology of Fishes

Boulenger, G. A., and Bridge, T. W. Fishes in the Cambridge Natural History, 1904.

Brehm's Tierleben, IV. Edition. Fische. Leipzig, 1914.

Darwin, Charles. Origin of Species. VI. Edition.

Day, Francis. The Fishes of Great Britain and Ireland. London, 1880–1884.

Dean, Bashford. Fishes, Living and Fossil. New York, 1895.

Geddes, P., and Thomson, J. Arthur. Evolution of Sex. London, 1901.

Goodrich, E. S. Fishes in Treatise on Zoology (E. Ray Lankester). London, 1909.

Houssay, Fr. Forme, puissance et stabilité des poissons. Paris, 1912.

Jordan, D. S. A Guide to the Study of Fishes. New York, 1905.

—— and Evermann, B. W. Fishes of North and Middle America, 1896–1900.

Kerr, J. Graham. Embryology of Vertebrates. London, 1919.

McIntosh, W. C. Resources of the Sea. London, 1899.

Murray, Sir John, and Hjort, Johan. The Depths of the Ocean. London, 1912.

Osborn, H. F. From the Greeks to Darwin. New York, 1896.

Regan, C. T. British Freshwater Fishes. London, 1911.

Roule, L. Traité de la pisciculture et des pêches. Paris, 1914.

SMITT, F. A. Scandinavian Fishes. 1893.

THOMPSON, D'ARCY W. Growth and Form. 1917.
THOMSON, J. ARTHUR. The Study of Animal Life. 1917.
—— Heredity.

WOODWARD, A. SMITH. Vertebrate Palæontology. Cambridge, 1898.

II. Books and Papers used in the Text

ABEL, O. Lehrbuch der Palæozoologie. Jena, 1920.
—— Ueber den wiederholten Wechsel der Körperform. Bijd. Dierk. Amsterdam, 1922.
ABRAHAM, N. On the Breeding Habits of Chromis philander. Ann. Mag. Nat. Hist., 1901.
ADAMETZ, L. Ueber die Beziehungen der Konstitution zu den endokrinen Drüsen. Zeitsch. f. Tierzücht. u. Züchtungsbiologie, 1924.
AFLALO, F. G. The Sea-fishing Industry of England and Wales. London, 1904.
ALLEN, E. J. On the Culture of the Plankton Diatom (Thalassiossira gravida Cleve) in Artificial Sea-water. Jour. Mar. Biol. Assoc. X., 1914.
—— Food from the Sea. Jour. Mar. Biol. Assoc. XI., 1917.
ALLIS, E. P. The Lateral Sensory Canals—of Mustelus levis. Quart. J. Mic. Science, 1901.
—— The Latero-sensory Canals and Related Bones in Fishes. Int. Monats. Anat. u. Phys. 21, 1904.
—— Certain Homologies of the Palato-quadrate of Selachians. Anat. Anzeiger.. 45, 1914.
ARNOLD, J. P. Trichogaster Labiosus, Day. Blätt. Aquar. Terrar. Kunde. 1911.
ATKINS, W. E. G. The Phosphate Content of Fresh and Salt Waters. Jour. Mar. Biol. Assoc., XIII. 1923.
—— On the Vertical Mixing of Sea-water and its Importance for the Algal Plankton. Jour. Mar. Biol. Assoc. 1924.

BAGLIONI, S. Zur Physiologie der Schwimmblase. Zeit. allg. Phys., 1908.
BALFOUR, F. M. A Treatise on Comparative Embryology. London, 1880–1881.
BALLOWITZ, E. Ueber den Bau des elektrischen Organes von Torpedo. Arch. Mikr. Anat. 1893.
BATESON, W. The Sense Organs and Perceptions of Fishes. Jour. Mar. Biol. Assoc. 1889.

BEARD, J. On Certain Problems of Vertebrate Embryology. Jena, 1896.

—— The Determination of Sex in Animal Development. Anat. Anz. 1902.

BEAUFORT, L. F. DE. Die Schwimmblase der Malacopterygii. Morph. Jahrb. 39. 1909.

BENNETT, F. Communication between Air-bladder and Cloaca in Herring. Jour. Anat. Phys. 1879.

BETHE, A. Ueber die Erhalting des Gleichgewichts. Biol. Centralbl. 1894.

BIANCO, LO S. Uove e larve di Trachypterus tænia Bl. Mitt. Zool. Stat. Neapel. 1908.

BIGELOW, H. B. The Sense of Hearing in the Goldfish Carassius auratus. Amer. Naturalist, 38. 1904.

BOAS, J. E. V. Lehrbuch der Zoologie. Jena, 1894.

BOEKE, J. Beiträge zur Entwicklungsgeschichte der Teleostier. Petrus Camper, 1902–1903.

BORLEY, J. O., and RUSSELL, E. S. Report on Herring Trawling. Fishery Investigations. Min. Agriculture and Fisheries, London, 1922.

BOTTARD, L. A. Poissons venimeux. Paris, 1889.

BOTTAZZI, F. Sulla regolazione della pressione osmotica negli organismi animali. Arch. Fisol. 1906.

BOULENGER, E. G. Notes on the Breeding of the "Millions" Fish (Girardinus pæciloides). Proc. Zool. Soc. London, 1912.

BOULENGER, G. A. Les poissons du Congo. Bruxelles, 1901.

BOUTAN, L. Sur le plan d'equilibre—des poissons Teleostéens à vessie natatoire. C. R. Acad. Sci. Paris, 1916.

BRAUER, A. Tiefseefische. Valdivia Expedition. Jena, 1908.

BUCHMANN, W. Naturwissenschaftliche Korrespondenz. 1924.

BUCHNER, P. Ueber den "tierische" Leuchten. Naturwissenschaften, 10. 1924.

BULLEN, F. T. Denizens of the Deep. London, 1904.

BURCKHARDT, C. R. On the Luminous Organs of Selachian Fishes. Ann. Mag. Nat. Hist. 1900.

BYRNE, L. W. On the Number and Arrangement of the Bony Plates of the Young John Dory. Biometrika. 1902.

CARBONNIER, P. Nidification du poisson arc-en-ciel de l'Inde. Bull. Soc. Acclim. Paris, 1876. Ann. Mag. Nat. Hist. 1876.

—— Le gourami et son nid. Ann. Mag. Nat. Hist. 1877.

CHUN, C. Aus den Tiefen des Weltmeeres. Jena, 1905

CLARK, R. S. Rays and Skates. No. 1. Egg-capsules and young. M. B. A., xii, 1922.

COLES, R. J. My Fight with the Devilfish. American Mus. Journ., XVI. 1916.

COUCH, J. British Fishes (1867). 1877.

COUTIÈRE, H. Sur la non-existence d'un appareil à venin chez la Murène hélène. C. R. Mém. S. B. Paris, 1902.

CUNNINGHAM, J. T. Researches on the Coloration of the Skins of Flatfishes. Jour. Mar. Biol. Assoc. 1893 and 1894.

—— Natural History of the Marketable Marine Fishes of the British Islands. London, 1896.

—— and MACMUNN, C. A. On the Coloration of the Skins of Fishes. Phil. Trans. Royal Soc. London, 1893.

CUVIER, G., and VALENCIENNES, A. Histoire naturelle des poissons. Paris, 1828.

DAHL, K. The Scales of the Herring as a means of determining Age, Growth and Migration. Norw. Fish. Inves. 1909.

DAKIN, W. J. The Osmotic Concentration of the Blood of Fishes. Biochem. Jour. III., 5 and 10. 1908.

—— Notes on the Biology of Fish Eggs and Larvæ. Inter. Rev. Hydrobiologie. 1910–1911.

DAMAS, D. Contribution à la biologie des Gadides. Rapp. Cons. Perm. Intern. Exp. Mer. 1909.

DEAN, BASHFORD. The Fin-fold Origin of the Paired Limbs. Anat. Anz. 1896.

—— Sharks as Ancestral Fishes. Nat. Science. London, 1896.

—— On the Embryology of Bdellostoma stouti. Festschrift v. Carl von Kupffer. Jena, 1899.

—— The Devonian "Lamprey," Palæospondylus gunni. Mem. N. Y. Acad. Sciences. 1899.

—— Reminiscence of Holoblastic Cleavage in—Heterodontus (Cestracion) japonicus. Ann. Zool. Japonenses. 1901.

—— Biometric Evidence in the Problem of the Paired Limbs of the Vertebrates. Amer. Naturalist. 1902.

—— Studies on Fossil Fishes. Mem. Amer. Mus. Nat. Hist. 1909.

DELSMAN, H. C. Fish Eggs and Larvæ from the Java Sea. Treubia, 1921.

—— Die larvale Entwicklung von Chirocentrus dorab. Bijd. t. d. Dierkunde. Leiden, 1922.

DENDY, A. The Pineal Gland. Science Progress. 1906.

DOLLO, L. Sur la phylogénie des Dipneustes. Bull. Soc. Belge Géol. 1896.

DUNCKER, G. Die Methode der Variationsstatistik. Arch. f. Entwicklungsmech. 1898.

—— Symmetrie und Asymmetrie bei bilateralen Tieren. Idem. 1904.

—— Syngnathiden-studien. Mitt. Zool. Mus. Hamburg. 1908.

—— Die Bestimmung d. Variation v. Merkmalen selektiv ausgemerzter Individuen. Idem. Hamburg, 1917.

EDDINGTON, A. On the Saprolegnia of Salmon Disease. Rep. Scot. Fish. Board, VII. 1889.

EHRENBAUM, E. Eier und Larven von Fischen der deutschen Bucht. Wiss. Meeresunt. (Helgoland.) 1897.

—— Eier und Larven von Fischen. Nordisches Plankton. 1909.

—— Ueber die Lebensverhältnisse unserer Fische. Der Fischerbote. 1909–1924.

—— The Mackerel and the Mackerel Fishery. Rapp. Proc. Verb. Cons. Intern. Copenhagen, 1912, 1914, 1923.

—— Der Wurmstar des Fischauges und seine Bekämpfung. Der Fischerbote. 1917.

—— and MARUKAWA, H. Ueber Altersbestimmung und Wachstum beim Aal. Zeitsch. f. Fischerei. Berlin, 1913.

EMERY, C. La specie del genere Fierasfer nel golfo di Napoli. Fauna u. Flora des Golfes von Neapel. 1880.

EVERMANN, B. W. A Report upon Salmon Investigations, Columbia River, Idaho. Bull. U.S. Fish. Commission, XVI. 1897.

—— and GOLDSBOROUGH, E. L. The Fishes of Alaska. Bull. Bur. Fisheries, Washington. 1906.

—— and SEALE, A. Fishes of the Philippine Islands. Bull. Bur. Fisheries. 1907.

EWART, J. COSSAR. The Electric Organ of the Skate. Phil. Trans. Royal Society. London, 1892 and 1893.

FABER, G. L. The Fisheries of the Adriatic. London, 1883.

FAGE, L. Shore Fishes. Danish Oceanographical Expeditions, 1908–1910, II., A3. Copenhagen, 1918.

FRANZ, V. Neuere Ergebnisse über Fischwanderungen in der Nord- und Ostsee. Intern. Rev. Hydrobiol. u. Hydrogr. 1908.

—— Ermittelungen zur natürlichen Elimination gescheckter Schollen im Kampf ums Dasein. Biolog. Zentralblatt. 1925.

FRISCH, K. v. Ueber farbige Anpassung bei Fischen. Zool. Jahrb. 1912.

—— Sinnesphysiologie der Wassertiere. Verh. Deut. Zool. Ges. Berlin, 1924.

FULTON, T. W. On the Growth and Maturation of the Ovarian Eggs of Teleostean Fishes. Scot. Fish. Board Rep. 1898.

—— Report on the Marking Experiments on Plaice. Idem. 1920.

GARSTANG, W. The Impoverishment of the Sea. Jour. Mar. Biol. Ass. 1900.

GAST, R. Ueber die Verwendung des Lichtes beim Fischen. Der Fischerbote, Hamburg, 1918.

GEMMILL, J. F. The Teratology of Fishes. Glasgow, 1912.

GILBERT, C. H. The Deep-sea Fishes of the Hawaiian Islands. Bull. U.S. Fish. Comm. 1905.

GILL, TH. Parental Care among Freshwater Fishes. Smithsonian Report for 1905. Washington, 1907.

GILSON, G. L'anguille, sa reproduction, ses migrations et son interêt économique en Belgique. Ann. Soc. Roy. Malacol. Belgique. 1908.

GOODE, G. B., and BEAN, T. H. Oceanic Ichthyology. Washington, 1895.

GOODRICH, E. S. Notes on the Development, Structure and Origin of the Median and Paired Fins of Fish. Quart. J.M.S. London, 1906.

—— On the Scales of Fish, Living and Extinct. Proc. Zool. Soc. London, 1908.

—— On a New Type of Teleostean Cartilaginous Pectoral Girdle found in Young Clupeoids. Jour. Linn. Soc. 1920.

GRAEFFE, ED. Uebersicht der Seethierfauna des Golfes vom Triest. Wien, 1888.

GRASSI, B. The Reproduction and Metamorphosis of the Common Eel. Proc. Roy. Soc. London, 1896.

—— Nuove ricerche sulla storia naturale dell' Anguilla. R. Com. Talass. Italiano. Venezia, 1919.

GREENE, C. W. Physiological Studies of the Chinook Salmon. Bull. Bur. of Fisheries. Washington, 1905.

—— The Migration of the Salmon in the Columbia River. Idem. 1911.

GUDERNATSCH, J. F. The Thyroid Gland of the Teleosts. Jour. Morphol. 1911.

GUDGER, E. W. The Breeding Habits and the Segmentation of the Egg of the Pipe-fish Siphostoma floridæ. Proc. U.S. Nat. Museum. 1905.

—— The Gaff-Topsail (Felichthys felis). Zoologica. New York, 1916.

—— On the Use of the Sucking-fish for Catching Fish and Turtles. The American Naturalist. 1919.

GÜNTHER, A. The Study of Fishes. Edinburgh, 1880.

HAEMPEL, O. Leitfaden der Biologie der Fische. Stuttgart, 1912.

HARDY, A. C. The Herring in Relation to its Animate Environment. Min. Agric. and Fisheries. Fishery Inves. London, 1924

HEINCKE, FR. Die Fische. Leipzig, 1882.
—— Naturgeschichte des Herings. Abh. Deutsch. Seefisch. Ver. Berlin, 1898.
—— Die Beteiligung Deutschland an der internationalen Meeresforschung, IV.–V. Jahresbericht. Berlin, 1908.
—— Untersuchungen über die Scholle. Generalbericht. Cons. perm. inter. Explor. de la mer. 1913.
—— and EHRENBAUM, E. Die Bestimmung der schwimmenden Fischeier. Wiss. Meeresun. (Helgoland.) 1900.
HENNIG, E. Ueber neuere Funde fossiler Fische aus Aequatorial und Südafrika. Sitz. Ges. Naturf. Freunde. Berlin, 1913.
—— Ueber die Bestimmung des Planktons. Ber. Comm. Unters. deutschen Meere. Kiel, 1887.
HENSEN, V. Ueber die Bestimmung des Fischbestandes im Meer. Wiss. Meeresun. Kiel, 1911.
HERDMAN, W. A. An Intensive Study of the Marine Plankton. Reports, Lancashire Sea-fish. Laboratory. 1914–1918.
—— Spolia Runiana, III. Jour. Linn. Soc. London, 1918.
HERRICK, C. J. The Organ and Sense of Taste in Fishes. Bull. U.S. Fish. Comm. (1902). 1904.
HERTWIG, R. Ueber den derzeitigen Stand des Sexualitätsproblems. Biol. Centr., 32. 1912.
HICKSON, S. J. A Naturalist in North Celebes. London, 1889.
HJORT, J. Reports Regarding the Herring. Cons. perm. inter. Explor. de la mer. 1908, 1909, 1910.
—— Fluctuations in the Great Fisheries of Northern Europe. Idem. 1914.
HOBER, R. Lehrbuch der Physiologie. 1920.
HOFER, B. Handbuch der Fischkrankheiten. München, 1904.
—— Studien über die Hautsinnesorgane der Fische. Ber. Kgl. Bayer. Biol. Versuchsstat. in München. Stuttgart, 1908.
—— Neue Forschungen über Alter und Wachstum der Scholle. Der Fischerbote. 1916.
HOFFMANN, H. Moderne Probleme der Tiergeographie. Die Naturwissenschaften, V. 1925.
HOLT, E. W. L. On the Breeding of the Dragonet, Callionymus lyra. Proc. Zool. Soc. London, 1898.
HUOT, A. Recherches sur les poissons lophobranches. Ann. des sci. nat., XIV. 1902.

JAGOW, K. Kulturgeschichte des Herings. 1920.
JÄRVI, T. H. Die kleine Maräne (Coregonus Albula L.) im Keitelsee. Ann. Acad. Scent. Fennicæ., XIV. 1920.

JENSEN, AD. S. The Selachians of Greenland. Mindeskrift for Japetus Steenstrup. Copenhagen, 1914.

IMMERMANN, F. Die innere Struktur der Schollen-Otolithen. Wiss. Meeresunt., Helgoland, VIII., 2. 1908.

JOHANSEN, A. C. On the Summer and Autumn-spawning Herrings of the North Sea. Medd. Komm. Havunders., VII. Copenhagen, 1924.

—— and JACOBSEN, J. P. Changes in Specific Gravity of Pelagic Fish Eggs. Medd. Komm. Havunders. 1908.

JOHNSTONE, J. The Dietetic Value of the Herring. Rep. Lancs. Sea-fish. Laboratory. Liverpool, 1918.

—— Malignant Tumours in Fishes. Jour. Mar. Biol. Assoc., XIII. 1924.

JORDAN, D. S., and SEALE, A. The Fishes of Samoa. Bull. Bur. Fish., XXV. Washington, 1905.

KERR, J. G. The Zoological Position of Palæospondylus. Proc. Phil. Soc. Cambridge, 1900.

KIDDER, J. H. Experiments upon the Animal Heat of Fishes. Proc. U.S. Nat. Museum. 1880.

KING, H. Studies in Sex-determination in Amphibians. Biol. Bull., XX. 1911.

KISHINOUYE, K. Contributions to the Comparative Study of the so-called Scombroid Fishes. Tokyo, 1923.

KRÜGER, A. Untersuchungen über das Pankreas der Knochenfische. Kiel, 1904.

KRUKENBERG, C. F. W. Versuche zur vergleichende Physiologie der Verdauung. Unters. Physiol. Inst. Heidelberg, 1877–1879.

KYLE, H. M. The Asymmetry, Metamorphosis and Origin of Flatfishes. Phil. Trans. Royal Soc. London, 1921.

—— The Classification and Phylogeny of the Teleostei Anteriores. Wiss. Meeresunt. (Helgoland.) 1923.

LEA, E. Further Studies concerning the methods of Calculating the Growth of Herrings. Publ. de Circonst., 66. Cons. p. i. Exp. mer. 1913.

LEBOUR, M. V. The Food of Young Clupeoids. Jour. Mar. Biol. Assoc. 1921.

—— Food of Young Herring. Idem. 1924.

LEE, R. M. An Investigation into the Methods of Growth Determination in Fishes. Cons.—Exp. de la mer. Pub. de Circ., 63. 1912.

LE DANOIS, E. Les conditions de la Pêche à la Morue sur les Bancs de Terre-Neuve. Notes et Mémoires, 35. Off. sci. et techn. des Pêches maritimes. Paris, 1924.

LE DANOIS, E., et BELLOC G., Recherches sur la Régime des Eaux Atlantiques et sur la biologie des Poissons comestibles. Idem., 34. 1923.

LENDENFELD, R. v. The Radiating Organs of the Deep-sea Fishes. Mem. Mus. Comp. Zool. Harvard, 1905.

LEONHARDT, E. E. Der Fisch. Stuttgart, 1913.

LISSNER, H. Das Gehirn der Knochenfische. Wiss. Meeresunt., XIV. Helgoland, 1922.

LOHBERGER, J. Ueber zwei riesige Embryonen von Lamna. Abh. d. Bayer. Akad. Wiss., IV. München, 1910.

LÜBBERT, H. Beobachtungen über den Schollenbestand . . . im Jahre 1922. Ber. deut. wiss. Komm. Meeresfor. 1925.

LUTHER, A. Stellt der " aculeiforme Anpassungstypus " (Abel) eine Anpassung an die planktonische Lebensweise dar? Int. Rev. Hydrobiologie, V. 1912.

McINTOSH, W. C. On the Paternal Instincts of Cyclopterus. Ann. Mag. Nat. Hist. 1886.

—— Scientific Work in the Sea-fisheries. The Zoologist. 1907.

—— and MASTERMAN, A. T. The Life-histories of the British Marine Food-fishes. London, 1897.

MAIER, H. N. Beiträge zur Altersbestimmung der Fische. Wiss. Meeresunt., Helgoland, VIII. 1906.

—— and SCHEURING, L. Entwicklung der Schwimmblase . . . bei Clupeiden, spez. beim Hering. Wiss. Meeresunt, XV. Helgoland. 1923.

MARSHALL, S. The Food of Calanus finmarchicus during 1923. Jour. Mar. Biol. Assoc. 1924.

MATSUBARA, S. Goldfish and their Culture in Japan. Proc. IV. Intern. Congress. Washington, 1908.

MATTHEWS, D. J. On the Amount of Phosphoric Acid in the Sea-water off Plymouth Sound. Jour. Mar. Biol. Assoc. 1916–1917.

MATTHEWS, J. D. On the Structure of the Herring and other Clupeoids. Rep. Scot. Fish. Board, V. 1886.

—— Note on the Ova, Fry and Nest of the Ballan Wrasse. Idem. 1886.

MEEK, A. Migrations of Fishes. London, 1916.

MELLEN, I. M. Tropical Toy Fishes. Zoolog. Society Bulletin. New York, 1921.

—— The Whitefishes. Zoologica. New York, 1923.

METZELAAR, J. On a Collection of Marine Fishes from the Lesser Antilles. Bijd. Dierkunde. Leiden, 1922.

MIELCK, W. Die Nahrung der Fischlarven. Festschrift f. Friedrich Heincke. Wiss. Meeresunt. Helgoland, 1922.

MILEWSKI, A. W. Zur Biologie, Morphologie und Anatomie der lebendgebärenden Zahnkarpfen. Int. Rev. Hydrobio., VIII. 1920.

MILROY, T. H. The Physical and Chemical Changes taking place in the Ova of Certain Marine Teleosteans. Rep. S. F. B. 1898.

MITSUKURI, K. The Cultivation of Marine and Fresh-water Animals in Japan. Bull. Bur. Fish. Washington (1904), 1905.

MOHR, E. Zur Naturgeschichte der Seezunge. Wiss. Meeresunt., XIV. Helgoland, 1918.

MONTI, R. La variabiltà della pressione osmotica. Atti Soc. Ital. Sc. Nat. Pavia, 1914.

MOREAU, E. Poissons de France. Paris, 1881.

MOSER, F. Beiträge zur vergleichenden Entwicklungsgeschichte der Schwimmblase der Fische. Arch. Mikr. Anat. 1903.

MOSSO, A. Ueber verschiedene Resistenz der Blutkörperchen bei verschiedenen Fischarten. Biol. Centralb. 1891.

MURRAY, SIR JOHN. The Distribution of Organisms in the Hydrosphere. Int. Rev. Hydrobiologie., I. 1908.

NEWBIGIN, M. I. Further Observations on the Source of the Pigment of Salmon Muscle. Scot. Fish. Board Rep. 1900.

NOMURA, K. A Note on the Two-year-old Fish of Ayu, Plecoglossus altinalis. Japanese Jour. of Zoology., I. 1922.

OOSTEN, J. v. A Study of the Scales of Whitefishes of Known Ages. Zoologica. New York, 1923.

ORTON, J. H. Sea-temperature, Breeding and Distribution in Marine Animals. Jour. Mar. Biol. Assoc., XII. 1920.

PANTIN, C. F. A. On the Physiology of Amœboid Movement. Jour. Mar. Biol. Assoc., XIII. 1923.

PARKER, G. H. The Optic Chiasma in Teleosts. Bull. Mus. Comp. Zool. Harvard, 1903.

—— Hearing and Allied Senses in Fishes. Bull. U.S. Fish. Comm. (1902). 1904.

—— The Function of the Lateral Line Organs in Fishes. Bull. Bur. Fisheries, XXIV. (1904). 1905.

PATON, D. N. Report of Investigations on the Life-history of the Salmon in Fresh Water. Rep. S. F. Board. 1898.

—— and NEWBIGIN, M. I. Further Investigations on the Life-history of the Salmon in Fresh Water. Idem. 1900.

PAVESI, P. L'Industria del Tonno. Roma, 1889.

PETERSEN, C. G. JOH. The Common Eel gets a Particular Breeding Dress before its Emigration to the Sea. Rep. Dan. Biol. Stat. 1896.

PETERSEN, C. G. JOH. The Yearly Immigration of young Plaice into the Limfjord from the German Sea. Idem. 1897.

—— Larval Eels of the Atlantic Coasts of Europe. Medd. Komm. Havunders. Fiskeri. 1905.

—— Om lysets indflydelse paa aalens vandringer. Rep. Dan Biol. Station. 1907.

—— The Sea Bottom and its Production of Fish-food. Rep. Danish Biol. Station. 1918.

—— On the Stock of Plaice in relation to the Intensive Fishing of the Present Times in the Beltsea and Other Waters. Idem. (1920). 1921.

PHILIPPI, E. Ein neuer Fall von Arrhenoidie. Sitzber. Ges. Naturfos. Freunde. Berlin, 1904.

—— Fortpflanzungsgeschichte der viviparen Teleostier . . . Zool. Jahrb. 1908.

PLEHN, M. Praktikum der Fischkrankheiten. Stuttgart, 1924.

PORTIER, P., et DUVAL, M. Variation de la pression osmotique du sang des poissons. Com. Ren. de l'Acad. des Sciences. Paris, 1922.

POUCHET, G. Des changements de coloration sous l'influence des nerfs. Jour. Anat. Physiol. Paris, 1876.

PRATJE, A. Das Leuchten der Organismen. Ergebnisse d. Physiologie. München, 1923.

PRINCE, E. E. On the Nest and Development of Gasterosteus spinachia. Ann. Mag. Nat. Hist. 1885.

RABEN, E. Quantitative Bestimmung der im Meerwasser gelösten Phosphorsäure. Wiss. Meeresunt. Kiel, 1916.

REDEKE, H. C. Ueber den gegenwärtigen Stand unserer Kenntnis von den Rassen der wichtigsten Nutzfische. Rapp. Proc. Verb. Cons. per. inter. Exp. de la mer. 1912.

REGAN, C. T. The Phylogeny of the Teleostomi. Ann. Mag. Nat. Hist. 1904.

—— The Classification of Teleostean Fishes. Idem. 1909.

—— The Caudal Fin of the Elopidæ and of some other Teleostean Fishes. Idem. 1910.

—— The Classification of the Percoid Fishes. Idem. 1913.

—— The Distribution of the Fishes of the Order Ostariophysi. Bijd. t.d. Dierkunde. Leiden, 1922.

—— Dwarfed Males Parasitic on the Females in Oceanic Angler-Fishes. Proc. Roy. Soc. London, 1925.

REIBISCH, J. Ueber die Eizahl bei Pleuronectes platessa und die Altersbestimmung dieser Form aus den Otolithen. Wiss. Meeresunt. Kiel, 1899.

RENNIE, J. The Epithelial Islets of the Pancreas in Teleostei. Quar. Jour. Micr. Sci. 1904.

RIDEWOOD, W. G. The Air-bladder and Ear of Osseous Fishes. Jour. Anat. Physiol. London, 1891.
—— On the Cranial Osteology of the Clupeoid Fishes. Proc. Zool. Soc. London, 1905.
ROMANES, G. J. Animal Intelligence. London, 1883.
ROULE, L. Sur l'influence exercée sur la migration de montée du saumon par la proportion d'oxygène dissous dans l'eau des fleuves. Com. ren. Acad. Sci. Paris, 1914.
RYDER, J. A. A Contribution to the Development and Morphology of the Lophobranchiates, Hippocampus antiquorum. Bull. U.S. Fish. Comm. 1882.
—— On the Development of Osseous Fishes. Rep. U.S. Fish. Comm. 1887.
—— On the Mechanical Genesis of the Scales of Fishes. Ann. Mag. Nat. Hist. 1893.
—— The Inheritance of Modifications due to Disturbances of the Early Stages of Development. Proc. Acad. Nat. Sci. Philadelphia, 1893.
—— The Vascular respiratory Mechanism of the Vertical Fins of the Viviparous Embiotocidæ. Idem. 1893.
SANZO, L. Larva di Stomias boa. Mem. R. Comit. Talassog. Ital. 1912.
SAEMUNDSSON, B. Zoologiske Meddelelser fra Island, XIV. Medd. Dansk. naturh. Foren. 1922.
SCHAEFFER, E. Aale auf dem Lande. Der Fischerbote. 1919.
SCHIEMENZ, F. Ueber den Farbensinn der Fische. Zeit. für vergl. Physiologie. Berlin, 1924.
SCHMALHAUSEN, J. J. Zur Morphologie der unpaaren Flossen. 1912–1913.
SCHMIDT, JOHS. Eel Investigations. Rapp. Cons. intern. Copenhagen, 1914. Medd. Komm. Havund., IV., 7. 1914. Phil. Trans. Roy. Soc. London, 1922.
—— Zoarces Investigations. Medd. Carlsberg Laboratoriet, XIII., 1917; XIV., 1921. Jour. of Genetics, VII., 1918; X., 1920. Com. ren. Labor. Carlsberg., XIV., 1919, 1920, 1922.
—— Lebistes Investigations. Meddel. Carlsberg Lab., XIV. 1919, 1920.
—— Diallel Crossings with Trout. Jour. of Genetics., IX. 1919. Com. ren. des travaux, labor. Carlsberg, XIV., 1919.
SCHNAKENBECK, W. Ueber Färbungsanomalien bei Pleuronectiden. Wiss. Meeresunt. Helgoland, 1923.
SCHULTZE, O. Zur Frage von dem feineren Bau der elektrischen Organe der Fische. Leipzig, 1906.

SCOTT, G. G. Effects of Change in the Density of Water upon the Blood of Fishes. Proc. IV. Inter. Fish. Cong. Washington, 1908.

SEDGWICK, A. British Association Meeting. Dover, 1899.

SELLIER, J. Sur la lipase du sang chez quelques groupes de poissons. Com. ren. Mem. Soc. Biol. Paris, 1902.

SEMPER, C. The Natural Conditions of Existence as they affect Animal Life. London, 1883.

SHELFORD, V. E. The Reaction of Herring and other Salt-water Fishes to Decomposition Products normal to Sea-water. Science. 1915.

SIMPSON, S. The Body Temperature of Fishes. Proc. Roy. Soc. Edinburgh, 1908.

—— Further Observations on the Body Temperature of Fishes. Jour. Anat. Physiol. London, 1908.

SOLLAS, W. J. and I. B. J. An Account of the Devonian Fish, Palæospondylus gunni. Phil. Trans. Roy. Soc. London, 1904.

STANSCH, K. Die exotischen Zierfische. Braunschweig, 1924.

STARKS, E. C. The Air-bladder in Clupea harengus. Science. 1911.

—— On a Posterior Communication of the Air-bladder with the Exterior in Fishes. Science. 1911.

STECHE, O. Die Leuchtorgane von Anomalops katoptron und Photoblepharon palpebratus. Zeitsch. Wiss. Zool. 1909.

STEMPELL, W., and KOCH, A. Elemente der Tierphysiologie. II. edit. Jena, 1924.

STENSIÖ, E. A : SON. Triassic Fishes from Spitzbergen. Vienna, 1921.

STIRLING, A. B. Notes on the Fungus Disease affecting Salmon. Proc. Roy. Edinburgh, 1878.

STIRLING, W. On the Chemistry and Histology of the Digestive Organs of Fishes. Report Scot. Fish., II. (1884). 1885.

STOCKARD, C. R. The Artificial Production of a Single Median Cyclopean Eye in the Fish Embryo. Arch. Ent.-Mech. Leipzig, 1907.

STRODTMANN, S. Die Anpassung der schwimmenden Fischeier an schwächere Salzgehalte. Der Fischerbote. 1915.

SULLIVAN, M. X. The Physiology of the Digestive Tract of Elasmobranchs. Bull. Bur. Fish. Washington (1907), 1908.

SUMNER, F. B. The Physiological Effect upon Fishes of Changes in the Density and Salinity of Water. Bull. Bur. Fish. Washington (1905). 1906.

SURBECK, G. Beitrag zur Kenntnis der Geschlechtsverteilung bei Fischen. Schweiz. Fisch.-Zeitung. 1913.

SURBECK, G. Ergebnisse der ersten Laichfischfang-Statistik Kanton Bern pro 1913–1914. Idem. 1914.

SWINNERTON, H. H. A Contribution to the Morphology of the Teleostean Head Skeleton (Gasterosteus aculeatus). Quar. Jour. Mic. Sci. 1902.

TÅNING, A. V. Mediterranean Scopelidæ. Rep. Danish Ocean. Exped., II., 5. 1918

THILO, O. Die Stacheln der Fische. "Nerthus," Altona. 1897.

—— Die Bedeutung der Weberschen Knöchelchen. Zool. Anz. 1908.

—— Luftdruckmesser an den Schwimmblasen der Fische. Int. Rev. Hydrobiol. and Hydrogr. 1908.

—— Die Vorfahren der Kugelfische. Biol. Centralbl. 1914.

THOMSON, J. S. The Periodic Growth of Scales in Gadidæ as an Index of Age. Jour. Mar. Biol. Assoc. 1904.

—— The Dorsal Vibratile Fin of the Rockling (Motella). Quar. Jour. Mic. Soc. London, 1912.

TÖRLITZ, H. Anatomische und Entwicklungsgeschichtliche Beiträge zur Artfrage unseres Flussaales. Zeitsch. Fisch. 1922.

TORNIER, G. Ueber experimentelles Hervorrufen . . . von Mopsköpfen, Cyclopen . . . bei Wirbeltieren. Sitzber. Ges. Naturf. Freunde. Berlin, 1908.

—— Ueber die Art, wie äussere Einflüsse den Aufbau des Tieres abändern. Verh. Deutsch. Zool. Ges. 1911.

TOWNSEND, C. H. The Power of the Shark-sucker's disk. Zool. Soc. Bull. New York, 1915.

—— The Swordfish and Thresher Shark Delusion. Idem. 1923.

—— Why Swordfish strike Ships. Idem. 1924.

TRACY, C. H. The Morphology of the Swim-bladder in Teleosts. Anat. Ang., 38. 1911.

TRAQUAIR, R. H. On the Fossil Fishes found at Achanarras Quarry, Caithness. Ann. Mag. Nat. Hist. 1890.

—— The Evolution of Fishes. Rep. Brit. Assoc. 1900.

TRESSLER, D. K. Marine Products of Commerce. New York, 1923.

TROJAN, E. The Structure of the Bud-like Organs of Malthopsis spinulosa. Mem. Mus. Comp. Zool. Harvard, XXX. 1905.

WALTER, E. Der Flussaal. Neudamm, 1910.

WEBER, MAX. Siboga Expeditie. Leiden, 1901.

—— Die Fische der Siboga-Expedition. Leyden, 1913.

WEGENER, A. Die Entstehung der Kontinente und Ozeane. Braunschweig (1915). 1922.

WEGENER, A. The Origin of Continents. 3rd Ed. Translation. London, 1925.

WELLS, The Chondrocranium of the Herring. Proc. Zool. Soc. London, 1921.

WELLS, M. M. The Reactions and Resistance of Fishes in their Natural Environment to Salts. Jour. Exper. Zoology. 1915.

WHITEHOUSE, R. H. The Caudal Fin of the Teleostomi. Proc. Zool. Soc. London, 1911.

WIEDERSHEIM, R. E. E. Vergleichende Anatomie der Wirbeltiere Jena, 1906.

WILLIAMSON, H. C. On the Anatomy of the Pectoral Arch of the Grey Gurnard with Special Reference to its Innervation. Rep. Fish. Board, Scot. 1893.

—— Notes on Some Points in Teleostean Development. Idem. 1897.

—— Report on Diseases and Abnormalities in Fishes. Idem. 1911.

WOODWARD, A. SMITH. The Problem of the Primeval Sharks. Natural Science. 1895.

—— The Use of Fossil Fishes in Stratigraphical Geology. Quar. Jour. Geol. Soc. London, 1915.

ZANDER, E. Das Kiemenfilter der Teleostier. Zeitsch. f. wissen. Zoologie. 1906.

ZENNECK, J. Reagieren die Fische auf Töne? Pflügers Arch. Physiol. 1903.

ZIEGLER, H. E. Der Begriff des Instinctes einst und jetzt. Supp. VII. d. Zool. Jahrb. Jena, 1904.

ZITTEL, K. A. VON. Grundzüge der Paläontologie. Berlin, 1923.

INDEX